Birkhäuser Advances in Infectious Diseases

Series editors
Stefan H.E. Kaufmann
Max Planck Institute for Infection Biology, Berlin, Germany

Andrew A. Mercer
Department of Microbiology and Immunology, University of Otago
Dunedin, New Zealand

Olaf Weber
Bonn, Germany

More information about this series at http://www.springer.com/series/5444

Carsten B. Schmidt-Weber

Editor

Allergy Prevention and Exacerbation

The Paradox of Microbial Impact
on the Immune System

 Springer

Editor
Carsten B. Schmidt-Weber
ZAUM
Technical University of Munich
Munich, Germany

ISSN 2504-3811 ISSN 2504-3838 (electronic)
Birkhäuser Advances in Infectious Diseases
ISBN 978-3-319-69967-7 ISBN 978-3-319-69968-4 (eBook)
https://doi.org/10.1007/978-3-319-69968-4

Library of Congress Control Number: 2017962326

Printed on acid-free paper

This Springer imprint is published by Springer Nature
The registered company is Springer International Publishing AG
The registered company address is: Gewerbestrasse 11, 6330 Cham, Switzerland

Contents

Overview: The Paradox of Microbial Impact on the Immune System in Allergy Prevention and Exacerbation

Carsten B. Schmidt-Weber

Abstract Allergy prevalence has been increasing to epidemic proportions. The term "allergy tsunami" was even coined after the prevalence of allergen sensitization in school-aged children jumped to more than 40%. Because of the huge economic impact on the health care system and the dramatic impact on quality of life, allergy prevention has become very relevant. In this context, it is interesting that infectious diseases are becoming less prevalent and increasingly under control, while noncommunicable diseases (including, but not limited to, allergy) increase. This also applies to autoimmune diseases, which are characterized by an overactive immune response.

The immune system's response to antigens and the microbiome is critical for the outcome of infections and allergies. Paradoxically, the microbiome can play a protective role in allergies, but it is also known to be the driver of exacerbations. The current chapter focuses on this paradox and therefore starts with the topic of autoallergy, in which microbial antigens are considered to be potential pathogens and disease initiators. Bacterial allergens are generally thought of as potential "virulence factors" and exacerbation factors in allergy. In contrast, the allergy-protective microbiome, which was discovered in protective agricultural environments, is the second key aspect of this chapter. Additional environmental elements that were lost due to lifestyle changes are also highlighted. Both farm and lifestyle effects contributed to the hygiene hypothesis, which is challenged by the author to show that "dirt" is not at all protective against allergies. Determinants and metabolites of bacteria and molds that co-evolutionized with man have been identified in rural environments, with stunning effects on our immune systems.

Allergy prevention at least partially develops in our gut through a complex interplay of microbiota with our immune system. Here, the persistence of type-2 allergen memory, such as interleukin-4 and immunoglobulin E–producing cells, is particularly important and relevant for food allergies. This chapter also describes a recently discovered mechanism that links the gut microbiome with pro-allergic

C.B. Schmidt-Weber
Center of Allergy and Environment (ZAUM), Technical University and Helmholtz Center Munich, Munich, Germany
e-mail: csweber@tum.de

© Springer International Publishing AG 2017
C.B. Schmidt-Weber (ed.), *Allergy Prevention and Exacerbation*, Birkhäuser
Advances in Infectious Diseases, https://doi.org/10.1007/978-3-319-69968-4_1

type-2 immunity. The mechanism of environmental sensing is discussed, using the example of the aryl hydrocarbon receptor to demonstrate the link between microbiota, their metabolites, and their recognition at cellular and molecular levels. For the prevention of microbiome-driven exacerbations, the chapter also discusses therapeutic options with novel diagnostics and biologics that are likely to influence the future management of allergies. (For a graphical chapter overview, see Fig. 1.)

1 Introduction: Allergy Exacerbation Triggers Are Also the Prevention Factors

Initial allergic sensitization is thought to originate as a loss of tolerance against otherwise harmless antigens, such as that contained in pollen. Signals that convey a danger signal, such as bacterial wall components, are thought to promote inflammation and downregulate inhibitory immune mechanisms. The result of this accidental co-recognition of harmless and potentially dangerous antigens is an unwanted immune response against the harmless antigen. This mechanism may not only apply to exogenous antigens from pollen, but also to autologous molecules of the organism and thus may cause autoimmune reactions (Fig. 1).

However, it is currently unclear why the allergen drives immunoglobulin (Ig) E/type-2 responses when autoimmunity is usually characterized by IgG/type-1 or 17 responses. Type-2 responses are characterized by interleukin (IL)-4, IL-5, and IL-13; here, IL-4 drives Th2 cells and the IgM to IgE switch of B cells, IL-5 promotes eosinophilic inflammation, and IL-4 and IL-13 promote epithelial and

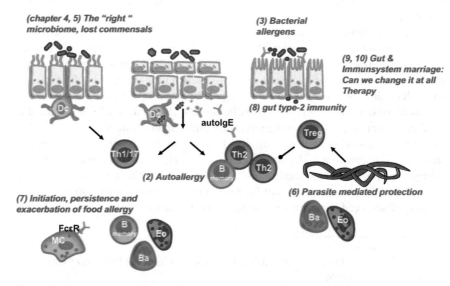

Fig. 1 Environmental challenges and in particular the microbiome can interact with the immune system on different levels. The paradox of triggering disease on one hand and promoting protection against allergies is illustrated and related book chapters indicated in brackets

macrophage differentiation of epithelial cells and macrophages. Type-1/17 responses promote a lytic immunity and are characterized by interferon (IFN)-γ and IL-17 expression, which promotes the antigen presentation of dendritic cells and the recruitment of neutrophils to the site of inflammation. An interesting exception to this distinction between types 1 and 2 is the phenomenon of autoantigens being recognized by IgE.

In the chapter titled "Microbial Triggers in Autoimmunity, Severe Allergy, and Autoallergy," the nature of autoallergy is described in detail; in particular, auto-IgE appears to recognize antigens with structures that are very similar between human and bacterial organisms because they are highly conserved evolutionarily. This "molecular mimicry" is one way that self-tolerance may be overcome accidentally (Virtanen et al. 1999). It is also speculated that this autoallergy against tissue antigens may drive transient allergies to become chronic allergies. However, it remains unclear why IgE is raised against these autoantigens, rather than IgG and type-1/17 responses as is normally observed for autoantigens.

A possible explanation is that epithelial cells provide this type-2 bias from the epithelial mediators IL-25, IL-33, and thymic stromal lymphopoietin (TSLP), which are also called "alarmins" (Bianchi 2007). These cytokines can activate a recently discovered group of cells called innate lymphoid cells (ILCs). As the name indicates, these cells cannot recognize antigens, but they release cytokines themselves, including IL-4 (Licona-Limon et al. 2013). The IL-4-secreting ILCs are called ILC2s; it was hoped that the initial source of IL-4 at the beginning of allergic sensitization was discovered in this group of cells. However, mice lacking a key ILC2 transcription factor have normal levels of Th2 cells (Verhoef et al. 2016); thus, ILC2s are currently thought to contribute as amplifiers to Th2 cell activity but not account for the root of type-2 immune responses. It was also shown that Th2-derived IL-4, but not ILC2-derived IL-4 and IL-13, were correlated with lung pathology in an experimental asthma model (Oeser et al. 2015). Other initial sources of IL-4 are still the subject of intensive research.

The question remained on how uncommitted, naive T cells can be driven or differentiated into cell that produce IL-4. Certain transcription factors, such as GATA3 and STAT5, were found to be important. However, weak T-cell receptor stimulations also favor Th2 commitment (for an overview, see Paul and Zhu 2010). Generally speaking, it seems that a unique Th2 path has not been discovered; possibly, multiple factors need to come together to generate type-2 immunity, break tolerance, and induce sensitization. Of note, an already committed Th2 cell may still be counteracted by regulatory immune responses and prevent greater support of B cells and IgE production. However, regulatory responses are diminished under conditions of inflammation, including type-2 cytokines (Mantel et al. 2006). Similarly type-1 and/or type-17 responses compete with and antagonize Th2 responses. These dynamics underlie the complex role of microbiota, which can trigger inflammation, exacerbation, tolerance, sensitization, and allergy. The spectrum of bacterial allergens is described in the chapter titled "Bacterial Allergens," which illustrates the importance of the *Staphylococcus* species and the specific virulence factors in allergy, while also highlighting protective factors such as *Staphylococcal pneumonia* PspC and PspA.

The identification of microbiotic factors that prevent allergies has been challenging and their nature is only partially understood. Early epidemiologic research suggested that agricultural environments may have a protective effect for hay fever; this was extrapolated in the "hygiene hypothesis". The history and recently identified key factors are presented in the chapter titled, "Good and Bad Farming: The Right Microbiome Protects from Allergy." A clear conclusion of the author is that the farm effect is far more specific than one would think in the context of the hygiene hypothesis. Farms keeping sheep only or those using pressed hay do not show equal levels of allergy protection; furthermore, allergies other than those directed against grass allergens are affected to a much lower degree.

Other hygiene effects, such as the presence of *Heliobacter pylori* (see the chapter titled, "The Lost Friend: *H. pylori*") are more controversially discussed and may explain the increase of allergies. In the 1950s, *H. pylori* colonization reached 80–90%; however, it has dropped to 2% in the population born after 2000 due to sanitary improvements. With these improvements, many parasites also disappeared as pathogens in developed countries (see the chapter titled, "Parasite-Mediated Protection Against Allergy"). Many parasites such as helminths elicit a type-2 nonlytic response, which allows the organism to survive and protects the host from major tissue damage. Particularly interesting for allergy is that helminths also developed immunosuppressive mechanisms that support their lifecycles, which are attenuated with deworming of the host. The immunoregulatory mechanisms of parasites are very specific: helminths, for example, induce both exacerbation and tolerance, whereas the whipworm *Trichuris suis* failed to improve allergic rhinitis. The developmental stage of the worm is relevant for these effects. In addition, the allergy-affected organ seems to be important: for example, worm-mediated protection works in the lung but not in the skin. The chapter titled "Parasite-Mediated Protection Against Allergy" provides a very specific and molecularly defined avenue for how parasites influence allergy.

2   Cellular Player: From Genetics to the Immune System

Some simple considerations that identify critical cellular mechanisms of allergy are likely to be important in the context of prevention and exacerbation. First, the fact that allergies occur every year and specifically for selected allergens highlights immune cells that mediate long-lived antigen-specific immunologic memory, such as T- and B-cells. Furthermore, as also discussed in the chapter titled, "Initiation, Persistence, and Exacerbation of Food Allergy," the IgE memory cell produces large amounts of the Ig (plasma cell), but these IgE-producing cells were found to be surprisingly short lived. Therefore, a reservoir of IgE plasma cells may reside in the IgG^+ memory B-cell compartment, which undergoes sequential class switching (from IgG to IgE).

T-cells require phagocytic antigen-presenting cells for their activation, multiple types and variants of which are present in different tissues. The induction of unresponsiveness (tolerance) is particularly well documented for antigens that are taken up orally. Specialized mechanisms mediate this tolerance, such as the transepithelial transport of antigens via microfold (M) cells (Suzuki et al. 2008) or other yet undiscovered mechanisms (Kraus et al. 2005). Novel mechanisms mediating tolerance, including specialized Tregs, are presented in the chapter titled, "The Role of the Gut in Type 2 Immunity."

Another important cellular player are epithelial cells because they directly face the environment and microbiota (e.g. skin, lung, gut). These cells are also home to genes such as *IL33* and *Filaggrin* mutations, which were discovered to be genetically associated with allergy (Campbell et al. 2015; Barnes 2006; Holgate 1999). Epithelial cells are in contact with multiple resident (dendritic cells, mast cells) and infiltrating immune competent cells (neutrophils, eosinophils, innate lymphoid cells) that interact closely upon microbial or parasite impact; therefore, they have different protective effects on allergy, as discussed in the chapter titled "Parasite Mediated Protection Against Allergy." These protective effects arise from species-specific evasion strategies. An example are *Schistosoma* components, which promote dendritic cells to a Th2-stimulating ability, and the nematode*Acanthocheilonema viteae*, which promotes anti-inflammatory IL-10 secretion by macrophages.

Common among T cells, innate lymphocytes, macrophages, and epithelial cells (Zissler et al. 2016) is that these cells can differentiate into a type-2 phenotype. This differentiation is usually initiated by IL-4 and antagonized by IFN-γ. This differentiation changes the functional profile of these cells, which is already well described for T cells; however, the meaning for tissue cells, such as epithelial cells, is yet unclear.

3 Protection Against Allergy

A key motivation for the concept of allergy protection comes from farm studies, where therapeutic vaccinations with allergens have been performed for more than 100 years. Immunological findings indicate that the immune system can actively and antigen-specifically suppress immune responses. In this situation, tolerogenic vaccination generates nonresponsiveness against allergens, which in turn generates allergen-specific IgG4 that competes and therefore neutralizes the inflammation caused by allergen-specific IgE. Furthermore, regulatory responses such as allergen-specific regulatory T cells and B cells are induced. However, can this approach also be used prior to an outbreak of allergy symptoms? This question was positively answered for peanuts through feeding studies of young subjects, with allergy prevalence shown to be lower than in the avoidance group (Du Toit et al. 2008). The risk of inducing an allergy rather than protecting against it is obviously the key obstacle; it currently prevents substantial engagement in this direction, although allergen avoidance at least is not recommended anymore.

Intentional exposure to a food allergen was shown to mediate allergen tolerance in children, although not without risks. Primary interventions and food tolerance are driven by the allergen itself, but the protection of farm effects as well as the parasite-mediated effects are not necessarily driven by the allergen itself. Thus, it is hoped that microbial- or parasite- derived factors can be identified, which could be used for therapeutic interventions (see the chapters titled "Bacterial Allergens" and "The Lost Friend: *H. pylori*"). An example of an allergen-independent mechanism is the route of the so-called aryl hydrocarbon receptor (AhR), which is present in many tissue and immune cells and is able to respond to microbial tryptophan metabolites. The AhR can induce immune mediators such as IL-22, which in turn act on epithelial cells to close or reinforce the epithelial barrier. Because the AhR can bind multiple ligands and also interacts with several adaptor proteins, both ligand specificity and function are still subject to intense research (see the chapter titled "The Role of the Gut in Type 2 Immunity").

4 Chronification and Exacerbation of Allergies

The cause of chronic disease progression is largely unknown, although it is important to note that allergen-specific immunotherapy can prevent the progression of allergic rhinitis into asthma. Viral exacerbations of airway diseases are well documented; they activate the innate immune system (e.g. the TLR3 receptor) and allow direct interventions (Silkoff et al. 2017). The previously mentioned autoallergery could be an explanation; at least for asthma, it has been observed that an exacerbation leads to another, more escalated disease severity level.

For both airway and skin manifestation, *Staphylococci* can cause exacerbations or predispose children to asthma if they are colonized in the hypopharyngeal region (Bisgaard et al. 2007). They are a source of enterotoxins (A, B, C, D), superantigens, and can even produce toxins that degranulate mast cells independently of IgE. *Staphylococci* antigens are targets of IgE (see the chapter titled "Bacterial Allergens") and can identify patients with chronic sinusitis and nasal polyposis (Tripathi et al. 2004; Van Zele et al. 2004). In atopic dermatitis, *Staphylococcus aureus* can cause a superinfection at sites of inflammation; this is thought to be supported by type-2 immunity, which inhibits the type-1 lytic immune response and would be more appropriate to defend the pathogen (Eyerich et al. 2009a). Therefore, *Staphylococci* may use type-2 immunity as an evasion strategy.

Another observation of T cells isolated from the airway and skin biopsies of patients with chronic allergies (Pennino et al. 2013; Eyerich et al. 2009b) is that these tissue-derived T cells show a very broad T-cell subset and uncharacteristic cytokine profile. For example, there is an unusual co-expression of IL-4 and IFN-γ that is not observed in the periphery, where IL-4 and IFN-γ fall into discrete and antagonistically regulated Th1- and Th2-subset phenotypes. Similar observations were reported for Th17 cells (Gelfand et al. 2017), which are activated in asthma exacerbation along with the inflammasome (Lee et al. 2014). The plasticity of tissue-derived T cells was proposed to be the result of repetitive or chronic stimulation of the T cells because it may occur under the crossfire of allergens and superantigens. This increased plasticity has become a major research focus because it is anticipated that these cells have a high potential to cause damage and worsen pathology of the disease. For example, the plastic Th17 cells switch to IFN-γ production and form the majority of cells found in synovial fluids of patients with rheumatoid arthritis (Nistala et al. 2010; Cosmi et al. 2011), the guts of patients with Crohn disease (Kleinschek et al. 2009), and cerebrospinal fluid (Kebir et al. 2009). It also appears that these cells are resistant to the suppression of regulatory T cells (Basdeo et al. 2017).

5   Concluding Remarks and Therapeutic Relevance

Vitamins B and K are well-known symbiotic elements of bacteria in our gut from which we benefit. These vitamins highlight the complex synthetic machinery of the 1.5 kg of bacteria carried by each person. Approximately 2–3 years after birth, an individual is living with a stable microbiota composition. Therefore, a question is raised: can the microbiome be manipulated to confer protection against allergy without creating other problems in the symbiotic relationship or turning symbionts into pathogens? This issue is currently being explored with some interesting therapeutic options arising, although it has been noted that frequent antibiotic intake can increase allergic sensitization (see the chapter titled "Aryl Hydrocarbon Receptor: An Environmental Sensor in Control of Allergy Outcomes").

The authors of this book draw a differentiated picture of the role of the microbiota and their impact on the exacerbation and prevention of allergies. Consequences for therapy will therefore only develop out of profound knowledge that considers the elements of the microbiota (ranging from superantigens to Ahr ligands) and cellular recipients (ranging from dendritic cells and epithelial cells to several lymphocyte populations). Equally important is the recipient, his or her genetic predisposition, and the diverse disease endotypes that are subject to current research. Biomarkers described in chapter titled "Specific Therapies for Asthma Endotypes: A New Twist in Drug Development" may support the endotype diagnosis so that appropriate therapies can be used to control disease along with the resulting microbiota (see the chapter titled "The Gut Microbiome and Its Marriage to the Immune System: Can We Change It All?").

References

Barnes KC (2006) Genetic epidemiology of health disparities in allergy and clinical immunology. J Allergy Clin Immunol 117(2):243–254. https://doi.org/10.1016/j.jaci.2005.11.030

Basdeo SA, Cluxton D, Sulaimani J, Moran B, Canavan M, Orr C, Veale DJ, Fearon U, Fletcher JM (2017) Ex-Th17 (nonclassical Th1) cells are functionally distinct from classical Th1 and

Th17 cells and are not constrained by regulatory T cells. J Immunol 198(6):2249–2259. https://doi.org/10.4049/jimmunol.1600737

Bianchi ME (2007) DAMPs, PAMPs and alarmins: all we need to know about danger. J Leukoc Biol 81(1):1–5. https://doi.org/10.1189/jlb.0306164

Bisgaard H, Hermansen MN, Buchvald F, Loland L, Halkjaer LB, Bonnelykke K, Brasholt M, Heltberg A, Vissing NH, Thorsen SV, Stage M, Pipper CB (2007) Childhood asthma after bacterial colonization of the airway in neonates. N Engl J Med 357(15):1487–1495. https://doi.org/10.1056/NEJMoa052632

Campbell DE, Boyle RJ, Thornton CA, Prescott SL (2015) Mechanisms of allergic disease – environmental and genetic determinants for the development of allergy. Clin Exp Allergy 45(5):844–858. https://doi.org/10.1111/cea.12531

Cosmi L, Cimaz R, Maggi L, Santarlasci V, Capone M, Borriello F, Frosali F, Querci V, Simonini G, Barra G, Piccinni MP, Liotta F, De Palma R, Maggi E, Romagnani S, Annunziato F (2011) Evidence of the transient nature of the Th17 phenotype of CD4+CD161+ T cells in the synovial fluid of patients with juvenile idiopathic arthritis. Arthritis Rheum 63(8):2504–2515. https://doi.org/10.1002/art.30332

Du Toit G, Katz Y, Sasieni P, Mesher D, Maleki SJ, Fisher HR, Fox AT, Turcanu V, Amir T, Zadik-Mnuhin G, Cohen A, Livne I, Lack G (2008) Early consumption of peanuts in infancy is associated with a low prevalence of peanut allergy. J Allergy Clin Immunol 122(5):984–991. https://doi.org/10.1016/j.jaci.2008.08.039

Eyerich K, Pennino D, Scarponi C, Foerster S, Nasorri F, Behrendt H, Ring J, Traidl-Hoffmann C, Albanesi C, Cavani A (2009a) IL-17 in atopic eczema: linking allergen-specific adaptive and microbial-triggered innate immune response. J Allergy Clin Immunol 123(1):59–66.e54. https://doi.org/10.1016/j.jaci.2008.10.031

Eyerich S, Eyerich K, Pennino D, Carbone T, Nasorri F, Pallotta S, Cianfarani F, Odorisio T, Traidl-Hoffmann C, Behrendt H, Durham SR, Schmidt-Weber CB, Cavani A (2009b) Th22 cells represent a distinct human T cell subset involved in epidermal immunity and remodeling. J Clin Invest 119(12):3573–3585. https://doi.org/10.1172/JCI40202

Gelfand EW, Joetham A, Wang M, Takeda K, Schedel M (2017) Spectrum of T-lymphocyte activities regulating allergic lung inflammation. Immunol Rev 278(1):63–86. https://doi.org/10.1111/imr.12561

Holgate ST (1999) Genetic and environmental interaction in allergy and asthma. J. Allergy Clin Immunol 104(6):1139–1146

Kebir H, Ifergan I, Alvarez JI, Bernard M, Poirier J, Arbour N, Duquette P, Prat A (2009) Preferential recruitment of interferon-gamma-expressing TH17 cells in multiple sclerosis. Ann Neurol 66(3):390–402. https://doi.org/10.1002/ana.21748

Kleinschek MA, Boniface K, Sadekova S, Grein J, Murphy EE, Turner SP, Raskin L, Desai B, Faubion WA, de Waal Malefyt R, Pierce RH, McClanahan T, Kastelein RA (2009) Circulating and gut-resident human Th17 cells express CD161 and promote intestinal inflammation. J Exp Med 206(3):525–534. https://doi.org/10.1084/jem.20081712

Kraus TA, Brimnes J, Muong C, Liu JH, Moran TM, Tappenden KA, Boros P, Mayer L (2005) Induction of mucosal tolerance in Peyer's patch-deficient, ligated small bowel loops. J Clin Invest 115(8):2234–2243. https://doi.org/10.1172/JCI19102

Lee TH, Song HJ, Park CS (2014) Role of inflammasome activation in development and exacerbation of asthma. Asia Pac Allergy 4(4):187–196. https://doi.org/10.5415/apallergy.2014.4.4.187

Licona-Limon P, Kim LK, Palm NW, Flavell RA (2013) TH2, allergy and group 2 innate lymphoid cells. Nat Immunol 14(6):536–542. https://doi.org/10.1038/ni.2617

Mantel PY, Ouaked N, Ruckert B, Karagiannidis C, Welz R, Blaser K, Schmidt-Weber CB (2006) Molecular mechanisms underlying FOXP3 induction in human T cells. J Immunol 176(6):3593–3602

Nistala K, Adams S, Cambrook H, Ursu S, Olivito B, de Jager W, Evans JG, Cimaz R, Bajaj-Elliott M, Wedderburn LR (2010) Th17 plasticity in human autoimmune arthritis is driven by the inflammatory environment. Proc Natl Acad Sci USA 107(33):14751–14756. https://doi. org/10.1073/pnas.1003852107

Oeser K, Maxeiner J, Symowski C, Stassen M, Voehringer D (2015) T cells are the critical source of IL-4/IL-13 in a mouse model of allergic asthma. Allergy 70(11):1440–1449. https://doi.org/ 10.1111/all.12705

Paul WE, Zhu J (2010) How are T(H)2-type immune responses initiated and amplified? Nat Rev Immunol 10(4):225–235. https://doi.org/10.1038/nri2735

Pennino D, Bhavsar PK, Effner R, Avitabile S, Venn P, Quaranta M, Marzaioli V, Cifuentes L, Durham SR, Cavani A, Eyerich K, Chung KF, Schmidt-Weber CB, Eyerich S (2013) IL-22 suppresses IFN-gamma-mediated lung inflammation in asthmatic patients. J Allergy Clin Immunol 131(2):562–570. https://doi.org/10.1016/j.jaci.2012.09.036

Silkoff PE, Flavin S, Gordon R, Loza MJ, Sterk PJ, Lutter R, Diamant Z, Turner RB, Lipworth BJ, Proud D, Singh D, Eich A, Backer V, Gern JE, Herzmann C, Halperin SA, Mensinga TT, Del Vecchio AM, Branigan P, San Mateo L, Baribaud F, Barnathan ES, Johnston SL (2017) Toll-like receptor 3 blockade in rhinovirus-induced experimental asthma exacerbations: a randomized controlled study. J Allergy Clin Immunol. https://doi.org/10.1016/j.jaci.2017.06.027

Suzuki H, Sekine S, Kataoka K, Pascual DW, Maddaloni M, Kobayashi R, Fujihashi K, Kozono H, McGhee JR, Fujihashi K (2008) Ovalbumin-protein sigma 1 M-cell targeting facilitates oral tolerance with reduction of antigen-specific CD4+ T cells. Gastroenterology 135(3):917–925. https://doi.org/10.1053/j.gastro.2008.05.037

Tripathi A, Conley DB, Grammer LC, Ditto AM, Lowery MM, Seiberling KA, Yarnold PA, Zeifer B, Kern RC (2004) Immunoglobulin E to staphylococcal and streptococcal toxins in patients with chronic sinusitis/nasal polyposis. Laryngoscope 114(10):1822–1826. https://doi. org/10.1097/00005537-200410000-00027

Van Zele T, Gevaert P, Watelet JB, Claeys G, Holtappels G, Claeys C, van Cauwenberge P, Bachert C (2004) Staphylococcus aureus colonization and IgE antibody formation to enterotoxins is increased in nasal polyposis. J Allergy Clin Immunol 114(4):981–983. https://doi.org/ 10.1016/j.jaci.2004.07.013

Verhoef PA, Constantinides MG, McDonald BD, Urban JF Jr, Sperling AI, Bendelac A (2016) Intrinsic functional defects of type 2 innate lymphoid cells impair innate allergic inflammation in promyelocytic leukemia zinc finger (PLZF)-deficient mice. J Allergy Clin Immunol 137(2): 591–600.e591. https://doi.org/10.1016/j.jaci.2015.07.050

Virtanen T, Zeiler T, Rautiainen J, Mantyjarvi R (1999) Allergy to lipocalins: a consequence of misguided T-cell recognition of self and nonself? Immunol Today 20(9):398–400

Zissler UM, Chaker AM, Effner R, Ulrich M, Guerth F, Piontek G, Dietz K, Regn M, Knapp B, Theis FJ, Heine H, Suttner K, Schmidt-Weber CB (2016) Interleukin-4 and interferon-gamma orchestrate an epithelial polarization in the airways. Mucosal Immunol 9(4):917–926. https:// doi.org/10.1038/mi.2015.110

Microbial Triggers in Autoimmunity, Severe Allergy, and Autoallergy

Fariza M.S. Badloe, Sherief R. Janmohamed, Johannes Ring, and Jan Gutermuth

Abstract The prevalence of immune-mediated diseases (allergies, autoimmune diseases, and autoinflammatory diseases) is rising world-wide and their management is difficult, since treatments usually are mainly symptomatic. Insight in the exact pathomechanisms is crucial for focused prevention or improvement of therapies.

Microbial factors including bacteria (mainly *Staphylococcus aureus*), viruses and fungi are important factors in atopic dermatitis. Allergic asthma is also heavily impacted by *Staphylococcus aureus*, but also by multiple viruses. Viral complications in severe allergies can lead to life-threatening disease states. In autoimmunity, many viruses and bacteria play a role in disease development due to molecular mimicry or bystander activation.

Furthermore, it is hypothesized that autoallergy can link allergy, chronification of disease, and autoimmunity. Here, fungal triggers are suspected to elicit autoreactive IgE and T helper 1 responses in atopic dermatitis, which can maintain a vicious inflammatory cycle due to constant autoallergen release by scratching. Taken together, multiple microbial factors elicit or aggravate immune mediated diseases, such as allergies and autoimmunity.

F.M.S. Badloe • S.R. Janmohamed • J. Gutermuth (✉)
Department of Dermatology, Vrije Universiteit Brussel (VUB), Universitair Ziekenhuis Brussel (UZB), Brussels, Belgium
e-mail: jan.gutermuth@uzbrussel.be

J. Ring
Department of Dermatology, Vrije Universiteit Brussel (VUB), Universitair Ziekenhuis Brussel (UZB), Brussels, Belgium

Department of Dermatology and Allergy, Biederstein, Technical University, Munich, Germany

© Springer International Publishing AG 2017   11
C.B. Schmidt-Weber (ed.), *Allergy Prevention and Exacerbation*, Birkhäuser
Advances in Infectious Diseases, https://doi.org/10.1007/978-3-319-69968-4_2

Abbreviations

α-NAC	Nascent polypeptide-associated complex subunit alpha
AD	Atopic dermatitis
APC	Antigen presenting cell
C. jejuni	*Campylobacter jejuni*
CVB4	Coxsackievirus B4
EBV	Epstein-Barr virus
GBS	Guillain-Barré syndrome
GD1a	Disialoganglioside
GM1	Monosialotetrahexosylganglioside
HLA	Human leucocyte antigen
H. pylori	*Helicobacter pylori*
HRV	Human rhinovirus
HSK	Herpetic stromal keratitis
IFV	Influenza virus
Ig	Immunoglobulin
IL	Interleukin
LPS	Lipopolysaccharide
MGL	*Malassezia globosa*
MHC	Major histocompatibility complex
MnSOD	Manganese superoxide dismutase
MRSA	Methicillin-resistant *Staphylococcus aureus*
NK cells	Natural killer cells
RSV	Respiratory syncytial virus
S. aureus	*Staphylococcus aureus*
S. epidermis	*Staphylococcus epidermidis*
SE	Staphylococcal enterotoxin
T1D	Type-1 diabetes
T_H	T helper cell
TLR	Toll-like receptor
TSLP	Thymic stromal lymphopoietin

1 Immune Dysfunction as Underlying Cause of Allergy, Autoimmunity and Autoallergy

Since the first appearance of life on earth about 4 billion years ago, protection has been necessary against threats from the outside world (environment), but also from interior self-derived cells and molecules. Homo sapiens, the most recent inhabitant of this planet developed around 200,000 years ago and bears an evolutionary optimized immune system to cope with all these environmental and self-derived threats. As outlined in this book however, sometimes things go astray.

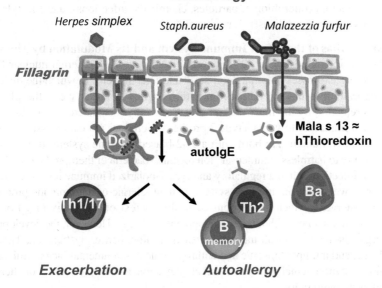

Fig. 1 Viruses, bacteriae and fungi can elicit autoreactive TH1/17 or TH2 biased immune responses. In case of autoallergy, either de novo autosensitization or cross-sensitization to exogenous, microbe-derived allergens can trigger B cell responses with autoreactive IgE or autoreactive T cells

Physical barriers form the first line of defence against biological external threats, such as bacteria or viruses, and include the skin but also mucous membranes of the lung, gut, and genitalia. The rapid-acting innate immune system with key players such as macrophages, neutrophils, NK cells and the complement system forms a second line of defence. Finally, the adaptive immune system constitutes the third line of defence, consisting of B cells and plasma cells, which produce antibodies (IgA, IgD, IgE, IgG, and IgM), and T cells, which target infection or tumour cells. The innate system reacts quickly within minutes but with little specificity. It is a short-term system without memory. In contrast, the adaptive immune system reacts slower (within days to weeks), but is highly specific and develops long-lasting memory. Both systems tackle invaders from outside (e.g. infection) or inside (e.g. cancer), but it is the adaptive immune system that leads in some individuals to allergy ('outside, foreign') or autoimmunity ('inside, self'). It is hypothesized that allergy can be complicated by autoimmunity in chronicity and severity, so-called "autoallergy" (Fig. 1). Indeed, allergy and autoimmunity share common features. Prevalence of both disease states have been rising over the last decades (allergy 25% versus 7% in autoimmunity). Allergy and Autoimmunity are characterized by T- and B cell reactions against harmless foreign proteins (allergy) or harmless self-proteins (autoimmunity), and both share similar genetic predispositions (HLA, cytokines, cytokine-receptors, TLR polymorphisms), environmental factors (microbiome, particles, chemicals, infections), and lifestyle factors (smoking, traffic exhaust) (Eaton et al. 2007).

T Helper 2 Bias of the Infant Immune System and Its Modulation by Microbes
The human foetus possesses 50% maternal DNA and 50% paternal nuclear DNA and therefore expresses paternal "non-self" antigens in utero, to which the maternal immune system is not tolerant. From a teleological point of view, the placenta prevents rejection by the maternal immune system by local secretion of IL-4. IL-4 shifts local immunity towards a type 2 profile, which does not cause tissue damage. In this context, humans are born with a T_H2-biased immune system. It is assumed that exposure to harmless commensal "tolerogenic" bacteria or their products and early microbial infections, elicit a regulatory and type 1-polarized immune responses, which results in a well-balanced immune system without allergic or autoimmune processes. Potential allergens, which are recognized in the context of regulatory (T_{reg}) or T_H1 responses, are tolerated without the development of allergy. However, the development of allergies is multi-factorial: individual genetic profile, barrier dysfunction, T_H1/T_H2 imbalance, and multiple biogenic and anthropogenic environmental factors (microbes, infections, environmental pollutants, etc.) all play a role in the development of allergies, as well as autoimmunity.

2 Autoimmunity and Microbial Triggers

2.1 Basic Pathogenesis of Autoimmunity

The pathogenesis of autoimmune diseases is multifactorial. The starting point is a genetically susceptible individual. During microbial infection, T cells are activated and in case of structural homology of microbial and human proteins (*molecular mimicry*), T cell receptors can cross-react between microbial and human epitopes, which results in self-tissue destruction. Moreover, in the context of an inflammatory process, autoreactive T- and B cells, which are rendered in an anergic state during immune homeostasis, can be activated as bystanders. Tissue damage by trauma can destroy immune barriers that normally keep highly tissue-specific antigens, e.g. in the eye or in the testes, from being recognized and targeted by T or B cells. Such release and priming towards *cryptic antigens* constitutes another important pathomechanism of autoimmunity and is of clinical relevance in encephalitis, male sterility and trauma-induced uveitis. Finally, *epitope spreading* is a process in which ongoing autoimmune processes are aggravated and perpetuated. Antigens from damaged cells are taken up and processed by antigen presenting cells (APC), leading to activation of additional autoreactive T cell clones during ongoing autoimmune disease. Multiple sclerosis is an example of a disease, in which relapse is dominated by a T cell clone that has been derived from a previous disease flare by epitope spreading (Ercolini and Miller 2009).

2.2 Microbial Triggers in Autoimmunity

There is a large number of bacteria, viruses, and fungi that can trigger autoimmune disorders or that are involved in their pathogenesis. The most important will be discussed below (Table 1).

2.2.1 Bacterial Triggers of Autoimmunity

Molecular mimicry of *Helicobacter pylori* (*H. pylori*) can induce gastric autoimmunity (Smyk et al. 2014; Hasni 2012). In autoimmune type A-gastritis for example, *H. pylori* infection leads to activation of gastric $CD4^+$ T cells that recognize cross-reactive epitopes shared by *H. pylori* and the human gastric H^+, K^+-ATPase (D'Elios et al. 2004).

Campylobacter jejuni (C. Jejuni) is associated with the development of Guillain-Barré syndrome (GBS). Lipopolysaccharide (LPS) on the outer layer of the *C. jejuni* serotype (O:4 and O:19) bacteria structurally mimics the human ganglioside (GM1 and GD1a) (Ercolini and Miller 2009). Similar to *C. jejuni*, *Haemophilus influenzae* infection can induce antibodies to LPS that are cross-reactive with human ganglioside (GM1) in neural tissue. The presence of a ganglioside-like structure on its LPS suggests that molecular mimicry may play a role in the induction of GBS (Ercolini and Miller 2009; Mori et al. 2000).

Table 1 Microbial triggers and their pathomechanisms in autoimmune diseases

	Disease	Mechanism
Bacteria		
H. pylori	Autoimmune gastritis	Molecular mimicry
C. jejuni	Guillain-Barré syndrome	Molecular mimicry
H. influenzae	Guillain-Barré syndrome	Molecular mimicry
Virus		
Epstein-Barr virus	Systemic lupus erythematosus, Rheumatoid arthritis, Multiple sclerosis	Molecular mimicry
Coxsackievirus B4	Type-1 diabetes	Molecular mimicry
Herpes simplex virus	Herpetic stromal keratitis (experimental model)	Molecular mimicry
		Bystander activation

2.2.2 Viral Triggers of Autoimmunity

Viruses have been suggested to being an important factor in the initiation of autoimmune processes. Murine herpetic stromal keratitis (HSK) provides an experimental model system for studying molecular mimicry and bystander activation elicited by viral infections and several viruses have been identified as important factors in the initiation of autoimmunity (Wickham and Carr 2004).

Epstein-Barr virus (EBV) is highly associated with the development of systemic lupus erythematosus, rheumatoid arthritis and multiple sclerosis. Prolonged infection with EBV can lead to constant activation and proliferation of B cells, which results in the production of mono- and polyclonal autoantibodies (Vojdani 2014).

Coxsackievirus B4 (CVB4) may cause type-I diabetes (T1D) by molecular mimicry. T cells isolated from T1D patients react with both glutamic acid decarboxylase (GAD-65) and protein 2C in CVB4 (Ercolini and Miller 2009).

3 Microbial Triggers in Severe Allergy

3.1 Atopic Dermatitis

Atopic dermatitis (AD; synonyms: eczema, atopic eczema, constitutional eczema and neurodermitis) is a chronic relapsing itchy inflammatory skin disease, which affects 10–20% of children and 1–3% of adults (Ring et al. 2012; Bieber 2008). The term AD was coined in 1933 by Fred Wise and Marion Sulzberger (Wallach and Taieb 2014; Ring 1985) but had already been mentioned in antiquity by Suetonius with his clear description of cutaneous and respiratory atopic symptoms in emperor Augustus (Wallach and Taieb 2014; Ring 1985). AD is characterized by dry skin, erythema, and itch and caused by genetic and environmental factors (Navarrete-Dechent et al. 2016; Park et al. 2016; Kaesler et al. 2014; Ring 2016). It causes a major health- and economic burden for patients and their parents with great impact on quality of life. AD has a polyetiologic pathophysiology with genetically determined barrier dysfunction within the epidermis and the innate and adaptive immune system, which results in cutaneous type 2 inflammation. Patients suffering from AD, especially severe AD, suffer from clinically relevant IgE mediated food- and respiratory allergies. Allergens such as house dust mite derived allergens, pollen, and food can elicit or aggravate AD (Werfel et al. 2015).

Exacerbations of AD can be triggered by several microbial interactions. Defects in the innate and adaptive immune system can result in an increased susceptibility to bacterial, viral, and fungal infections. Colonization and infection with *Staphylococcus aureus (S. aureus)* and *Malassezia* spp. cause disease exacerbation (Baker 2006). This paragraph reviews microbial triggers of AD and their pathomechanisms (Table 2).

Table 2 Microbial triggers and mechanisms of action in atopic dermatitis and allergic asthma

	Atopic dermatitis	Allergic asthma
Bacterial		
Staphylococcus epidermidis	Upregulation IL-17 expression	T_H2 response
Staphylococcus aureus	Enterotoxins superantigens	
	T-cell mediated inflammation	
MRSA	Higher number of superantigens	
Mycoplasma pneumoniae		Induction of T_H2 cytokines, IgE production
Haemophilus influenzae		T_H2 response
Viral		
Herpes simplex virus	α-toxin	
Vaccinia virus	α-toxin	
Respiratory syncytial virus		T_H2 response
Human rhinovirus		IL-33, TSLP, and IL-25. T_H2 response
Influenza virus		T_H2 response
Yeast		
Malassezia spp.	MnSOD cross-reactivity, auto-reactive T-cells, Histamine release, Cytokines, IgE antibodies	

MRSA methicillin-resistant *Staphylococcus aureus, TSLP* thymic stromal lymphopoietin, *MnSOD* manganese superoxide dismutase

3.1.1 Staphylococci

Staphylococcus epidermidis (*S. epidermidis*) is the most abundant commensal bacteria detected in healthy human skin. A recent study showed that a decrease in cutaneous bacterial diversity (microbiome) leads to dominance of *S. aureus* and *S. epidermidis*, in flares of AD (Kong et al. 2012). *S. epidermidis* differentially upregulates the expression of IL-17 in the skin, suggesting a specific role in skin immunity (Ong 2014; Naik et al. 2012).

S. aureus is an important bacterial factor in the exacerbation of AD. *S. aureus* colonizes the skin of the majority (90%) of AD patients, at both lesional and non-lesional sites. As mentioned above, analyses of the cutaneous microbiome revealed that flares of AD are characterized by almost complete colonization of lesional skin with *S. aureus*, while remissions and the skin of healthy individuals show a diversity in colonization with commensals (Park et al. 2013, 2016; Wollina 2017; Breuer et al. 2002; Kong et al. 2012). The colonization grade is associated with the severity of AD (Jun et al. 2017; Tauber et al. 2016). *S. aureus*-colonization is a result of skin barrier dysfunction caused by genetics and immune responses triggered by allergens, scratching, or both, and is increased in the presence of IL-4 and IL-13. Interleukin-4 and IL-13 enhance *S. aureus* binding to AD-skin by inducing *S. aureus*-adhesins and reduce *S. aureus*-killing by inhibiting the production of keratinocytes derived antimicrobial peptides required for control of *S. aureus*-abundance (Huang et al. 2017).

S. aureus is a gram-positive bacterium that produces exotoxins with superantigenic properties and can cause life-threatening infections. Approximately 50–80% of the colonizing *S. aureus* in AD patients produce these enterotoxins (Park et al. 2013, 2016; Wollina 2017; Breuer et al. 2002; Ong and Leung 2016). Staphylococcus enterotoxins (SE) (A, B, C, D) are considered superantigens because of their ability to bind to class II MHC molecules on APCs and stimulate large populations of T cells. SEs contribute to an increase of allergic skin inflammation in AD by penetrating the skin barrier (Lin et al. 2007). Antigen presenting cells opsonize SEs to CD4$^+$ T cells, which polyclonally activate approximately 20% T cells via the variable β-chain of the T-cell receptor. This results in T cell-mediated inflammation in AD-lesions (Ong and Leung 2016).

Methicillin-Resistant *Staphylococcus aureus* (MRSA)
The prevalence of MRSA-colonisation in AD-patients is 4–10 times higher than the rate of MRSA-colonisation in the general population, and the presence of MRSA is significantly higher in moderate to severe AD, as compared to mild AD. MRSA produces significantly higher number of superantigens than S. aureus, which augments its potential to cause infection and severe cutaneous inflammation in AD (Ong and Leung 2016).

3.1.2   Viral Infections

Herpes Simplex Virus and Vaccinia Virus
S. aureus produces α-toxins, which may be particularly virulent to filaggrin-deficient keratinocytes that lack sphingomyelinase (Fig. 1), an enzyme that is required to cleave α-toxin receptor. α-toxin may also increase the risk of viral infections including herpes simplex virus and vaccinia virus in AD, resulting in life threatening complications (eczema herpeticum and eczema vaccinatum) (Ong 2014; Wollenberg 2012).

Coxsackievirus
Infection with Coxsackievirus A6 typically leads to hand-foot-mouth disease in children but can also lead to other atypical manifestations, including Giannotti-Crosti syndrome and eczema coxsackium, which is concentrated in areas of active AD (Mathes et al. 2013; Feder et al. 2014). It is not clear if in the latter the virus is an aetiological factor or a complication of AD. Of note, the first episode of AD is often seen during viral infections in childhood, such as varicella.

3.1.3   Fungi

Malassezia Species
The interaction between multiple *Malassezia* species and the immune system is linked to skin inflammation in AD. *Malassezia sympodialis* allergens Mala s

11 (manganese-dependent superoxide dismutase), and Mala s 13, (thioredoxin; Fig. 1), are structurally homologue to their corresponding human proteins. This leads to cross-reactivity of autoreactive T cells between the human enzyme, contributing to skin inflammation in AD (Balaji et al. 2011; Schmid-Grendelmeier et al. 2005).

Malassezia globosa protein MGL_1304, can be detected in the sweat of AD-patients and induces histamine release from basophils (Hiragun et al. 2013). This protein is an important allergen in AD and can cause type-I hypersensitivity. The severity of AD can also correlate to the degree of MGL_1304-sensitization (Hiragun et al. 2014).

Malassezia spp. produce a variety of immunogenic proteins that elicit the production of specific IgE-antibodies and may induce the release of pro-inflammatory cytokines (Glatz et al. 2015). These cytokines and the IgE-antibodies may contribute to skin inflammation in AD.

3.2 Allergic Asthma

Following food allergies with or without AD, asthma is the second manifestation of the atopic march. Episodic wheeze occurs in about 30% of all children, while persistent asthma occurs in about 10% of all children and 5% of adults (Thomsen 2015). Asthma was already described in ancient Egyptian times and probably before that. The Georg Ebers Papyrus, discovered in Egypt in the 1870s, contains prescriptions written in hieroglyphics for over 700 remedies (Wahn 2014). One of the ancient Egyptian remedies was to heat a mixture of herbs on bricks and inhale their fumes. The term Asthma is delineated from the Greek verb *"aazein,"* to pant or to exhale with the open mouth. It is first mentioned in The Iliad (Homerus), about the siege of Troy. As a medical term, it was already used by Hippocrates in the Corpus Hippocraticum, but it could have been both a clinical entity, or a symptom. At the present time however, asthma is considered a heterogeneous disease with different endotypes: early-onset allergic asthma, adult-onset (severe) eosinophilic asthma, late onset T_H2 asthma, late-onset non-allergic asthma of the elderly, virus-induced asthma, exercise induced asthma, obesity-related asthma, non-eosinophilic asthma, brittle asthma, menstrual-linked asthma, and smoking associated asthma (Koczulla et al. 2017). This deliberate endotyping of asthma compared to AD allows the development of tailored guidelines and treatment approaches. Valid biomarkers would be extremely helpful keys to disease endotyping and pose a current research topic (Chaker et al. 2017; Zissler et al. 2016). This paragraph will review the mechanism of microbial triggers in allergic asthma (see Table 2).

3.2.1 Bacterial Triggers in Allergic Asthma

Several bacteria have been linked to the exacerbation of asthma (Earl et al. 2015; Lan et al. 2016; Green et al. 2014; Papadopoulos et al. 2011; Guilbert and Denlinger

2010). Infections with *Chlamydophila pneumoniae* and *Mycoplasma pneumoniae* have been reported to have an association with asthma exacerbation (Endo et al. 2017; Iramain et al. 2016; Lieberman et al. 2003; Hong 2012). Moreover, colonization or infection with *S. aureus, Haemophilus influenzae, Streptococcus pneumoniae* and *Moraxella pneumoniae* have also been linked to trigger asthma exacerbation. It has been hypothesized that the effect of bacteria on T_H2 response could play an important role (Lan et al. 2016; Deo et al. 2010).

3.2.2 Viral Triggers in Allergic Asthma

Respiratory syncytial virus (RSV), human rhinovirus (HRV), and influenza virus (IFV) in infancy are associated with the development of allergic asthma and are major causes of asthma exacerbation (Busse et al. 2010; Iikura et al. 2015; Pelaia et al. 2006). RSV, HRV, and IFV may cause asthma exacerbation due to their induction of epithelial cell-derived alarmines such as IL-33, thymic stromal lymphopoietin (TSLP), and IL-25. These cytokines are strong inducers of type-2 immune responses via innate lymphoid type 2 and dendritic cells (Lan et al. 2016; Beale et al. 2014).

4 Microbial Triggers in Autoallergy

An especially difficult to treat population are patients with chronic and severe AD. Already in 1926 Van Leeuwen et al. raised the theory that auto-sensitivity against human skin dander plays a role in atopic patients. These findings were reproduced by Hampton and Cooke in the 1940s and since the late 1990s Valenta and others demonstrated the presence of autoreactive (auto)-IgE in the sera of AD patients. Auto-IgE titers correlate with AD severity (Natter et al. 1998; Valenta et al. 1998). It was postulated that IgE-autosensitisation and autoreactive T cells can perpetuate and aggravate AD via autoimmune inflammation and such an "autoallergy" might constitute a novel endotype of AD (Bieber et al. 2017; Hradetzky et al. 2015) (Fig. 2).

Thus, the term "autoallergy" describes an autoimmune process accompanying atopic dermatitis, with auto-reactive IgE as a hallmark (see Table 3). Microbial factors are suspected as trigger factors, including IgE-cross-reactivity of human Manganese superoxide dismutase (MnSOD) or Thioredoxin with IgE directed against the respective homologue proteins derived from *Aspergillus fumigatus* and *M. sympodialis* (Hom s 2 and Hom s 3) (Hradetzky et al. 2015; Schmid-Grendelmeier et al. 2005; Crameri et al. 1996; Balaji et al. 2011). The mechanism of molecular mimicry and cross-reactivity between human and microbial proteins were described in the previous section. Moreover, a cellular T_H1 immune response plus an IL17/IL22-dominated immune response can be elicited by Hom s 2. In the context of inflammation and itch, auto-allergens may be continuously released from

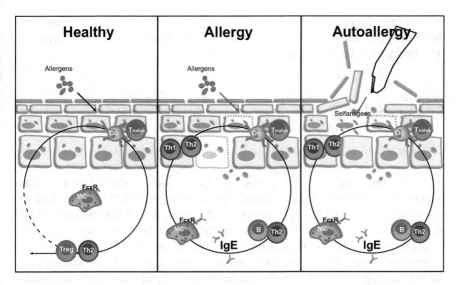

Fig. 2 Vicious cycle of autoallergy. The auto-allergen may replace the need for allergens to drive tissue inflammation and convert seasonal allergies into chronic disorders. *Left panel*: in non-atopic healthy subjects, allergens are phagocytised by Langerhans cells (a dendritic cell subset located in the epidermis), and a regulatory T cell response is induced in the draining lymph nodes, inhibiting allergic TH2 sensitization. *Middle panel*: in patients suffering from seasonal atopic diseases, a lack of regulatory T cell response leads to the formation of allergen-specific TH2 cells and plasma cells that produce allergen-specific IgE-molecules. Because self-antigens are presented under inflammatory conditions, auto-IgE can be induced. *Right panel*: in autosensitized patients, autoallergy with self-antigen-specific IgE and TH1/TH2 cells maintain inflammation, which is constantly triggered by the traumatic release of self-antigens via scratching (Picture courtesy of C. Schmidt-Weber)

Table 3 Examples of IgE-reactive auto-allergens. Adapted from Hradetzky et al. 2015

Auto-allergens with cross-reactivity to exogenous allergens	Auto-allergens without known cross-reactivity to exogenous allergens	"Classical" auto-antigens with IgE reactivity
Profilin (Pan-allergen)	Hom s 1 (SART-1)	Collagen (Rheumatoid arthritis)
Manganese superoxide dismutase (A. fumigatus)	Hom s 2 (α-NAC)	Thyreoperoxidase (Hashimoto's thyroiditis)
Thioredoxin (M. sympodialis)	Hom s 3 (BCL7B, Oncogen)	BP180, BP230 (Bullous pemphigoid)
Acidic ribosomal P2 protein (A. fumigatus)	Hom s 4 (Calcium-Binding)	Desmoglein 3 (Pemphigus vulgaris)
Serum albumin (high conservation among mammals)	Hom s 5 (Cytokeratin II)	Nuclear antigens (Systemic lupus erythematosus)

the skin cells, such as keratinocytes and fibroblasts, by scratching or wounding and thereby maintaining a vicious autoallergic cycle of inflammation in autoallergic AD (Hradetzky et al. 2015; Mittermann et al. 2008).

5 Conclusion

Severe reactions in allergic asthma and the type 2 dominated inflammation in atopic dermatitis can be triggered or aggravated by multiple microbial factors, including bacteria, viruses, and fungi. Of special concern is *S. aureus*, which is of significant importance in AD and allergic asthma, especially strains of MRSA. In asthma the aggravating impact of multiple bacteria and viruses is well-documented, and in atopic dermatitis viral complications can be life threatening, such as eczema herpecticum. In head-and-neck eczema the colonization with malassezia furfur is a therapeutic target. In autoimmunity, microbial infections can be the starting point for the development of disease by various molecular pathways, including mainly molecular mimicry or bystander activation. Here frequent pathogens, such as EBV and Coxsackie viruses can induce the pathophysiologic catastrophy of "break of self tolerance." Current research is geared towards detection of protective microbial species and the physiologic time window of commensal microbial colonization. Adequate anti-microbial therapies and prevention strategies thus are important for prevention or amelioration of allergic- and autoimmune diseases. Finally, it is hypothesized that autoallergy could initially be initiated by mechanisms such as molecular mimicry and may thereby lead into a chronic disorder that has become secondarily allergen independent. Here again fungal malassezia species are suspected as eliciting pathogens. Taken together, microbes of all classes, bacteria, viruses and fungi, have a strong impact in severe allergic disease and therefore need to be considered as targets in prevention and therapy of allergic diseases. Future research needs to address beneficial and harmful effects of microbes in the origination, elicitation and chronification of severe allergic conditions.

Conflicts of Interest None

Funding Sources None

References

Baker BS (2006) The role of microorganisms in atopic dermatitis. Clin Exp Immunol 144(1):1–9. https://doi.org/10.1111/j.1365-2249.2005.02980.x
Balaji H, Heratizadeh A, Wichmann K, Niebuhr M, Crameri R, Scheynius A, Werfel T (2011) Malassezia sympodialis thioredoxin-specific T cells are highly cross-reactive to human thioredoxin in atopic dermatitis. J Allergy Clin Immunol 128(1):92–99.e94. https://doi.org/10.1016/j.jaci.2011.02.043

Beale J, Jayaraman A, Jackson DJ, Macintyre JD, Edwards MR, Walton RP, Zhu J, Ching YM, Shamji B, Edwards M, Westwick J, Cousins DJ, Hwang YY, McKenzie A, Johnston SL, Bartlett NW (2014) Rhinovirus-induced IL-25 in asthma exacerbation drives type 2 immunity and allergic pulmonary inflammation. Sci Trans Med 6(256):256ra134. https://doi.org/10.1126/scitranslmed.3009124

Bieber T (2008) Atopic dermatitis. N Engl J Med 358(14):1483–1494. https://doi.org/10.1056/NEJMra074081

Bieber T, D'Erme AM, Akdis CA, Traidl-Hoffmann C, Lauener R, Schappi G, Schmid-Grendelmeier P (2017) Clinical phenotypes and endophenotypes of atopic dermatitis: where are we, and where should we go? J Allergy Clin Immunol 139(4s):S58–s64. https://doi.org/10.1016/j.jaci.2017.01.008

Breuer K, HA S, Kapp A, Werfel T (2002) Staphylococcus aureus: colonizing features and influence of an antibacterial treatment in adults with atopic dermatitis. Br J Dermatol 147 (1):55–61

Busse WW, Lemanske RF Jr, Gern JE (2010) Role of viral respiratory infections in asthma and asthma exacerbations. Lancet 376(9743):826–834. https://doi.org/10.1016/s0140-6736(10)61380-3

Chaker AM, Zissler UM, Wagenmann M, Schmidt-Weber C (2017) Biomarkers in allergic airway disease: simply complex. ORL J Oto-rhino-laryngol Relat Spec 79(1–2):72–77. https://doi.org/10.1159/000455725

Crameri R, Faith A, Hemmann S, Jaussi R, Ismail C, Menz G, Blaser K (1996) Humoral and cell-mediated autoimmunity in allergy to Aspergillus fumigatus. J Exp Med 184(1):265–270

D'Elios MM, Bergman MP, Amedei A, Appelmelk BJ, Del Prete G (2004) Helicobacter pylori and gastric autoimmunity. Microbes Infect 6(15):1395–1401. https://doi.org/10.1016/j.micinf.2004.10.001

Deo SS, Mistry KJ, Kakade AM, Niphadkar PV (2010) Role played by Th2 type cytokines in IgE mediated allergy and asthma. Lung India 27(2):66–71. https://doi.org/10.4103/0970-2113.63609

Earl CS, An SQ, Ryan RP (2015) The changing face of asthma and its relation with microbes. Trends Microbiol 23(7):408–418. https://doi.org/10.1016/j.tim.2015.03.005

Eaton WW, Rose NR, Kalaydjian A, Pedersen MG, Mortensen PB (2007) Epidemiology of autoimmune diseases in Denmark. J Autoimmun 29(1):1–9

Endo Y, Shirai T, Saigusa M, Mochizuki E (2017) Severe acute asthma caused by Chlamydophila pneumoniae infection. Respirol Case Rep 5(4):e00239. https://doi.org/10.1002/rcr2.239

Ercolini AM, Miller SD (2009) The role of infections in autoimmune disease. Clin Exp Immunol 155(1):1–15. https://doi.org/10.1111/j.1365-2249.2008.03834.x

Feder HM Jr, Bennett N, Modlin JF (2014) Atypical hand, foot, and mouth disease: a vesiculobullous eruption caused by Coxsackie virus A6. Lancet Infect Dis 14(1):83–86. https://doi.org/10.1016/s1473-3099(13)70264-0

Glatz M, Bosshard PP, Hoetzenecker W, Schmid-Grendelmeier P (2015) The role of Malassezia spp. in atopic dermatitis. J Clin Med 4(6):1217–1228. https://doi.org/10.3390/jcm4061217

Green BJ, Wiriyachaiporn S, Grainge C, Rogers GB, Kehagia V, Lau L, Carroll MP, Bruce KD, Howarth PH (2014) Potentially pathogenic airway bacteria and neutrophilic inflammation in treatment resistant severe asthma. PLoS one 9(6):e100645. https://doi.org/10.1371/journal.pone.0100645

Guilbert TW, Denlinger LC (2010) Role of infection in the development and exacerbation of asthma. Expert Rev Respir Med 4(1):71–83. https://doi.org/10.1586/ers.09.60

Hasni SA (2012) Role of Helicobacter pylori infection in autoimmune diseases. Curr Opin Rheumatol 24(4):429–434. https://doi.org/10.1097/BOR.0b013e3283542d0b

Hiragun T, Ishii K, Hiragun M, Suzuki H, Kan T, Mihara S, Yanase Y, Bartels J, Schroder JM, Hide M (2013) Fungal protein MGL_1304 in sweat is an allergen for atopic dermatitis patients. J Allergy Clin Immunol 132(3):608–615.e604. https://doi.org/10.1016/j.jaci.2013.03.047

Hiragun M, Hiragun T, Ishii K, Suzuki H, Tanaka A, Yanase Y, Mihara S, Haruta Y, Kohno N, Hide M (2014) Elevated serum IgE against MGL_1304 in patients with atopic dermatitis and cholinergic urticaria. Allergol Int 63(1):83–93. https://doi.org/10.2332/allergolint.13-OA-0611

Hong SJ (2012) The role of Mycoplasma pneumoniae infection in asthma. Allergy, Asthma Immunol Res 4(2):59–61. https://doi.org/10.4168/aair.2012.4.2.59

Hradetzky S, Werfel T, Rosner LM (2015) Autoallergy in atopic dermatitis. Allergo J Int 24(1):16–22. https://doi.org/10.1007/s40629-015-0037-5

Huang YJ, Marsland BJ, Bunyavanich S, O'Mahony L, Leung DY, Muraro A, Fleisher TA (2017) The microbiome in allergic disease: current understanding and future opportunities-2017 PRACTALL document of the American Academy of Allergy, Asthma & Immunology and the European Academy of Allergy and Clinical Immunology. J Allergy Clin Immunol 139 (4):1099–1110. https://doi.org/10.1016/j.jaci.2017.02.007

Iikura M, Hojo M, Koketsu R, Watanabe S, Sato A, Chino H, Ro S, Masaki H, Hirashima J, Ishii S, Naka G, Takasaki J, Izumi S, Kobayashi N, Yamaguchi S, Nakae S, Sugiyama H (2015) The importance of bacterial and viral infections associated with adult asthma exacerbations in clinical practice. PLoS one 10(4):e0123584. https://doi.org/10.1371/journal.pone.0123584

Iramain R, De Jesus R, Spitters C, Jara A, Jimenez J, Bogado N, Cardozo L (2016) Chlamydia pneumoniae, and mycoplasma pneumoniae: are they related to severe asthma in childhood? J Asthma 53(6):618–621. https://doi.org/10.3109/02770903.2015.1116085

Jun SH, Lee JH, Kim SI, Choi CW, Park TI, Jung HR, Cho JW, Kim SH, Lee JC (2017) Staphylococcus aureus-derived membrane vesicles exacerbate skin inflammation in atopic dermatitis. Clin Exp Allergy 47(1):85–96. https://doi.org/10.1111/cea.12851

Kaesler S, Volz T, Skabytska Y, Koberle M, Hein U, Chen KM, Guenova E, Wolbing F, Rocken M, Biedermann T (2014) Toll-like receptor 2 ligands promote chronic atopic dermatitis through IL-4-mediated suppression of IL-10. J Allergy Clin Immunol 134(1):92–99. https://doi.org/10.1016/j.jaci.2014.02.017

Koczulla AR, Vogelmeier CF, Garn H, Renz H (2017) New concepts in asthma: clinical phenotypes and pathophysiological mechanisms. Drug Discov Today 22(2):388–396. https://doi.org/10.1016/j.drudis.2016.11.008

Kong HH, Oh J, Deming C, Conlan S, Grice EA, Beatson MA, Nomicos E, Polley EC, Komarow HD, Murray PR, Turner ML, Segre JA (2012) Temporal shifts in the skin microbiome associated with disease flares and treatment in children with atopic dermatitis. Genome Res 22(5):850–859. https://doi.org/10.1101/gr.131029.111

Lan F, Zhang N, Gevaert E, Zhang L, Bachert C (2016) Viruses and bacteria in Th2-biased allergic airway disease. Allergy 71(10):1381–1392. https://doi.org/10.1111/all.12934

Lieberman D, Lieberman D, Printz S, Ben-Yaakov M, Lazarovich Z, Ohana B, Friedman MG, Dvoskin B, Leinonen M, Boldur I (2003) Atypical pathogen infection in adults with acute exacerbation of bronchial asthma. Am J Respir Crit Care Med 167(3):406–410. https://doi.org/10.1164/rccm.200209-996OC

Lin YT, Wang CT, Chiang BL (2007) Role of bacterial pathogens in atopic dermatitis. Clin Rev Allergy Immunol 33(3):167–177. https://doi.org/10.1007/s12016-007-0044-5

Mathes EF, Oza V, Frieden IJ, Cordoro KM, Yagi S, Howard R, Kristal L, Ginocchio CC, Schaffer J, Maguiness S, Bayliss S, Lara-Corrales I, Garcia-Romero MT, Kelly D, Salas M, Oberste MS, Nix WA, Glaser C, Antaya R (2013) "Eczema coxsackium" and unusual cutaneous findings in an enterovirus outbreak. Pediatrics 132(1):e149–e157. https://doi.org/10.1542/peds.2012-3175

Mittermann I, Reininger R, Zimmermann M, Gangl K, Reisinger J, Aichberger KJ, Greisenegger EK, Niederberger V, Seipelt J, Bohle B, Kopp T, Akdis CA, Spitzauer S, Valent P, Valenta R (2008) The IgE-reactive autoantigen Hom s 2 induces damage of respiratory epithelial cells and keratinocytes via induction of IFN-gamma. J Investig Dermatol 128(6):1451–1459. https://doi.org/10.1038/sj.jid.5701195

Mori M, Kuwabara S, Miyake M, Noda M, Kuroki H, Kanno H, Ogawara K, Hattori T (2000) Haemophilus influenzae infection and Guillain-Barre syndrome. Brain 123(Pt 10):2171–2178

Naik S, Bouladoux N, Wilhelm C, Molloy MJ, Salcedo R, Kastenmuller W, Deming C, Quinones M, Koo L, Conlan S, Spencer S, Hall JA, Dzutsev A, Kong H, Campbell DJ, Trinchieri G, Segre JA, Belkaid Y (2012) Compartmentalized control of skin immunity by resident commensals. Science 337(6098):1115–1119. https://doi.org/10.1126/science.1225152

Natter S, Seiberler S, Hufnagl P, Binder BR, Hirschl AM, Ring J, Abeck D, Schmidt T, Valent P, Valenta R (1998) Isolation of cDNA clones coding for IgE autoantigens with serum IgE from atopic dermatitis patients. FASEB J 12(14):1559–1569

Navarrete-Dechent C, Perez-Mateluna G, Silva-Valenzuela S, Vera-Kellet C, Borzutzky A (2016) Humoral and cellular autoreactivity to epidermal proteins in atopic dermatitis. Arch Immunol Ther Exp 64(6):435–442. https://doi.org/10.1007/s00005-016-0400-3

Ong PY (2014) New insights in the pathogenesis of atopic dermatitis. Pediatr Res 75 (1-2):171–175. https://doi.org/10.1038/pr.2013.196

Ong PY, Leung DY (2016) Bacterial and viral infections in atopic dermatitis: a comprehensive review. Clin Rev Allergy Immunol 51(3):329–337. https://doi.org/10.1007/s12016-016-8548-5

Papadopoulos NG, Christodoulou I, Rohde G, Agache I, Almqvist C, Bruno A, Bonini S, Bont L, Bossios A, Bousquet J, Braido F, Brusselle G, Canonica GW, Carlsen KH, Chanez P, Fokkens WJ, Garcia-Garcia M, Gjomarkaj M, Haahtela T, Holgate ST, Johnston SL, Konstantinou G, Kowalski M, Lewandowska-Polak A, Lodrup-Carlsen K, Makela M, Malkusova I, Mullol J, Nieto A, Eller E, Ozdemir C, Panzner P, Popov T, Psarras S, Roumpedaki E, Rukhadze M, Stipic-Markovic A, Todo Bom A, Toskala E, van Cauwenberge P, van Drunen C, Watelet JB, Xatzipsalti M, Xepapadaki P, Zuberbier T (2011) Viruses and bacteria in acute asthma exacerbations—a GA(2) LEN-DARE systematic review. Allergy 66(4):458–468. https://doi.org/10.1111/j.1398-9995.2010.02505.x

Park HY, Kim CR, Huh IS, Jung MY, Seo EY, Park JH, Lee DY, Yang JM (2013) Staphylococcus aureus colonization in acute and chronic skin lesions of patients with atopic dermatitis. Ann Dermatol 25(4):410–416. https://doi.org/10.5021/ad.2013.25.4.410

Park KD, Pak SC, Park KK (2016) The pathogenetic effect of natural and bacterial toxins on atopic dermatitis. Toxins 9(1):3. https://doi.org/10.3390/toxins9010003

Pelaia G, Vatrella A, Gallelli L, Renda T, Cazzola M, Maselli R, Marsico SA (2006) Respiratory infections and asthma. Respir Med 100(5):775–784. https://doi.org/10.1016/j.rmed.2005.08.025

Ring J (1985) 1st description of an "atopic family anamnesis" in the Julio-Claudian imperial house: Augustus, Claudius, Britannicus. Der Hautarzt; Zeitschrift fur Dermatologie, Venerologie, und verwandte Gebiete 36(8):470–471

Ring J (2016) Atopic dermatitis – eczema. Springer, Cham. https://doi.org/10.1007/978-3-319-22243-1

Ring J, Alomar A, Bieber T, Deleuran M, Fink-Wagner A, Gelmetti C, Gieler U, Lipozencic J, Luger T, Oranje AP, Schafer T, Schwennesen T, Seidenari S, Simon D, Stander S, Stingl G, Szalai S, Szepietowski JC, Taieb A, Werfel T, Wollenberg A, Darsow U (2012) Guidelines for treatment of atopic eczema (atopic dermatitis) part I. J Eur Acad Dermatol Venereol 26 (8):1045–1060. https://doi.org/10.1111/j.1468-3083.2012.04635.x

Schmid-Grendelmeier P, Fluckiger S, Disch R, Trautmann A, Wuthrich B, Blaser K, Scheynius A, Crameri R (2005) IgE-mediated and T cell-mediated autoimmunity against manganese superoxide dismutase in atopic dermatitis. J Allergy Clin Immunol 115(5):1068–1075. https://doi.org/10.1016/j.jaci.2005.01.065

Smyk DS, Koutsoumpas AL, Mytilinaiou MG, Rigopoulou EI, Sakkas LI, Bogdanos DP (2014) Helicobacter pylori and autoimmune disease: cause or bystander. World J Gastroenterol 20 (3):613–629. https://doi.org/10.3748/wjg.v20.i3.613

Tauber M, Balica S, Hsu CY, Jean-Decoster C, Lauze C, Redoules D, Viode C, Schmitt AM, Serre G, Simon M, Paul CF (2016) Staphylococcus aureus density on lesional and nonlesional skin is strongly associated with disease severity in atopic dermatitis. J Allergy Clin Immunol 137(4):1272–1274.e1271-1273. https://doi.org/10.1016/j.jaci.2015.07.052

Thomsen SF (2015) Epidemiology and natural history of atopic diseases. Eur Clin Respir J 2:24642. https://doi.org/10.3402/ecrj.v2.24642

Valenta R, Natter S, Seiberler S, Wichlas S, Maurer D, Hess M, Pavelka M, Grote M, Ferreira F, Szepfalusi Z, Valent P, Stingl G (1998) Molecular characterization of an autoallergen, Hom s 1, identified by serum IgE from atopic dermatitis patients. J Investig Dermatol 111 (6):1178–1183. https://doi.org/10.1046/j.1523-1747.1998.00413.x

Vojdani A (2014) A potential link between environmental triggers and autoimmunity. Autoimmune Dis 2014:437231. https://doi.org/10.1155/2014/437231

Wahn U (2014) K.C. Bergmann, J. Ring (Ed.): History of allergy. Pediatr Allergy Immunol 25(6): i–i. https://doi.org/10.1111/pai.12292

Wallach D, Taieb A (2014) Atopic dermatitis/atopic eczema. Chem Immunol Allergy 100:81–96. https://doi.org/10.1159/000358606

Werfel T, Heratizadeh A, Niebuhr M, Kapp A, Roesner LM, Karch A, Erpenbeck VJ, Losche C, Jung T, Krug N, Badorrek P, Hohlfeld JM (2015) Exacerbation of atopic dermatitis on grass pollen exposure in an environmental challenge chamber. J Allergy Clin Immunol 136 (1):96–103.e109. https://doi.org/10.1016/j.jaci.2015.04.015

Wickham S, Carr DJ (2004) Molecular mimicry versus bystander activation: herpetic stromal keratitis. Autoimmunity 37(5):393–397. https://doi.org/10.1080/08916930410001713106

Wollenberg A (2012) Eczema herpeticum. Chem Immunol Allergy 96:89–95. https://doi.org/10.1159/000331892

Wollina U (2017) Microbiome in atopic dermatitis. Clin Cosmet Investig Dermatol 10:51–56. https://doi.org/10.2147/ccid.s130013

Zissler UM, Esser-von Bieren J, Jakwerth CA, Chaker AM, Schmidt-Weber CB (2016) Current and future biomarkers in allergic asthma. Allergy 71(4):475–494. https://doi.org/10.1111/all.12828

Bacterial Allergens

Gómez-Gascón Lidia and Barbara M. Bröker

Abstract According to the hygiene hypothesis, bacterial infection as well as exposure to a complex microbial environment early in life confer protection from allergy. On the flip side of the coin, colonization and infection by certain bacterial species have been associated with an increased risk of allergy development as well as with exacerbations of allergic symptoms. Moreover, bacteria themselves and their products can become targets of allergic immune responses, eliciting the generation of specific Th2 cells and IgE. Some bacterial factors are even able to trigger allergic inflammation in animal models, suggesting that they are true allergens. This review summarizes the state of the art regarding pro-allergenic properties of bacteria.

1 Introduction

Allergies or hypersensitivity reactions are complex diseases featuring tissue inflammation triggered by exaggerated responses of the immune system when contacting certain environmental substances, called allergens. The most common allergens groups are pollens, environmental fungi, dust mites, animal dander as well as certain foods or drugs (Ipci et al. 2016). Allergies are characterized by type 2 inflammation, including increased numbers of Th2 cells, which upon allergen exposure release IL-4, IL-5 and IL-13, as well as by allergen-specific IgE (Wills-Karp et al. 2012; Barnes 2009). Asthma, an allergic reaction of the respiratory system, is one of the most common chronic diseases throughout the world. About 300 million people currently suffer from asthma (Masoli et al. 2004). Two forms of asthma can be distinguished, the allergic (exogenous) asthma and the so-called non-allergic (endogenous) asthma, which is also known as idiopathic or intrinsic asthma. Both forms are mediated by IgE and characterized by the release of mast cell products which results in bronchial constriction and impaired

G.-G. Lidia (✉) • B.M. Bröker
Department of Immunology, University Medicine Greifswald, Greifswald, Germany
e-mail: lidia.gomezgascon@uni-greifswald.de; broeker@uni-greifswald.de

© Springer International Publishing AG 2017   27
C.B. Schmidt-Weber (ed.), *Allergy Prevention and Exacerbation*, Birkhäuser
Advances in Infectious Diseases, https://doi.org/10.1007/978-3-319-69968-4_3

respiration. Allergic asthma is triggered and maintained by typical inhalation allergens, such as pollens, animal and herbal dusts, animal proteins and chemicals. On the contrary, non-allergic asthma is not caused by known allergens. Infections of the respiratory system, intolerance to drugs, and exposure to poisons are discussed as causes for this asthma form. Nevertheless, in at least 10% of the endogenous, non-allergic asthma cases the causative agent is still unknown. These are referred to as cases of idiopathic or intrinsic asthma (Barnes 2009, 2012).

Cystic fibrosis (CF), is an autosomal-recessive disorder caused by mutations in the CF transmembrane regulator gene, which encodes a chloride channel. This causes a defect in the function of ciliated epithelial cells impairing airway clearance mechanisms, which predisposes CF patients to high density colonization and infection by a number of bacteria, in particular *Staphylococcus aureus* and *Pseusomonas aeruginosa*. Several groups suggest that the T cells responses are dysregulated in CF patients which may play an important role in inducing and maintaining the lung inflammation in these individuals (Upritchard et al. 2008; Tiringer et al. 2013; Hector et al. 2015).

Usually bacteria are connected with protection from allergy. Infections with pathogens such as *Mycobacterium tuberculosis* or respiratory viruses enhance Th1 responses and limit Th2 responses, counteracting allergy development. A well-known example of this is the protection from asthma and allergy that is conferred by vaccination with Bacillus Calmette- Guérin (BCG), a mycobacterial organism. In 1997, an inverse correlation between tuberculin skin responses and atopic diseases was reported, and numerous studies have since confirmed that tuberculosis as well as vaccination with BCG or other mycobacteria reduces the prevalence of allergy, both in animal and human studies (Shirakawa et al. 1997; Umetsu et al. 2002; Kim et al. 2014; Choi and Koh 2002, 2003; Choi 2014). Such findings inspired Jean François Bach to propose that "the main factor in the increased prevalence of these diseases [allergies and autoimmune diseases, see also chapter "Microbial Triggers in Autoimmunity, Severe Allergy, and Autoallergy"] in industrialized countries is the reduction in the incidence of infectious diseases in those countries over the past three decades", which is often referred to as the original hygiene hypothesis (Bach 2002).

Mycobacteria can modulate the immune system. On the one hand, they are potent inducers of Th1 responses and INF-γ, which is associated with the suppression of asthma at least initially (Yoshida et al. 2002); and on the other hand, they elicit regulatory T cell (Treg) responses, probably the main mechanism of protection from allergies (Kim et al. 2014; Choi and Koh 2002; Choi 2014). Tregs release the anti-inflammatory cytokines IL-10 and TGF-β, and they may also suppress inflammatory effector T cells (Teffs) by cell-cell contact through CTLA-4 or other surface receptors (Kwon et al. 2010; Read et al. 2000). More recently, the role of the commensal microflora in inflammatory homeostasis and immune regulation moved into the focus of attention. The hygiene hypothesis was modified: Not just infections but also exposure to innocuous exogenous and endogenous microorganisms early in life protects from allergy (Schaub et al. 2006). The proposed mechanisms of protection are similar to those discussed in relation to infection: a shift in the Th1–Th2 immune balance and the induction of Treg-mediated tolerance to allergens (Umetsu et al. 2002; Seroogy and Gern 2005; Bacher et al. 2016). Hence, variations of the microbiome, both in number and diversity of bacteria, may significantly affect in the incidence of allergic manifestations, including allergic airway diseases like asthma (Ramsey and Celedon 2005; Hilty et al. 2010; Edwards et al. 2012; Atkinson 2013; Ribet and Cossart 2015; Ipci et al. 2016). Recently, Birzele and colleagues reported a highly significant inverse association between bacterial diversity in the mattress dust and asthma that was even more pronounced than the correlation with the nasal microbiome. They suggest that colonization of the intestine or inhalation of microbial metabolites with immunomodulatory properties could contribute to the inflammatory homeostasis and asthma protection (Birzele et al. 2017). In consequence, probiotics, selected microbes of the commensal gut microflora, are being tested for their capacity of enhancing immune tolerance (Umetsu et al. 2002; Kalliomäki et al. 2001; Kwon et al. 2010).

However, common bacterial infections can also have an important role in asthma induction and exacerbation (Welliver and Duffy 1993; Emre et al. 1995; Seggev et al. 1996; Edwards et al. 2012). Already in the 1980s, some authors demonstrated a correlation between bacterial colonization and allergic diseases. Specific antibacterial serum IgE was also studied as well as the ability of the bacteria to induce histamine release from human basophil leukocytes and mast cells, (Pauwels et al. 1980; Tee and Pepys 1982; Welliver and Duffy 1993; Emre et al. 1995; Seggev et al., 1996; Larsen et al. 1998; Kjaergard et al. 1996). Once asthma is present, bacterial infection and colonization may exacerbate disease symptoms alone or in conjunction with viruses such as human rhinovirus or respiratory syncytial virus (Barnes 2009; Edwards et al. 2012; Darveaux and Lemanske 2014).

The underlying mechanisms comprise the induction of lung inflammation as well as tissue injury and repair. The bacteria can infect airway epithelial cells causing cell death and/or increasing epithelial permeability. The damage to the integrity of the airways facilitates microbial invasion as well as exposure of the immune system to environmental pollutants and allergens. On the other hand, respiratory pathogens can induce mediators that are required for airway repair resulting in airway remodeling with thickening of the airway walls and impairment of lung function. Fibroblast growth factors and vascular endothelial growth factors are involved in angiogenesis, airway smooth muscle proliferation and hypertrophy, collagen and fibronectin deposition, as well as in the generation of new lymphatic vessels (Edwards et al. 2012). Besides such indirect effects, bacteria and their products are also discussed as allergens, eliciting specific IgE antibodies that bind to epitopes expressed by the infectious agents (Hales et al. 2012, 2008; Barnes 2009; Ikezawa 2001; Stentzel et al. 2016). Moreover, bacteria have been shown to promote the degranulation of mast cells and basophils, either mediated by IgE or IgE independent (Ye et al. 2014; Kjaergard et al. 1996; Nakamura et al. 2013; Ahren et al. 2003; Clementsen et al. 1990). These mechanisms are depicted in Fig. 1.

The bacterial species *Haemophilus influenzae, Streptococcus pneumoniae, Streptococcus pyogenes* and *Staphylococcus aureus* have been associated with an increased risk of asthma, and also *Mycoplasma pneumoniae, Chlamydia pneumonia*

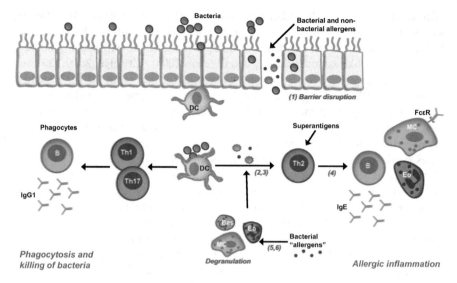

Fig. 1 Roles of bacteria in allergic diseases. Typically bacteria induce immune responses that are dominated by Th1 and Th17. These T cell subtypes facilitate phagocytosis and killing of the bacteria, crucial mechanisms of host defence against bacterial infections. Under some conditions, however, bacteria and their compounds can promote Th2 instead of Th1 immune responses, leading to allergic inflammation. Bacteria may benefit from this "allergic deviation", because it counteracts the anti-bacterial Th1 and Th17 immune responses. The following pro-allergenic activites have been described in bacteria: (1) Damage to the integrity of the epithelial airway barriers facilicing microbial invasion and intensifying the contact of inhalant allergens with the immune system; (2) Induction of differentiation of naïve T cells into Th2 cells; (3) Induction of Th2 cytokine release; (4) Elicitation of specific IgE antibodies; (5) Recruitment and activation of eosinophils and basophils; (6) Promotion of degranulation of mast cells and basophils by IgE-dependent and independent mechanisms. Abbreviations: B, B cell; Bas, basophil; DC, dendritic cell; Eo, eosinophil; FCεR, High affinity Immunoglobulin E receptor; MC, mast cell; Th, T helper cell

or *Chlamydia trachomatis* may play a role in asthma onset and exacerbations (Korppi 2009; Hilty et al. 2010; Edwards et al. 2012; Atkinson 2013; Darveaux and Lemanske 2014). Bacteria may benefit from an "allergic deviation" of the immune response, because it counteracts the development of Th1 and Th17 responses, which are effectively promoting bacterial elimination. In this sense, type 2 reactions contribute to bacterial immune evasion.

The aim of this review is to discuss the available data on the relationship between bacteria and allergic host immune responses with an emphasis on bacterial compounds that may act as causative agents, especially in allergic airway diseases (summarized in Table 1). The allergen platforms http://www.allergen.org/ and http://www.allergome.org/ as well as Medline (PubMed) and the search engine http://sci-hub.cc/ were searched for relevant articles.

Table 1 Bacteria involved in the initiation and exacerbations of Th2 immune responses

Bacteria	Associated allergic diseases	Evidence for a role in pathogenesis	Putative allergens	References
C. trachomatis	Chronic allergic airway diseases	Colonization associated with higher risk of asthma Specific-IgE in sera and BAL (patients)	MOMP, CrpA, POMP, HSP60 Unidentified proteins (250 KDa, 64 KDa)	Emre et al. (1995) Patel et al. (2012)
C. pneumoniae	Asthma Wheezing episodes	Colonization associated with higher risk of asthma Specific-IgE in sera and BAL(patients) Increased levels of IL-4 gene expression in BAL cells (murine model) Induction levels of IgE and IL-4 (human PBMCs from asthmatic patients)	MOMP, CrpA, POMP, HSP60, LBP Unidentified proteins (98 KDa, 78 KDa, 58–60 KDa, 36 KDa)	Emre et al. (1995) Larsen et al.(1998) Ikezawa et al. (2001) Wark et al. (2002) Webley et al. (2009) Chen et al. (2009) Hahn et al. (1991, 2008, 2012) Patel et al. (2012) Smith-Norowitz et al. (2016)
M. pneumoniae	Asthma	Colonization associated with higher risk of asthma Specific-IgE in sera and BAL(patients) High numbers of eosinophils and basophils (human blood and BAL from asthmatic children) Increased levels of IL-4 (human BAL and sera) IFNγ suppression (murine model) Low Th1/Th2 ratio (human blood) High levels of IL-5 (human sera)	P1, CARDS toxin	Yano et al. (1994) Koh et al. (2001) Martin et al. (2001) Hou et al. (2003) Medina et al. (2012) Watanabe et al. (2014) Ye et al. (2014) Wang et al. (2015) Giavina-Bianchi and Kalil (2016) Yeh et al. (2016)
S. aureus	Asthma Wheezing episodes AD	Colonization associated with higher risk of asthma Specific-IgE in sera and BAL (asthma patients) Th1, Th2 cytokines release following stimulation with SEs	Extracellular vesicles, SEs (A-U), TSST-1, FBP, δ-toxin, Spls (A-E)	Bachert et al. (2003) Bachert and Zhang (2012) Tripathi et al. (2004) Barnes (2009) Hong et al. (2011) Kowalski et al. (2011)

(continued)

Table 1 (continued)

Bacteria	Associated allergic diseases	Evidence for a role in pathogenesis	Putative allergens	References
		Mast cell degranulation (in vitro and in vivo) High levels of IgE, IgG4 and Th2 cytokines (human sera, PBMCs, murine model)		Reginald et al. (2011a, b) Huvenne et al. (2013) Nakamura et al. (2013) Spaulding et al. (2013) Liu et al. (2014) Davis et al. (2015) Stentzel et al. (2016)
S. epidermidis	AD	Mast cell degranulation induced by bacterial culture supernatant		Nakamura et al. (2013)
H. influenza	Asthma Wheezing episodes	Histamine release, both IgE-dependent and independent Specific-IgE in sera (patients) High levels of IgE, IgG4 and Th2 cytokines (Human sera and PMBCs)	P4 P6	Clementsen et al. (1990) Pauwels et al. (1980) Brarda et al. (1996) Kjaerdgard et al. (1996) Ahren et al. (2003) King et al. (2003) Hales et al. (2008, 2009, 2012) Hollams et al. (2010) Larsen et al. (2014)
S. pneumoniae	Asthma Wheezing episodes	Colonization associated with higher risk of asthma Specific-IgE in sera (patients) Histamine release IgE-dependent High eosinophil counts (patients' blood) High levels of Th2 cytokines (PMBCs)	PscP	Brarda et al. (1996) Kjaerdgard et al. (1996) Bisgaard et al. (2007, 2010) Hollams et al. (2010) Darveaux and Lemanske (2014) Larsen et al. (2014)
S. pyogenes	Chronic allergic airways diseases	Specific-IgE in sera (patients) Th1, Th2 cytokine release following stimulation with SPEs	SPEs (A,C and G to M), SSA, $SMEZ_n$)	Bachert et al. (2001, 2003) Tripathi et al. (2004) Spaulding et al. (2013)

(continued)

Table 1 (continued)

Bacteria	Associated allergic diseases	Evidence for a role in pathogenesis	Putative allergens	References
M. catarrhalis	Asthma Wheezing episodes	Colonization associated with higher risk of asthma Specific-IgE in sera (patients) High levels of Th2 cytokines pattern (PMBCs)		Brarda et al. (1996) Bisgaard et al. (2007, 2010) Larsen et al. (2014)
P. aeruginosa	CF/Allergy	High numbers of Th2 cells, IL-4 and Il-13 (sera and BAL from patients) TNFα and IL-12 suppression and High levels of IgG1 (murine model)	OdDHL	Teleford et al. (1998) Moser et al. (2000) Ritchie et al. (2003, 2007) Brazova et al. (2005) Hartl et al. (2006) Upritchard et al. (2008) Tiringer et al. (2013) Hector et al. (2015)
B. burgdorferi	Lyme disease	Specific-IgE in sera (patients) Increase of CD8+ and CD60+ (patients)	p31, p34, p41, p45 and p60	Bluth et al. (2007)
S. saprophyticus	AD	Mast cell degranulation induced by bacterial culture supernatant		Nakamura et al. (2013)

2 Atypical Bacterial Pathogens

2.1 Chlamydia trachomatis

C. trachomatis is an atypical bacterium causing pneumonia, bronchiolitis and wheezing illness in infants (Korppi 2009; Webley et al. 2009; Patel et al. 2012). Persistent infections with these bacteria may play a role in the pathogenesis of asthma (Emre et al. 1995; Patel et al. 2012; Atkinson 2013). In children with chronic respiratory disease, most of them diagnosed with asthma, total serum IgE levels were higher in patients who were positive for *Chlamydia* compared to *Chlamydia*-negative individuals. Moreover, IgE with specificity for six *C. trachomatis* antigens were detected by Western-blot in patients' bronchoalveolar lavage (BAL) samples and sera. Band patterns on the blots differed significantly between individuals and kind of samples, but the dominant allergens were lipo-polysaccharide (LPS), major outer membrane protein (MOMP), a 250 kDa protein, a 64–66 kDa protein, cysteine-rich membrane protein (CrpA), polymorphic outer membrane protein (POMP) and chlamydial heat shock protein (HSP 60). The most immunoreactive proteins were the 250 kDa protein in both kind of samples as well as MOMP and POMP in BAL and serum, respectively. This suggests that *C. trachomatis* induces specific IgE antibodies and may play a direct role in the onset or exacerbation of chronic allergic airway diseases (Patel et al. 2012; Emre et al. 1995).

2.2   Chlamydia pneumoniae

C. pneumoniae is widely associated with wheezing episodes and asthma exacerba-tions in children and adults as well as with the incidence of asthma and lung remodeling (Johnston and Martin 2005; Hahn et al. 1991, 2012; Hahn and Peeling 2008; Korppi, 2009). This pathogen has been identified frequently in BAL, nasal washes and sera from asthmatic patients (Webley et al. 2009; Emre et al. 1995; Wark et al. 2002). Already in the 1990s, Emre et al. observed that in asthmatic children production of *C. pneumoniae*-specific IgE was more frequent in infected than non-infected or healthy individuals. On immunoblots the dominant IgE-binding bands were 98 KDa, 58–60 kDa, 36 kDa and 78 KDa (Emre et al. 1995). These initial findings have been corroborated by several groups. Studying children with *C. pneumoniae* respiratory tract infection, Ikewaza et al. observed *C. pneumoniae*-specific serum IgE significantly more frequently in the sera from the patients with wheezing compared to those without wheezing, while total IgE levels did not differ (Ikezawa 2001). Patel et al. studied sera and BAL in asthmatic children and found that during infection with *C. pneumoniae* or *C. trachomatis* the most immunoreactive allergens were similar, MOMP in BAL and LPS in serum (Patel et al. 2012). Moreover, Hahn et al. reported a significant association between *C. pneumoniae*-IgE and asthma diagnosis. *C. pneumoniae*-IgE was detectable in a large proportion of asthma patients and its amount was positively correlated with disease severity. The allergens most commonly recognized were largely overlapping with those identified by the group of Patel; CrpA, MOMP, HSP 60, LPS as well as lectin binding protein (LBP) (Hahn et al. 2012). In contrast to these reports of an association between allergy and *C. pneumoniae*, Larsen et al. did not find differences between asthmatic adults and healthy blood donors when measuring histamine release from basophil leukocytes sensitized with *C. pneumoniae*-specific serum IgE (Larsen et al. 1998).

Turning to the cellular immune response, Chen et al. observed induction of IL-4 gene expression in BAL fluid cells in mice with chronic and recurrent *C. pneumoniae* infection, which was correlated with an increase of thickness of subepithelial basement membrane suggestive of airway remodeling (Chen et al. 2009). Smith-Norowitz et al. studied T helper (Th) responses in peripheral blood mononuclear cells (PBMCs) from patients infected with *C. pneumoniae*. They found that PBMCs from asthmatic patients produced more IL-4 in response to exposure to *C. pneumoniae* than those from the control group. Moreover, higher concentrations of total IgE were detected in culture supernatants of PBMCs from asthmatic people (Smith-Norowitz et al. 2016). Treatment of the chlamydial infection with doxycycline dampened the IgE and IL-4 responses in PBMCs of asthma patients infected with *C. pneumoniae*. (Dzhindzhikhashvili et al. 2013; Smith-Norowitz et al. 2016). These results provide evidence for a relationship between bacterial infection and allergic response in patients with asthma.

2.3 Mycoplasma pneumoniae

M. pneumoniae causes respiratory disease or asymptomatic infections, which are associated with exacerbations of bronchial asthma. The bacteria are frequently found in asthma patients, and elevated total IgE serum concentrations are linked to the development of *M. pneumoniae*-specific IgE and/or IgE binding to common allergens in asthmatics patients during *M. pneumoniae* infection. Asthmatic children with *M. pneumonie* have elevated numbers of basophils in the peripheral blood and eosinophilia in the BAL suggestive of exacerbations of bronchial asthma (Huhti et al. 1974; Seggev et al. 1996; Yano et al. 1994; Tang et al. 2009; Ye et al. 2014). The association of this bacteria with initial onset of bronchial asthma was shown by Yano et al., in the case of one patient who developed asthma upon recovery from a *M. pneumoniae* infection. This was associated with a positive skin test for *M. pneumoniae* antigens reflecting the generation of a specific IgE response. This indicates that mycoplasma can induce allergic sensitization and/or contribute to immediate hypersensitivity reactions (Yano et al. 1994). On the other hand, Yeh et al. (2016), reported that patients with *M. pneumoniae* infection had a higher incidence of asthma than those without (hazard ratio of 3.35) (Yeh et al. 2016).

Patients infected with *M. pneumoniae* have significantly higher levels of IL-4 and elevated IL-4/IFN-γ ratios in the BAL and the serum compared with healthy controls, corresponding to a predominant Th2-like cytokine response (Koh et al. 2001; Ye et al. 2014). Moreover, elevated serum levels of IL-5, a cytokine released by Th2 cells and mast cells to promote the eosinophil growth and differentiation, have been measured in *M. pneumoniae* infected asthmatic but also non-asthmatic patients (Wang et al. 2015). Acute respiratory mycoplasma infection may cause bronchial hyperresponsiveness in a murine model probably due to IFN-γ suppression (Martin et al. 2001). In addition, Hou et al., observed a higher percentage of CD3+, CD4+ T lymphocyte and lower percentage of Th1, NK cells in PBMCs, linked to a lower ratio of Th1/Th2 cells in 3–13 year old children infected with *M. pneumoniae* (Hou et al. 2003).

Chu et al studied the effects of *M. pneumoniae* airway infection in a murine allergic asthma model before or after allergen sensitization. In accordance with the hygiene hypothesis, they observed protection from allergy when animals were infected before sensitization. Bronchial hyperresponsiveness, lung inflammation and IL-4 concentrations were reduced in the BAL, whereas INF-γ levels and Th1/Th2 ratios were increased. In contrast, *M. pneumoniae* infection of sensitized animals in the effector phase after challenge resulted in enhanced IL-4 release, lung inflammation and bronchial hypersensitivity. Moreover, increased airway collagen deposition by *M. pneumoniae* infection was observed in this murine allergic-asthma model (Chu et al. 2003, 2005).

In search of *M. pneumoniae* allergens, the group of Medina tested recombinant CARDS toxin, an ADP-ribosylating and vacuolating toxin termed the community-acquired respiratory distress syndrome toxin. This toxin binds to human surfactant protein A and shares regions of sequence identity with the pertussis toxin S1 protein (Becker et al. 2015; Medina et al. 2012). Single intranasal exposure to this toxin in the absence of adjuvant induced allergic pulmonary inflammation in naïve mice, which was characterized by airway hyperreactivity, eosinophilia and metaplasia of the mucosal tissue. Th2 type cytokines such as IL-13 or IL-4 and the chemokines CCL17 and CCL22, which can recruit Th2 cell to the site of inflammation, were also found elevated. Prolonged presence of CARDS toxin protein has been detected in the airway secretions from therapy refractory asthma patients (Peters et al. 2011). This suggests that *M. pneumoniae* may trigger and maintain asthma (Medina et al. 2012; Peters et al. 2011).

3   Common Bacterial Inhabitants of the Human Respiratory Tract

3.1   Staphylococcus aureus

S. aureus is often found as member of the normal microflora of the upper respiratory tract, but the microorganism is also a notorious human pathogen, causing a broad range of infections in- and outside hospitals (Lowy 1998; Wertheim et al. 2005). Davis et al. observed an association between *S. aureus* nasal colonization and the prevalence wheezing and asthma in children and young adults (Davis et al. 2015).

S. aureus can produce numerous virulence factors, among them superantigens, many of which also act as enterotoxins (SEs) causing food poisoning (Foster 2005; Thammavongsa et al. 2015). Superantigens comprise a group of proteins called staphylococcal enterotoxins A (SEA), B (SEB), C, D, E (up to U), and toxic shock syndrome toxin-1 (TSST-1)(Grumann et al. 2014). These proteins have the capacity to stimulate T cell proliferation directly without processing by antigen presenting cells (APC). Superantigens bind to the major histocompatibility complex (MHC) II and to the T-cell receptor (TCR) outside the typical antigen binding site and can thereby trigger a large fraction of T cells (Fraser et al. 2000; Fraser and Proft 2008). Memory T cells respond with the release of their effector cytokines, both of Th1 and Th2 profile. In rare instances this may culminate in a cytokine storm, known as toxic shock syndrome (Proft and Fraser 2003; Spaulding et al. 2013). The presence of staphylococci in the upper airways and the adaptive immune response directed against superantigens have been associated with allergic airway diseases. SE-specific IgE has been found in serum and in the airways of patients with allergy, mainly rhinitis, sinusitis, and asthma (Proft and Fraser 2003; Tripathi et al. 2004; Barnes 2009; Kowalski et al. 2011; Reginald et al. 2011a; Bachert et al. 2003; Pastacaldi et al. 2011; Spaulding et al. 2013; Huvenne et al. 2013; Liu et al. 2014). SE-specific IgE serum concentrations were correlated with asthma severity. Especially intrinsic asthma, defined as bronchial asthma in the absence of an allergic reaction to common aeroallergens (Barnes 2009), is frequently associated with an IgE response directed at *S. aureus* superantigens (Bachert et al. 2003; Bachert and Zhang 2012). Liu et al. observed that specific anti-SE IgE was significantly more prevalent in sera from asthma patients compared to chronic rhinosinusitis. Prevalence was lowest in healthy subjects. Moreover, they observed higher levels of specific IgE to indoor allergens such as dust mite and total IgE in patients with high levels of SE-specific IgE (Liu et al. 2014). This was corroborated by Kowalski et al. who reported that total and SE-specific serum IgE as well as eosinophil cationic protein concentrations were correlated with clinical and immunological parameters of asthma severity, suggesting a role of SEs in asthma pathogenesis (Kowalski et al. 2011).

Some Gram-positive bacteria, including *S. aureus*, produce extracellular vesicles (EVs) similar to Gram-negative bacteria (Lee et al. 2009; Gurung et al. 2011). Hong et al. evaluated whether *S. aureus*-derived EVs were related to the pathogenesis of AD. They did not find the superantigens SEA and SEB in *S. aureus* derived EVs. Nevertheless the vesicles induced an AD-like skin inflammation, including epidermal thickening and infiltration of the dermis by inflammatory cells (mainly mast cells and eosinophils) in mice. This inflammatory response was associated with the production of Th1/Th17/Th2 cytokines. Long-term exposure to *S. aureus* derived-EVs triggered the production of EV-specific IgE and a Th17-cell response in the animals' skin-draining lymph nodes. Turning to AD patients, the authors observed that their EV-specific IgE levels were significantly higher than those in healthy controls. Taken together, these data support the hypothesis that *S. aureus* EVs are involved in the pathogenesis of AD (Hong et al. 2011). Kim et al., studied the role of *S. aureus* EVs in the pathogenesis of pulmonary inflammation. They observed that in mice inhalation of *S. aureus* EVs induced the production of pro-inflammatory mediators, such as TNF-α and Th1/Th17 cytokines as well as neutrophil infiltration into the lung. However, in contrast to AD, Th2 cytokines or IgE production were not observed (Kim et al. 2012).

The quest for *S. aureus* virulence factors that may cause allergies by initiating or promoting a type 2 immune response is ongoing. IgE binding to *S. aureus* protein extracts has been reported, but the nature of the recognized proteins remained elusive (Reginald et al. 2011b). The group of Reginald then identified fibronectin-binding protein (FBP) as a target of IgE and IgG4 in sera of AD patients with atopic dermatitis. They demonstrated that this protein induced a Th2-driven IgE-mediated allergic immune response in mice, indicating that *S. aureus* FBP is an allergen

(Reginald et al. 2011a). Stenzel et al. conducted a systematic search for IgG4 reactive extracellular proteins of *S. aureus* in sera of healthy subjects and identified staphylococcal serine protease-like proteins (SplA-E), group of six serine proteases endoced in one operon, as prominent IgG4 targets. These proteins elicited a type 2 cytokine response in human T cells and also high levels of specific IgE in asthmatic patients. In mouse models intra-tracheal application of SplD in the absence of adjuvant induced a strong allergic lung inflammation accompanied by bronchial hyperresponsiveness. Infiltration of eosinophils, neutrophils and T- cells in the BAL and Spl-specific serum IgE were observed. Stimulation of immune cells from the lung-draining lymph nodes with anti-CD3- and anti-CD28 antibodies induced the release of Th2 cytokines in the SplD exposed group. This shows that *S. aureus* proteins can play a direct role in the pathophysiology of the allergic diseases (Stentzel et al. 2016).

Type 2 deviation of the immune response appears to be a concerted action in *S. aureus*. Besides superantigens, FBP and Spls, *S. aureus* δ-toxin contributes to this by directly triggering mast cell degranulation, independently of IgE and antigen (Nakamura et al. 2013). This was important in the induction of allergic skin inflammation in mice, because, on the one hand, δ-toxin alone could trigger allergic skin disease through activation of mast cells, while, on the other, δ-toxin-deficient *S. aureus* strains induced less skin inflammation and lower amounts of total serum IgE, IgG1 and IL-4 (Nakamura et al. 2013).

3.2   Staphylococcus epidermidis

S. epidermidis belongs to the coagulase-negative Staphylococci (CoNS). The Gram-positive commensal bacteria are commonly present on the human skin, in the nasal cavity and the respiratory tract. As nosocomial pathogens they are frequently involved in infections of indwelling devices such as catheters and implants (Otto 2014; Becker et al. 2014). Nakamura et al. observed that supernatant of these bacteria elicited mast cell degranulation, being the first to relate this bacterial species with allergy. Also in *S. epidermidis* δ-toxin triggers mast cell degranulation, however, in contrast to *S. aureus*, mutation of the δ-toxin gene did not fully abolish this effect, indicating that other factors may also be involved (Nakamura et al. 2013).

3.3   Haemophilus influenza

H. influenzae is a common inhabitant of the upper respiratory tract. Its presence is associated with induction and exacerbation of asthma, chronic obstructive pulmonary diseases and also with recurrent wheezing early in life (Kjaergard et al. 1996; Korppi 2009; Barnes 2009; Bisgaard et al. 2007, 2010; Hilty et al., 2010;

Hales et al., 2012). Kjaergard and colleagues proposed that the bacteria as well as *S. pneumoniae* could drive acute exacerbations of chronic bronchitis, because IgE directed against these bacteria could trigger histamine release (Kjaergard et al. 1996). In patients with bronchial asthma anti-*H. influenzae* IgE was also observed (Pauwels et al. 1980). Patients with bronchiectasis and *H. influenzae* infection developed a Th2 cytokine pattern and increased serum concentrations of specific IgG1, IgG3 and IgG4 (King et al. 2003). Blood cells of 6 month old children who developed asthma by 7 years of age elaborated higher levels of IL-5 and IL-13 when exposed to the bacteria than blood cells of children who remained asthma-free (Larsen et al. 2014). Hales and colleagues observed anti-P6 (an outer membrane protein) IgG4 and IgE antibodies in allergic patients (symptoms of rhinitis or asthma and allergen specific IgE against mite, cat and/or grass pollen), both children and adults (Hales et al. 2008). During convalescence from asthma exacerbations an strong increase of anti-P6 IgE was observed in allergic children, reaching titers similar to those induced by major allergens (Hales et al. 2008, 2009). Moreover, the bacteria can activate eosinophils and potentiate the release of inflammatory mediators in basophils and eosinophils when these are triggered by IgE-dependent or -independent mechnisms (Clementsen et al. 1990; Ahren et al. 2003).

However, there are also findings indicating that exposure to *H. influenzae* may protect from allergy, even when an anti-bacterial immune response of type 2 can be measured. Epton and colleagues reported that T cells produced mainly Th1 cytokines in response to *H. influenzae* P6 and D15 proteins, and specific IgE antibodies were below the threshold of detection in allergic and non-allergic individuals (Epton et al. 2002). Hollams et al. reported that in teenagers specific IgE antibodies against P6 were inversely correlated with asthma risk (Hollams et al. 2010). Similarly, when the immune response against two *H. influenzae* proteins, P4 and P6, was studied in children (1–5 years) with a high risk of atopy, asthma was associated with low titres of specific IgG1 as well as IgE. In fact, there was an inverse correlation between specific IgE binding to *H. influenzae* P6 on the one hand, and total serum IgE concentrations, anti-house dust mite-IgE titers and positive skin prick tests to common allergens on the other (Hollams et al. 2010; Hales et al. 2012).

Hence, with regard to allergy development, *H. influenzae* has to be considered a double edged sword. It remains to be elucidated, when exposure to the microorganisms protects from allergy and when, to the contrary, it facilitates and promotes allergic inflammation. Genetic predisposition of the host as well as his age and the circumstances of encounter with *H. influenzae* may be decisive factors.

3.4 Streptococcus pneumoniae

The Gram-positive bacterium *Streptococcus pneumoniae*, commonly called pneumococcus, is a major cause of pneumonia worldwide. However the bacteria can also behave as commensal inhabitants of the upper respiratory tract, and they are found in healthy carriers of all age groups (Korppi 2009; Olaya-Abril et al. 2015). Several studies demonstrate an association between colonization with pneumococci and the incidence of asthma as well as exacerbations of the disease. Bacterial colonization in asymptomatic neonates indicated an increased risk of allergy development, such as recurrent wheeze and asthma. At 4 years of age these *S. pneumoniae* carriers had higher eosinophil counts and total IgE titres in the blood, while there was no evidence of an IgE response to *S. pneumoniae* itself (Bisgaard et al. 2007, 2010; Darveaux and Lemanske 2014). Similarly, when PBMCs of 6 months old *S. pneumoniae* carriers were exposed to the bacteria, cells produced more IL-5 and IL-13 in subjects who developed asthma by 7 years of age than in those who did not (Larsen et al. 2014). Kjaergard et al. demonstrated anti-bacterial IgE in sera of chronic bronchitis patients infected with *S. pneumoniae*. Moreover, IgE-dependent histamine release could be induced in the patients' basophils by incubation with the infecting *S. pneumoniae* strain, indicating allergic sensitization (Kjaergard et al. 1996). *S. pneumoniae*-specific IgE antibodies were also found in sera from asthmatic children (Brarda et al. 1996).

As in the case of *H. influenzae*, there is also evidence for allergy protection conferred by *S. pneumoniae*. In 2010, Hollams and colleagues measured serum titers of IgE against the surface protein C (PscP) of pneumococci in teenagers. Specific IgE was frequent, however, it was inversely correlated with asthma risk (Hollams et al. 2010). In children between 1–5 years of age with high risk of atopy high titres of IgG1 specific for two *S. pneumoniae* proteins, PspC and PspA, were associated with a lower incidence of asthma (Hales et al. 2012).

Hence, as with *H. influenzae*, there is evidence for both promotion of allergy and protection from it. While numerous studies showed an association of *S. pneumoniae* with asthma, there is to date no experimental evidence for a causative role of the bacteria or their products in asthma development.

3.5   Streptococcus pyogenes

Streptococcus pyogenes is a group A streptococcal pathogen which causes a broad range of infections from sore throat, pharyngitis or skin lesions to fatal necrotising fasciitis, pneumonia and streptococcal toxic shock syndrome. Moreover *S. pyogenes* can asymptomatically colonize humans (Spaulding et al. 2013; Gur et al. 2002; Juhn et al. 2012). Like *S. aureus*, group A streptococci can produce enterotoxins called streptococcal pyrogenic exotoxins (SPEs) that act as superantigens and cause serious human illness such as devastating skin and soft tissue infection as well as toxic shock syndrome (Spaulding et al. 2013; Proft and Fraser 2003). The streptococcal superantigens include SPE (A, B, C, and G to M), streptococcal superantigen (SSA) and streptococcal mitogenic exotoxin Z_n (SMEZ$_n$). They have the capacity to stimulate a massive T cell response with resultant of Th1 and Th2 cytokine release. The superantigen can circumvent the need of processing by APCs by directly crosslinking MHC II to T cell receptors, binding outside of the typical antigen binding sites (Spaulding et al. 2013; Proft and Fraser 2003; Tripathi et al. 2004). IgE antibodies specific for SPEs have been found in patients with chronic sinusitis /nasal polyposis indicating their potential role in allergy, which is similar to the situation in *S. aureus* (Tripathi et al. 2004).

3.6 Moraxella catarrhalis

Moraxella catarrhalis is found as an commensal in the upper respiratory tract. The bacteria can cause sinusitis, otitis media, diseases of the lower respiratory tract as well as pneumonia (Korppi 2009; Hilty et al. 2010; Bisgaard et al. 2010). *M. catarrhalis*-specific IgE antibodies were found in sera from asthmatic children (Brarda et al. 1996), and there was a significant association between bacterial infection of the airways and acute wheezing episodes in young children (Bisgaard et al. 2010). Similar to *S. pneumoniae*, colonization with *M. catarrhalis* in asymptomatic neonates was associated with later development of recurrent wheeze, asthma, and allergy. Children colonized neonatally had higher eosinophil counts and total IgE in blood at 4 years of age than those who were non-carriers (Bisgaard et al. 2007). Moreover, children who develop asthma by 7 years of age showed a Th2 type in vitro response to *M. catarrhalis* already at the age of 6 months (Larsen et al. 2014). Apparently, an allergic predisposition may manifest itself very early in life by a type 2 reaction to commensal bacteria.

3.7 Pseudomonas aeruginosa

P. aeruginosa is a prominent cause of infection in CF patients of whom approximately the 80% are infected by this microorganism (Hartl et al. 2006; Upritchard et al. 2008; Hector et al. 2015, 2016; Tiringer et al. 2013).

It has been proposed that impairment of Treg function combined with a dysbalance between Th2/Th17 and Th1 cells, is important in inducing and maintaining lung inflammation in CF, and *P. aeruginosa* infection was found to be associated with all these features of Th2 immune deviation (Upritchard et al. 2008; Tiringer et al. 2013; Hector et al. 2015). Hartl and colleagues analyzed the BAL of *P. aeruginosa* infected CF patients and observed large numbers of Th2 cells and increased levels IL-4 and IL-13 but only low concentrations of INF-γ. Th2-cytokine concentrations were correlated with the severity of the disease. Similar observations were made in peripheral blood, implying that the Th2 bias was systemic in CF patients with chronic *P. aeruginosa* lung infections (Hartl et al. 2006; Brazova et al. 2005; Moser et al. 2000). In contrast, an immunological Th1 response profile was associated with a better pulmonary outcome. (Hartl et al. 2006; Upritchard et al. 2008; Hector et al. 2015; Bisgaard et al. 2007). It remains to be shown whether the Th2 deviation of the immune response precedes and facilitates *P. aeruginosa* infection in CF patients or whether its results from the host's chronic interaction with the bacteria.

Already in the 90s, Teleford and co-workers reported that the protein N-(3-Oxododecanoyl)-L-homoserine lactone (OdDHL), a molecule involved in quorum sensing (a bacterial intracellular communication in *P. aeruginosa*), can modulate the immune cell activity in cell culture. In murine peritoneal macrophages incubation with OdDHL reduced the production of TNF-α and IL-12, cytokines which direct the generation of a Th1 immune response. Moreover, in spleen cells from keyhole limpet hemocyanin primed mice OdDHL suppressed the overall production of specific antibodies following antigen stimulation but increased the relative amount of IgG1, an immunoglobulin isotype, which is associated with a Th2 response in mice. In human immune cells OdDHL reduced IL-12 synthesis and increased the production of IgE antibodies (Telford et al. 1998). In contrast, Smith and coworkers observed a predominant Th1 immune response when murine naive T cells were activated with antigen and APCs in the presence of OdDHL (Smith et al. 2002), whereas the group of Ritchie reported that OdDHL inhibited the differentiation of both, Th1 and Th2 cells. Some of these conflicting data may be reconciled by Richie's observation that upon secondary stimulation, IFN-γ production was more sensitive to OdDHL inhibition than IL-4, leading to a Th2 bias (Ritchie et al. 2003, 2007). Further studies are needed to clarify the relationship between *P. aeruginosa* infections, allergic immune responses and CF in general and to reconcile the conflicting information about the immunomodulatory activity of OdDHL in particular.

4 Bacteria Not Related to the Respiratory Tract

4.1 Borrelia burgdorferi

The spirochete *B. burgdorferi* is the causal agent of the tick-borne Lyme disease, characterized by flu-like symptoms, skin lesions or erythema migrans. Antigen-specific *B. burgdorferi*-IgE antibodies (proteins p31, p34, p41, p45 and p60) and increased numbers of CD8+ and CD60+ T cells, which are related to regulation of human allergic responses, have been observed in blood of infected subjects, however, the clinical implications of these findings are not yet clear (Bluth et al. 2007).

4.2 Staphylococcus saprophyticus

S. saprophyticus are coagulase-negative staphylococci (CoNS), Gram-positive commensals often colonizing the human rectum and genitourinary tract. The microorganisms are causative agents of lower urinary tract infections, which usually have a bland disease course but may be complicated by acute pyelonephritis and nephrolithiasis, epididymitis, and prostatitis. The bacteria have also been indentified in septicemia associated with catheters as well as in endocarditis (Becker et al. 2014). Nakamura et al. (2013), observed that, similar to *S. aureus* and *S. epidermidis*, cell culture supernatants of *S. saprophyticus* elicit mast cell degranulation. This is the first report linking this microorganism to allergy (Nakamura et al. 2013).

5 Conclusions

The presented data provide clear evidence that bacterial colonization and infection may increase the risk of allergy development and exacerbate allergic inflammation. The reasons why some species are more closely associated with allergy than others remains to be elucidated. Regarding the mechanisms, it is plausible that in an allergic individual bacterial antigens may become antigenic targets of Th2 cells and IgE in the process of epitope spreading. Once atopic memory has been established to them, the bacterial antigens will trigger allergic responses similar to other inhalant allergens. Moreover, the fact that certain bacterial factors can cause allergic lung inflammation in naïve experimental animals suggests that some bacterial species are capable of triggering allergic inflammation de novo. Bacteria may benefit from an allergic deviation of the immune response toward a Th2 profile because this counteracts anti-bacterial clearance mechanisms, which are mainly orchestrated by Th1 and Th17 cells. Hence, pro-allergenic mechanisms can be considered as means of immune evasion, which may explain why some species, such as *S. aureus*, have developed a whole arsenal of virulence factors favoring or even triggering allergic inflammation. To date, only few bacterial allergens have been described. Efforts are now required to find out whether such bacterial factors may explain allergies of hitherto unknown cause.

References

Ahren IL, Eriksson E, Egesten A, Riesbeck K (2003) Nontypeable *Haemophilus influenzae* activates human eosinophils through beta-glucan receptors. Am J Respir Cell Mol Biol 29(5):598–605. https://doi.org/10.1165/rcmb.2002-0138OC

Atkinson TP (2013) Is asthma an infectious disease? New evidence. Curr Allergy Asthma Rep 13(6):702–709. https://doi.org/10.1007/s11882-013-0390-8

Bach JF (2002) The effect of infections on susceptibility to autoimmune and allergic diseases. N Engl J Med 347(12):911–920. https://doi.org/10.1056/NEJMra020100

Bacher P, Heinrich F, Stervbo U, Nienen M, Vahldieck M, Iwert C et al (2016) Regulatory T cell specificity directs tolerance versus allergy against aeroantigens in humans. Cell 167(4): 1067–1078.e1016. https://doi.org/10.1016/j.cell.2016.09.050

Bachert C, Zhang N (2012) Chronic rhinosinusitis and asthma: novel understanding of the role of IgE 'above atopy'. J Intern Med 272(2):133–143. https://doi.org/10.1111/j.1365-2796.2012. 02559.x

Bachert C, Gevaert P, Holtappels G, Johansson SG, van Cauwenberge P (2001) Total and specific IgE in nasal polyps is related to local eosinophilic inflammation. J Allergy Clin Immunol 107(4):607–614. https://doi.org/10.1067/mai.2001.112374

Bachert C, Gevaert P, Howarth P, Holtappels G, van Cauwenberge P, Johansson SG (2003) IgE to *Staphylococcus aureus* enterotoxins in serum is related to severity of asthma. J Allergy Clin Immunol 111:1131–1132

Barnes PJ (2009) Intrinsic asthma: not so different from allergic asthma but driven by super-antigens? Clin Exp Allergy 39(8):1145–1151. https://doi.org/10.1111/j.1365-2222.2009. 03298.x

Barnes PJ (2012) Severe asthma: advances in current management and future therapy. J Allergy Clin Immunol 129(1):48–59. https://doi.org/10.1016/j.jaci.2011.11.006

Becker A, Kannan TR, Taylor AB, Pakhomova ON, Zhang Y, Somarajan SR et al (2015) Structure of CARDS toxin, a unique ADP-ribosylating and vacuolating cytotoxin from *Mycoplasma pneumoniae*. Proc Natl Acad Sci USA 112(16):5165–5170. https://doi.org/10.1073/pnas. 1420308112

Becker K, Heilmann C, Peters G (2014) Coagulase-negative staphylococci. Clin Microbiol Rev 27(4):870–926. https://doi.org/10.1128/cmr.00109-13

Birzele LT, Depner M, Ege MJ, Engel M, Kublik S, Bernau C et al (2017) Environmental and mucosal microbiota and their role in childhood asthma. Allergy 72(1):109–119. https://doi.org/ 10.1111/all.13002

Bisgaard H, Hermansen MN, Buchvald F, Loland L, Halkjaer LB, Bonnelykke K et al (2007) Childhood asthma after bacterial colonization of the airway in neonates. N Engl J Med 357(15):1487–1495. https://doi.org/10.1056/NEJMoa052632

Bisgaard H, Hermansen MN, Bonnelykke K, Stokholm J, Baty F, Skytt NL et al (2010) Associ-ation of bacteria and viruses with wheezy episodes in young children: prospective birth cohort study. BMJ 341:c4978. https://doi.org/10.1136/bmj.c4978

Bluth MH, Robin J, Ruditsky M, Norowitz KB, Chice S, Pytlak E et al (2007) IgE anti-Borrelia burgdorferi components (p18, p31, p34, p41, p45, p60) and increased blood CD8+CD60+ T cells in children with Lyme disease. Scand J Immunol 65(4):376–382. https://doi.org/10.1111/ j.1365-3083.2007.01904.x

Brarda OA, Vanella LM, Boudet RV (1996) Anti-*Staphylococcus aureus*, anti-*Streptococcus pneumoniae* and anti-*Moraxella catarrhalis* specific IgE in asthmatic children. J Investig Aller-gol Clin Immunol 6(4):266–269

Brazova J, Sediva A, Pospisilova D, Vavrova V, Pohunek P, Macek M Jr et al (2005) Differential cytokine profile in children with cystic fibrosis. Clin Immunol 115(2):210–215. https://doi.org/ 10.1016/j.clim.2005.01.013

Chen CZ, Yang BC, Lin TM, Lee CH, Hsiue TR (2009) Chronic and repeated *Chlamydophila pneumoniae* lung infection can result in increasing IL-4 gene expression and thickness of airway subepithelial basement membrane in mice. J Formos Med Assoc 108(1):45–52. https:// doi.org/10.1016/s0929-6646(09)60031-0

Choi IS (2014) Immunomodulating approach to asthma using mycobacteria. Allergy Asthma Immunol Res 6(3):187–188. https://doi.org/10.4168/aair.2014.6.3.187

Choi IS, Koh YI (2002) Therapeutic effects of BCG vaccination in adult asthmatic patients: a randomized, controlled trial. Ann Allergy Asthma Immunol 88(6):584–591. https://doi.org/10. 1016/S1081-1206(10)61890-X

Choi IS, Koh YI (2003) Effects of BCG revaccination on asthma. Allergy 58(11):1114–1116

Chu HW, Honour JM, Rawlinson CA, Harbeck RJ, Martin RJ (2003) Effects of respiratory *Mycoplasma pneumoniae* infection on allergen-induced bronchial hyperresponsiveness and lung inflammation in mice. Infect Immun 71(3):1520–1526

Chu HW, Rino JG, Wexler RB, Campbell K, Harbeck RJ, Martin RJ (2005) *Mycoplasma pneumoniae* infection increases airway collagen deposition in a murine model of allergic airway inflammation. Am J Physiol Lung Cell Mol Physiol 289(1):L125–L133. https://doi.org/10.1152/ajplung.00167.2004

Clementsen P, Milman N, Kilian M, Fomsgaard A, Baek L, Norn S (1990) Endotoxin from *Haemophilus influenzae* enhances IgE-mediated and non-immunological histamine release. Allergy 45(1):10–17

Darveaux JI, Lemanske RF Jr (2014) Infection-related asthma. J Allergy Clin Immunol Pract 2(6):658–663. https://doi.org/10.1016/j.jaip.2014.09.011

Davis MF, Peng RD, McCormack MC, Matsui EC (2015) *Staphylococcus aureus* colonization is associated with wheeze and asthma among US children and young adults. J Allergy Clin Immunol 135(3):811–813. e815. https://doi.org/10.1016/j.jaci.2014.10.052

Dzhindzhikhashvili MS, Joks R, Smith-Norowitz T, Durkin HG, Chotikanatis K, Estrella E et al (2013) Doxycycline suppresses *Chlamydia pneumoniae*-mediated increases in ongoing immunoglobulin E and interleukin-4 responses by peripheral blood mononuclear cells of patients with allergic asthma. J Antimicrob Chemother 68(10):2363–2368. https://doi.org/10.1093/jac/dkt179

Edwards MR, Bartlett NW, Hussell T, Openshaw P, Johnston SL (2012) The microbiology of asthma. Nat Rev Microbiol 10(7):459–471. https://doi.org/10.1038/nrmicro2801

Emre U, Sokolovskaya N, Roblin PM, Schachter J, Hammerschlag MR (1995) Detection of anti-*Chlamydia pneumoniae* IgE in children with reactive airway disease. J Infect Dis 172(1):265–267

Epton MJ, Hales BJ, Thompson PJ, Thomas WR (2002) T cell cytokine responses to outer membrane proteins of *Haemophilus influenzae* and the house dust mite allergens Der p 1 in allergic and non-allergic subjects. Clin Exp Allergy 32(11):1589–1595

Foster TJ (2005) Immune evasion by staphylococci. Nature Rev Microbiol 3:948–958

Fraser J, Arcus V, Kong P, Baker E, Proft T (2000) Superantigens - powerful modifiers of the immune system. Mol Med Today 6(3):125–132

Fraser JD, Proft T (2008) The bacterial superantigen and superantigen-like proteins. Immunol Rev 225:226–243. https://doi.org/10.1111/j.1600-065X.2008.00681.x. IMR681 [pii]

Giavina-Bianchi P, Kalil J (2016) Mycoplasma pneumoniae infection induces asthma onset. J Allergy Clin Immunol 137(4):1024–1025. https://doi.org/10.1016/j.jaci.2015.11.011

Grumann D, Nübel U, Bröker BM (2014) *Staphylococcus aureus* toxins-their functions and genetics. Infect Genet Evol 21:583–592. https://doi.org/10.1016/j.meegid.2013.03.013. S1567-1348(13)00086-5 [pii]

Gur D, Ozalp M, Sumerkan B, Kaygusuz A, Toreci K, Koksal I et al (2002) Prevalence of antimicrobial resistance in *Haemophilus influenzae, Streptococcus pneumoniae, Moraxella catarrhalis* and *Streptococcus pyogenes*: results of a multicentre study in Turkey. Int J Antimicrob Agents 19(3):207–211

Gurung M, Moon DC, Choi CW, Lee JH, Bae YC, Kim J et al (2011) *Staphylococcus aureus* produces membrane-derived vesicles that induce host cell death. PLoS One 6(11):e27958. https://doi.org/10.1371/journal.pone.0027958

Hahn DL, Peeling RW (2008) Airflow limitation, asthma, and *Chlamydia pneumoniae*-specific heat shock protein 60. Ann Allergy Asthma Immunol 101(6):614–618. https://doi.org/10.1016/s1081-1206(10)60224-4

Hahn DL, Dodge RW, Golubjatnikov R (1991) Association of *Chlamydia pneumoniae* (strain TWAR) infection with wheezing, asthmatic bronchitis, and adult-onset asthma. JAMA 266(2):225–230

Hahn DL, Schure A, Patel K, Childs T, Drizik E, Webley W (2012) *Chlamydia pneumoniae*-specific IgE is prevalent in asthma and is associated with disease severity. PLoS One 7(4):e35945. https://doi.org/10.1371/journal.pone.0035945

Hales BJ, Pearce LJ, Kusel MM, Holt PG, Sly PD, Thomas WR (2008) Differences in the antibody response to a mucosal bacterial antigen between allergic and non-allergic subjects. Thorax 63(3):221–227. https://doi.org/10.1136/thx.2006.069492

Hales BJ, Martin AC, Pearce LJ, Rueter K, Zhang G, Khoo SK et al (2009) Anti-bacterial IgE in the antibody responses of house dust mite allergic children convalescent from asthma exacerbation. Clin Exp Allergy 39(8):1170–1178. https://doi.org/10.1111/j.1365-2222.2009.03252.x

Hales BJ, Chai LY, Elliot CE, Pearce LJ, Zhang G, Heinrich TK et al (2012) Antibacterial antibody responses associated with the development of asthma in house dust mite-sensitised and non-sensitised children. Thorax 67(4):321–327. https://doi.org/10.1136/thoraxjnl-2011-200650

Hartl D, Griese M, Kappler M, Zissel G, Reinhardt D, Rebhan C et al (2006) Pulmonary T(H)2 response in *Pseudomonas aeruginosa*-infected patients with cystic fibrosis. J Allergy Clin Immunol 117(1):204–211. https://doi.org/10.1016/j.jaci.2005.09.023

Hector A, Schafer H, Poschel S, Fischer A, Fritzsching B, Ralhan A et al (2015) Regulatory T-cell impairment in cystic fibrosis patients with chronic pseudomonas infection. Am J Respir Crit Care Med 191(8):914–923. https://doi.org/10.1164/rccm.201407-1381OC

Hector A, Frey N, Hartl D (2016) Update on host-pathogen interactions in cystic fibrosis lung disease. Mol Cell Pediatr 3(1):12. https://doi.org/10.1186/s40348-016-0039-5

Hilty M, Burke C, Pedro H, Cardenas P, Bush A, Bossley C et al (2010) Disordered microbial communities in asthmatic airways. PLoS One 5(1):e8578. https://doi.org/10.1371/journal.pone.0008578

Hollams EM, Hales BJ, Bachert C, Huvenne W, Parsons F, de Klerk NH et al (2010) Th2-associated immunity to bacteria in teenagers and susceptibility to asthma. Eur Respir J 36(3):509–516. https://doi.org/10.1183/09031936.00184109

Hong SW, Kim MR, Lee EY, Kim JH, Kim YS, Jeon SG et al (2011) Extracellular vesicles derived from *Staphylococcus aureus* induce atopic dermatitis-like skin inflammation. Allergy 66(3):351–359. https://doi.org/10.1111/j.1398-9995.2010.02483.x

Hou AC, Lu Y, Sha L, Liu LG, Shen J, Xu Y (2003) T(H1) and T(H2) cells in children with mycoplasma pneumonia. Zhonghua Er Ke Za Zhi 41(9):652–656

Huhti E, Mokka T, Nikoskelainen J, Halonen P (1974) Association of viral and mycoplasma infections with exacerbations of asthma. Ann Allergy 33(3):145–149

Huvenne W, Hellings PW, Bachert C (2013) Role of staphylococcal superantigens in airway disease. Int Arch Allergy Immunol 161(4):304–314. https://doi.org/10.1159/000350329

Ikezawa S (2001) Prevalence of *Chlamydia pneumoniae* in acute respiratory tract infection and detection of anti-*Chlamydia pneumoniae*-specific IgE in Japanese children with reactive airway disease. Kurume Med J 48(2):165–170

Ipci K, Altintoprak N, Muluk NB, Senturk M, Cingi C (2016) The possible mechanisms of the human microbiome in allergic diseases. Eur Arch Otorhinolaryngol 274(2):617–626. https://doi.org/10.1007/s00405-016-4058-6

Johnston SL, Martin RJ (2005) *Chlamydophila pneumoniae* and *Mycoplasma pneumoniae*: a role in asthma pathogenesis? Am J Respir Crit Care Med 172(9):1078–1089. https://doi.org/10.1164/rccm.200412-1743PP

Juhn YJ, Frey D, Li X, Jacobson R (2012) *Streptococcus pyogenes* upper respiratory infection and atopic conditions other than asthma: a retrospective cohort study. Prim Care Respir J 21(2):153–158. https://doi.org/10.4104/pcrj.2011.00110

Kalliomäki M, Salminen S, Arvilommi H, Kero P, Koskinen P, Isolauri E (2001) Probiotics in primary prevention of atopic disease: a randomised placebo-controlled trial. Lancet 357(9262):1076–1079. https://doi.org/10.1016/S0140-6736(00)04259-8

Kim MR, Hong SW, Choi EB, Lee WH, Kim YS, Jeon SG et al (2012) *Staphylococcus aureus*-derived extracellular vesicles induce neutrophilic pulmonary inflammation via both Th1 and Th17 cell responses. Allergy 67(10):1271–1281. https://doi.org/10.1111/all.12001

Kim YJ, Kim HJ, Kang MJ, Yu HS, Seo JH, Kim HY et al (2014) Bacillus Calmette-Guérin suppresses asthmatic responses via CD4(+)CD25(+) regulatory T cells and dendritic cells. Allergy Asthma Immunol Res 6(3):201–207. https://doi.org/10.4168/aair.2014.6.3.201

King PT, Hutchinson PE, Johnson PD, Holmes PW, Freezer NJ, Holdsworth SR (2003) Adaptive immunity to nontypeable *Haemophilus influenzae*. Am J Respir Crit Care Med 167(4):587–592. https://doi.org/10.1164/rccm.200207-728OC

Kjaergard LL, Larsen FO, Norn S, Clementsen P, Skov PS, Permin H (1996) Basophil-bound IgE and serum IgE directed against *Haemophilus influenzae* and *Streptococcus pneumoniae* in patients with chronic bronchitis during acute exacerbations. APMIS 104(1):61–67

Koh YY, Park Y, Lee HJ, Kim CK (2001) Levels of interleukin-2, interferon-gamma, and interleukin-4 in bronchoalveolar lavage fluid from patients with *Mycoplasma pneumonia*: implication of tendency toward increased immunoglobulin E production. Pediatrics 107(3): E39

Korppi M (2009) Management of bacterial infections in children with asthma. Expert Rev Anti Infect Ther 7(7):869–877. https://doi.org/10.1586/eri.09.58

Kowalski ML, Cieslak M, Perez-Novo CA, Makowska JS, Bachert C (2011) Clinical and immunological determinants of severe/refractory asthma (SRA): association with Staphylococcal superantigen-specific IgE antibodies. Allergy 66(1):32–38. https://doi.org/10.1111/j.1398-9995.2010.02379.x

Kwon HK, Lee CG, So JS, Chae CS, Hwang JS, Sahoo A et al (2010) Generation of regulatory dendritic cells and CD4+Foxp3+ T cells by probiotics administration suppresses immune disorders. Proc Natl Acad Sci USA 107(5):2159–2164. https://doi.org/10.1073/pnas.0904055107

Larsen FO, Norn S, Mordhorst CH, Skov PS, Milman N, Clementsen P (1998) *Chlamydia pneumoniae* and possible relationship to asthma. Serum immunoglobulins and histamine release in patients and controls. APMIS 106(10):928–934

Larsen JM, Brix S, Thysen AH, Birch S, Rasmussen MA, Bisgaard H (2014) Children with asthma by school age display aberrant immune responses to pathogenic airway bacteria as infants. J Allergy Clin Immunol 133(4):1008–1013. https://doi.org/10.1016/j.jaci.2014.01.010

Lee EY, Choi DY, Kim DK, Kim JW, Park JO, Kim S et al (2009) Gram-positive bacteria produce membrane vesicles: proteomics-based characterization of *Staphylococcus aureus*-derived membrane vesicles. Proteomics 9(24):5425–5436. https://doi.org/10.1002/pmic.200900338

Liu JN, Shin YS, Yoo HS, Nam YH, Jin HJ, Ye YM et al (2014) The prevalence of serum specific IgE to superantigens in asthma and allergic rhinitis patients. Allergy Asthma Immunol Res 6(3):263–266. https://doi.org/10.4168/aair.2014.6.3.263

Lowy F (1998) *Staphylococcus aureus* infections. N Engl J Med 339:520–529

Martin RJ, Chu HW, Honour JM, Harbeck RJ (2001) Airway inflammation and bronchial hyperresponsiveness after *Mycoplasma pneumoniae* infection in a murine model. Am J Respir Cell Mol Biol 24(5):577–582. https://doi.org/10.1165/ajrcmb.24.5.4315

Masoli M, Fabian D, Holt S, Beasley R, Program, G. I. F. A. G (2004) The global burden of asthma: executive summary of the GINA Dissemination Committee report. Allergy 59(5): 469–478. https://doi.org/10.1111/j.1398-9995.2004.00526.x

Medina JL, Coalson JJ, Brooks EG, Winter VT, Chaparro A, Principe MF et al (2012) *Mycoplasma pneumoniae* CARDS toxin induces pulmonary eosinophilic and lymphocytic inflammation. Am J Respir Cell Mol Biol 46(6):815–822. https://doi.org/10.1165/rcmb.2011-0135OC

Moser C, Kjaergaard S, Pressler T, Kharazmi A, Koch C, Hoiby N (2000) The immune response to chronic *Pseudomonas aeruginosa* lung infection in cystic fibrosis patients is predominantly of the Th2 type. APMIS 108(5):329–335

Nakamura Y, Oscherwitz J, Cease KB, Chan SM, Munoz-Planillo R, Hasegawa M et al (2013) Staphylococcus delta-toxin induces allergic skin disease by activating mast cells. Nature 503 (7476):397–401. https://doi.org/10.1038/nature12655

Olaya-Abril A, Jimenez-Munguia I, Gomez-Gascon L, Obando I, Rodriguez-Ortega MJ (2015) A pneumococcal protein array as a platform to discover serodiagnostic antigens against infection. Mol Cell Proteomics 14(10):2591–2608. https://doi.org/10.1074/mcp.M115.049544

Otto M (2014) *Staphylococcus epidermidis* pathogenesis. Methods Mol Biol 1106:17–31. https://doi.org/10.1007/978-1-62703-736-5_2

Pastacaldi C, Lewis P, Howarth P (2011) Staphylococci and staphylococcal superantigens in asthma and rhinitis: a systematic review and meta-analysis. Allergy 66(4):549–555. https:// doi.org/10.1111/j.1398-9995.2010.02502.x

Patel KK, Anderson E, Salva PS, Webley WC (2012) The prevalence and identity of Chlamydia-specific IgE in children with asthma and other chronic respiratory symptoms. Respir Res 13:32. https://doi.org/10.1186/1465-9921-13-32

Pauwels R, Verschraegen G, van der Straeten M (1980) IgE antibodies to bacteria in patients with bronchial asthma. Allergy 35(8):665–669

Peters J, Singh H, Brooks EG, Diaz J, Kannan TR, Coalson JJ et al (2011) Persistence of community-acquired respiratory distress syndrome toxin-producing *Mycoplasma pneumoniae* in refractory asthma. Chest 140(2):401–407. https://doi.org/10.1378/chest.11-0221

Proft T, Fraser JD (2003) Bacterial superantigens. Clin Exp Immunol 133(3):299–306

Ramsey CD, Celedon JC (2005) The hygiene hypothesis and asthma. Curr Opin Pulm Med 11(1): 14–20

Read S, Malmström V, Powrie F (2000) Cytotoxic T lymphocyte-associated antigen 4 plays an essential role in the function of CD25(+)CD4(+) regulatory cells that control intestinal inflammation. J Exp Med 192(2):295–302

Reginald K, Westritschnig K, Linhart B, Focke-Tejkl M, Jahn-Schmid B, Eckl-Dorna J et al (2011a) *Staphylococcus aureus* fibronectin-binding protein specifically binds IgE from patients with atopic dermatitis and requires antigen presentation for cellular immune responses. J Allergy Clin Immunol 128(1):82–91. e88. https://doi.org/10.1016/j.jaci.2011.02.034

Reginald K, Westritschnig K, Werfel T, Heratizadeh A, Novak N, Focke-Tejkl M et al (2011b) Immunoglobulin E antibody reactivity to bacterial antigens in atopic dermatitis patients. Clin Exp Allergy 41(3):357–369. https://doi.org/10.1111/j.1365-2222.2010.03655.x

Ribet D, Cossart P (2015) How bacterial pathogens colonize their hosts and invade deeper tissues. Microbes Infect 17(3):173–183. https://doi.org/10.1016/j.micinf.2015.01.004

Ritchie AJ, Yam AO, Tanabe KM, Rice SA, Cooley MA (2003) Modification of in vivo and in vitro T- and B-cell-mediated immune responses by the *Pseudomonas aeruginosa* quorum-sensing molecule N-(3-oxododecanoyl)-L-homoserine lactone. Infect Immun 71(8): 4421–4431

Ritchie AJ, Whittall C, Lazenby JJ, Chhabra SR, Pritchard DI, Cooley MA (2007) The immuno-modulatory *Pseudomonas aeruginosa* signalling molecule N-(3-oxododecanoyl)-L-homoserine lactone enters mammalian cells in an unregulated fashion. Immunol Cell Biol 85(8):596–602. https://doi.org/10.1038/sj.icb.7100090

Schaub B, Lauener R, von Mutius E (2006) The many faces of the hygiene hypothesis. J Allergy Clin Immunol 117(5):969–977.; quiz 978. https://doi.org/10.1016/j.jaci.2006.03.003

Seggev JS, Sedmak GV, Kurup VP (1996) Isotype-specific antibody responses to acute *Mycoplasma pneumoniae* infection. Ann Allergy Asthma Immunol 77(1):67–73. https://doi.org/10. 1016/s1081-1206(10)63482-5

Seroogy CM, Gern JE (2005) The role of T regulatory cells in asthma. J Allergy Clin Immunol 116(5):996–999. https://doi.org/10.1016/j.jaci.2005.07.015

Shirakawa T, Enomoto T, Shimazu S, Hopkin JM (1997) The inverse association between tuberculin responses and atopic disorder. Science 275(5296):77–79

Smith RS, Harris SG, Phipps R, Iglewski B (2002) The *Pseudomonas aeruginosa* quorum-sensing molecule N-(3-oxododecanoyl)homoserine lactone contributes to virulence and induces inflammation in vivo. J Bacteriol 184(4):1132–1139

Smith-Norowitz TA, Chotikanatis K, Erstein DP, Perlman J, Norowitz YM, Joks R et al (2016) *Chlamydia pneumoniae* enhances the Th2 profile of stimulated peripheral blood mononuclear cells from asthmatic patients. Hum Immunol 77(5):382–388. https://doi.org/10.1016/j. humimm.2016.02.010

Spaulding AR, Salgado-Pabon W, Kohler PL, Horswill AR, Leung DY, Schlievert PM (2013) Staphylococcal and streptococcal superantigen exotoxins. Clin Microbiol Rev 26(3):422–447. https://doi.org/10.1128/CMR.00104-12

Stentzel S, Teufelberger A, Nordengrün M, Kolata J, Schmidt F, van Crombruggen K et al (2016) Staphylococcal serine protease-like proteins are pacemakers of allergic airway reactions to *Staphylococcus aureus*. J Allergy Clin Immunol 139(2):492–500.e8. https://doi.org/10.1016/j.jaci.2016.03.045

Tang LF, Shi YC, Xu YC, Wang CF, Yu ZS, Chen ZM (2009) The change of asthma-associated immunological parameters in children with *Mycoplasma pneumoniae* infection. J Asthma 46(3):265–269. https://doi.org/10.1080/02770900802647557

Tee RD, Pepys J (1982) Specific serum IgE antibodies to bacterial antigens in allergic lung disease. Clin Allergy 12(5):439–450

Telford G, Wheeler D, Williams P, Tomkins PT, Appleby P, Sewell H et al (1998) The *Pseudomonas aeruginosa* quorum-sensing signal molecule N-(3-oxododecanoyl)-L-homoserine lactone has immunomodulatory activity. Infect Immun 66(1):36–42

Thammavongsa V, Kim HK, Missiakas D, Schneewind O (2015) Staphylococcal manipulation of host immune responses. Nat Rev Microbiol 13(9):529–543. https://doi.org/10.1038/nrmicro3521

Tiringer K, Treis A, Fucik P, Gona M, Gruber S, Renner S et al (2013) A Th17- and Th2-skewed cytokine profile in cystic fibrosis lungs represents a potential risk factor for *Pseudomonas aeruginosa* infection. Am J Respir Crit Care Med 187(6):621–629. https://doi.org/10.1164/rccm.201206-1150OC

Tripathi A, Conley DB, Grammer LC, Ditto AM, Lowery MM, Seiberling KA et al (2004) Immunoglobulin E to staphylococcal and streptococcal toxins in patients with chronic sinusitis/nasal polyposis. Laryngoscope 114(10):1822–1826. https://doi.org/10.1097/00005537-200410000-00027

Umetsu DT, McIntire JJ, Akbari O, Macaubas C, DeKruyff RH (2002) Asthma: an epidemic of dysregulated immunity. Nat Immunol 3(8):715–720. https://doi.org/10.1038/ni0802-715

Upritchard HG, Cordwell SJ, Lamont IL (2008) Immunoproteomics to examine cystic fibrosis host interactions with extracellular *Pseudomonas aeruginosa* proteins. Infect Immun 76(10):4624–4632

Wang L, Chen Q, Shi C, Lv H, Xu X, Yu L (2015) Changes of serum TNF-alpha, IL-5 and IgE levels in the patients of mycoplasma pneumonia infection with or without bronchial asthma. Int J Clin Exp Med 8(3):3901–3906

Wark PA, Johnston SL, Simpson JL, Hensley MJ, Gibson PG (2002) *Chlamydia pneumoniae* immunoglobulin A reactivation and airway inflammation in acute asthma. Eur Respir J 20(4):834–840

Watanabe H, Uruma T, Nakamura H, Aoshiba K (2014) The role of Mycoplasma pneumoniae infection in the initial onset and exacerbations of asthma. Allergy Asthma Proc 35(3):204–210. https://doi.org/10.2500/aap.2014.35.3742

Webley WC, Tilahun Y, Lay K, Patel K, Stuart ES, Andrzejewski C, Salva PS (2009) Occurrence of Chlamydia trachomatis and *Chlamydia pneumoniae* in paediatric respiratory infections. Eur Respir J 33(2):360–367. https://doi.org/10.1183/09031936.00019508

Welliver RC, Duffy L (1993) The relationship of RSV-specific immunoglobulin E antibody responses in infancy, recurrent wheezing, and pulmonary function at age 7–8 years. Pediatr Pulmonol 15(1):19–27

Wertheim HFL, Melles DC, Vos MC, van Leeuwen W, van Belkum A, Verbrugh HA, Nouwen JI (2005) The role of nasal carriage in *Staphylococcus aureus* infections. Lancet Infect Dis 5:751–762

Wills-Karp M, Rani R, Dienger K, Lewkowich I, Fox JG, Perkins C et al (2012) Trefoil factor 2 rapidly induces interleukin 33 to promote type 2 immunity during allergic asthma and hookworm infection. J Exp Med 209(3):607–622. https://doi.org/10.1084/jem.20110079

Yano T, Ichikawa Y, Komatu S, Arai S, Oizumi K (1994) Association of *Mycoplasma pneumoniae* antigen with initial onset of bronchial asthma. Am J Respir Crit Care Med 149(5):1348–1353. https://doi.org/10.1164/ajrccm.149.5.8173777

Ye Q, Xu XJ, Shao WX, Pan YX, Chen XJ (2014) *Mycoplasma pneumoniae* infection in children is a risk factor for developing allergic diseases. ScientificWorldJournal 2014:986527. https://doi.org/10.1155/2014/986527

Yeh JJ, Wang YC, Hsu WH, Kao CH (2016) Incident asthma and *Mycoplasma pneumoniae*: a nationwide cohort study. J Allergy Clin Immunol 137(4):1017–1023.e1011–1016. https://doi.org/10.1016/j.jaci.2015.09.032

Yoshida M, Leigh R, Matsumoto K, Wattie J, Ellis R, O'Byrne PM, Inman MD (2002) Effect of interferon-gamma on allergic airway responses in interferon-gamma-deficient mice. Am J Respir Crit Care Med 166(4):451–456. https://doi.org/10.1164/rccm.200202-095OC

Good and Bad Farming: The Right Microbiome Protects from Allergy

Markus Johannes Ege

Abstract The hygiene hypothesis has provided a useful framework for the assessment of the asthma and atopy epidemic in relation to changes in lifestyle. As a model for a traditional environment rich in microbial exposure, the farm effect has advanced the development of the hygiene hypothesis. Exposure to farming, however, is ambiguous: a strong exposure such as in occupational settings can also increase the risk of allergies and respiratory disease, whereas in childhood most though not all farm-related exposures are protective for asthma and allergies.

First analyses relying on generic markers of microbial exposures supported the notion that microorganisms were involved in the beneficial effects of farm exposure; and direct evidence was provided by molecular techniques targeting microbial DNA.

Exposure to endotoxin and Gram-negative bacteria was shown to protect from atopy most likely through innate immune signalling. Risk of asthma, but not atopy was reduced by a high microbial diversity in the environment. A high diversity might enhance the probability of encountering asthma-protective groups of microorganisms. Candidates were found in the mould genera *Eurotium* and *Penicillium* and among bacteria. Microbial diversity in the environment was also reflected in the mucosal microbiome of the nasal cavity, suggesting a direct influence on the upper airways and perhaps indirectly on the epithelium of the lower airways as postulated by the united airways hypothesis.

At the mucosal surface the microorganisms interact with the host immune system; a broad microbial diversity may prevent outgrowth of pathogens. Restoration of a balanced microbiome may thus enhance prevention from allergy.

M.J. Ege
LMU Munich, Dr. von Hauner Children's Hospital, Munich, Germany
e-mail: markus.ege@med.lmu.de

© Springer International Publishing AG 2017   51
C.B. Schmidt-Weber (ed.), *Allergy Prevention and Exacerbation*, Birkhäuser
Advances in Infectious Diseases, https://doi.org/10.1007/978-3-319-69968-4_4

1   The Hygiene Hypothesis and the Farming Studies

In November 1989 Prof. David Strachan reported an inverse association of hay fever prevalence and number of older siblings (Strachan 1989). He saw the missing link between hay fever and sibship size in pre- or early postnatal infections 'transmitted by unhygienic contact' and postulated an allergy-protective effect by cross infections in young families. This is nowadays considered the foundation of the hygiene hypothesis, which amalgamated various pre-existing ideas about lack of infections and disturbed immunity (Ege and Rompa 2016).

Hardly a month later, Dr. Gassner-Bachmann published data from a survey in adolescents of a rural community in Switzerland and found a strong inverse association of hay fever prevalence and living on a farm with an odds ratio of 0.12 [0.03–0.49], $p < 0.001$ (Gassner-Bachmann 1989). At that time, he attributed the extremely low hay fever prevalence among farm children (1.5%) to the permanent exposure with grass pollen. This might now be seen as the birth of the so-called farm effect—or at least its rebirth, since the impressively low prevalence of hay fever in farmers was already described in 1860 by Édouard Cornaz in Switzerland and in 1873 by Charles Blackley in England as Dr. Gassner-Bachmann acknowledged himself (Gassner-Bachmann and Wüthrich 2000).

Both scientific concepts developed independently during subsequent years. The idea of favourable side-effects of unhygienic contacts was soon supported by reports on lower risk of atopic sensitization in individuals having experienced orofecal infections with hepatitis A virus, *Helicobacter pylori*, and *Toxoplasma gondii* (Matricardi et al. 1997, 2000).

The farm effect took some more loops and winding paths. In search of effects by indoor pollution, an unanticipated finding was made in Bavaria: children living in homes with open coal or wood heating had a lower risk of asthma, hay fever, and sensitization to pollen (von Mutius et al. 1996). This paradoxical discovery led the authors to speculate that coal and wood heating was a proxy for living on a farm, which was soon thereafter confirmed by three consecutive studies in Switzerland, Austria, and Germany (Braun-Fahrlander et al. 1999; Riedler et al. 2000; Von Ehrenstein et al. 2000). Subsequently the authors of the latter studies jointly performed the allergy and endotoxin (ALEX, Table 1) study, which focused on the effects of lipopolysaccharides, cell wall components of Gram-negative bacteria, which are also called endotoxins for their role in septic shock. This study confirmed the farm effect on asthma and atopy and revealed two major farm-related exposures, i.e. regular stays in animal sheds and consumption of unprocessed cow's milk directly obtained from a farm (Riedler et al. 2001). Though the latter exposure was not related to an increased bacterial count (Loss et al. 2011) or contamination with endotoxin (Gehring et al. 2008), the exposure to animal sheds obviously was. Actually exposure to endotoxin was strongly linked to a lower prevalence of atopy, hay fever, and atopic asthma (Braun-Fahrlander et al. 2002). As a consequence, the old idea of permanent pollen exposure was gradually replaced by the paradigm of farming as an environment rich in microbial exposure. This was

Table 1 The European farm children studies

Study acronym[a]	Year	Design	Countries	Population	Findings on microbial exposures
ALEX	1999	Cross-sectional	AT, CH, DE	n = 812	Endotoxin and atopy
PARSIFAL	2000–2002	Cross-sectional	AT, CH, DE, SE, NL	n = 8263	Bacterial diversity and asthma
GABRIELA	2006–2007	Cross-sectional	AT, CH, DE, PL	Phase 3 (n = 444)	Asthma and *Eurotium/ Penicillium* Atopy and Gram-negative bacteria
				– Austrian subset (n = 86)	Bacterial diversity in mattress dust and nasal swabs
				– Bavarian subset (n = 325)	Asthma and diversity in nasal but not throat swabs
PASTURE	Since 2002	Birth cohort	AT, CH, DE, FR, SF	n = 1133, 0–11 years	Gut microbiome (prospective)

[a]Full names: *ALEX* Allergy and endotoxin study (Riedler et al. 2001), *PARSIFAL* Prevention of Allergy Risk factors for Sensitization In children related to Farming and Anthroposophic Lifestyle (Alfven et al. 2006), *GABRIELA* Multidisciplinary Study to Identify the Genetic and Environmental Causes of Asthma in the European Community Advanced Study (Genuneit et al. 2011), *PASTURE* Protection against Allergy—Study in Rural Environments (von Mutius et al. 2006)

impressively illustrated by higher gene expression levels of toll-like receptor (TLR) 2 and CD14 in farm children of the ALEX study (Lauener et al. 2002). These two receptors of the innate immunity recognize microbial cell wall components such as lipopeptides and lipopolysaccharides and can essentially be seen as monitors of sustained microbial exposure.

With its new focus on microbial exposure the farm effect on asthma and allergies virtually boosted the development of the hygiene hypothesis. Vice-versa the hygiene hypothesis with its emerging immunologic foundation in the Th1/Th2 paradigm lent theoretical support to the farm effect.

2 Good and Bad Farming

The ALEX study revealed two separate effects of animal shed exposure and farm milk consumption (Riedler et al. 2001). The high sample size of the subsequent PARSIFAL study (Table 1), however, offered the opportunity to address many

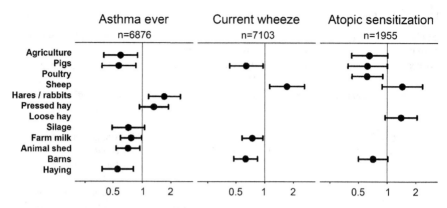

Fig. 1 Protective and risk factors among farm exposures. Odds ratios with 95%-confidence intervals are derived from multivariable models following backward elimination based on Akaike's information criterion. Reprint from Journal of Allergy and Clinical Immunology, Vol 119: 1140–7, Copyright 2007, with permission from Elsevier

more farm exposures and time windows in detail. Among others, this cross-sectional study included 8263 children from rural areas of five European countries (Alfven et al. 2006). A differential assessment of asthma, wheeze, and atopic sensitization revealed distinct protective exposures for these outcomes (Fig. 1). Also, unfavourable farm-related exposures were found, e.g. sheep farming or usage of certain types of animal fodder such as pressed hay, which was also inversely related to expression of TLR5 at school age (Ege et al. 2007). TLR5 is a receptor for bacterial flagellin and underexpressed in the lungs of severe asthmatics (Shikhagaie et al. 2014) thereby supporting the detected association with pressed hay.

Moreover, the PARSIFAL study revealed that gene expression of TLR2, TLR4, and CD14 at school age was already determined by maternal exposure to animal sheds during pregnancy (Ege et al. 2006). Particularly a dose-response relationship of gene expression and the number of animal species the mother had contact to emerged (Fig. 2). Traditional, family-run farms keep a variety of different animals, so this relationship may reflect effects of traditional farming as it was still common in Alpine regions around the turn of the millennium. However, other explanations are conceivable. The respective animal species are colonized by characteristic microbial communities, so the number of farm animal species may stand for diversity of microbial exposure.

Traditional farms differ from industrialized farms in the intensity of exposures, e.g. very high endotoxin levels as emitted by hog farms may cause asthma and wheeze particularly in individuals residing in the close neighbourhood or in chronically exposed farm workers (Mapp et al. 2005). However, chronic-obstructive pulmonary disease has particularly been attributed to traditional farming (Marescaux et al. 2016).

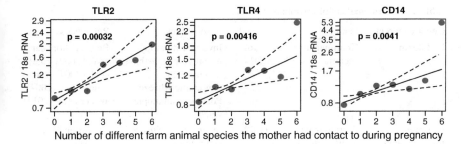

Number of different farm animal species the mother had contact to during pregnancy

Fig. 2 Gene expression of receptors of the innate immunity at school-age. Geometric means of gene expression standard to a house keeping gene (18s rRNA) and regression lines with 95%-confidence intervals are given. Reprint from Journal of Allergy and Clinical Immunology, Vol 117: 817–823, Copyright 2006, with permission from Elsevier

The question remains whether there is good and bad farming or, more precisely, farming with favourable and unfavourable health effects. Or does farm exposure follow Paracelsus' rule (Langman and Kapur 2006): *sola dosis facit venenum*? Children raised on farms may experience just the right dosage of exposure, not too high as in farm workers and not too low as in non-farm children.

3 Generic Markers of Microbial Exposure

Besides endotoxin various other generic markers of microbial exposure have been explored (Table 2). They represent different components of different microorganisms; thus a diverse picture of associations with allergy-related outcomes is not unexpected. Though inverse associations with allergic outcomes have repeatedly been found, it remains unclear whether the generic markers truly reflect microbial exposure or whether the associations are spurious. The effect on atopic asthma in the ALEX study was stronger for endotoxin *load*, i.e. endotoxin units per square meter of mattress surface, with an odds ratio of 0.48 [0.28–0.81] as compared to endotoxin *level* (OR = 0.73 [0.44–1.19]), which is calculated as endotoxin units per dust mass (Braun-Fahrlander et al. 2002). This discrepancy suggests that a relevant share of the association is rather due to differences in the dust mass collected from a definite surface, which again might stand for any substance contained in dust. So it remains unclear whether microorganisms are effectively involved in the asthma- and allergy protective effects observed in the ALEX study. In other words: is the farm effect actually related to microbial exposure?

Table 2 Generic markers of microbial exposure

Generic marker	Source	Ligands	Associations	References
Endotoxin	Gram-negative bacteria	TLR4, CD14	Inverse with atopy Negative with asthma Inverse with non-atopic wheeze Positive with atopic wheeze	Braun-Fahrlander et al. (2002), Ege et al. (2007), van Strien et al. (2004), Weber et al. (2015)
N-acetyl-muramic acid	Peptidoglycans of all bacteria, particularly Gram-positive bacteria	TLR2	Inverse with wheeze and asthma	van Strien et al. (2004)
Extracellular polysaccharides (EPS)	Indoor moulds, mostly based on EPS from *Penicillium* sp. and *Aspergillus* sp.		Inverse with asthma	Ege et al. (2007)
β (1→3) glucans	Cell wall constituents of most fungi, but also of some bacteria and many plants		Inverse with atopy (nonsignificant)	Ege et al. (2007)
Ergosterol	Marker of fungal biomass		No clear association	Karvonen et al. (2014)

4 Microbial DNA

Microorganisms leave their fingerprints anywhere like burglars; hence they can be convicted by the same methods. There are rather specific segments in the genetic code of bacteria and fungi, which can be used to trace down the culprits to species level. The DNA coding for the smaller subunit of bacterial ribosomes (16s rRNA) is characteristic for bacteria and contains a highly variable fragment, which reveals the identity of the bacterial genus if not species (Legatzki et al. 2014). For fungi a similar DNA fragment between the genes of their rRNA subunits, the internal transcribed spacer (ITS) region provides the same features. So, the amplification and subsequent sequencing of these molecular fingerprints offers a rather specific toolkit for the detection of microorganisms. As long as parallel sequencing was unaffordable, gel-based techniques such as polymerase chain reaction—single strand conformation polymorphism (PCR-SSCP) were used to separate DNA fragments based on their primary structure, i.e. the nucleotide sequence (Korthals et al. 2008). Only bands of interest were then identified by Sanger sequencing.

When applying this technique to mattress dust samples of the PARSIFAL study, about 70 individual bands were found with the vast majority being more frequently detected in samples from farm children as compared to rural reference children (Ege et al. 2011). In combination with the above-mentioned association of gene

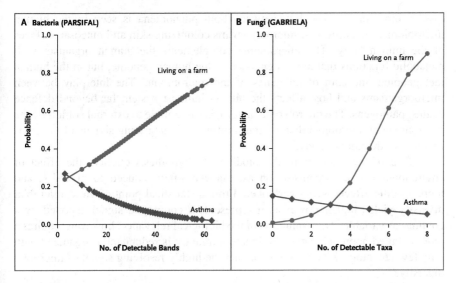

Fig. 3 Microbial diversity and asthma. From Exposure to environmental microorganisms and childhood asthma, Ege, M.J., Mayer, M., Normand, A.C., Genuneit, J., Cookson, W.O., Braun-Fahrlander, C., Heederik, D., Piarroux, R. & von Mutius, E., 364, 701–9. Copyright (2011) Massachusetts Medical Society. Reprinted with permission from Massachusetts Medical Society

expression of the innate immunity with the diversity of animal species during prenatal exposure (Fig. 2) this finding suggested a higher diversity in bacterial exposure in farm children. As illustrated by Fig. 3, this assumption was clearly confirmed. Independently of this association, also the asthma risk dropped by almost 40% for every 10 additional PCR-SSCP bands (Fig. 3). In addition, a quarter of the farm effect on asthma was explained by bacterial diversity. For the first time, a part of the farm effect on asthma could be attributed to microbial exposure as determined by bacterial DNA.

5   Microbial Diversity and Hot Spots of Microbial Exposure

The interpretation of microbial diversity, however, is difficult. A higher diversity may enhance the stochastic probability of exposure to advantageous microorganisms. On the other hand, an ample variety of microorganisms may keep detrimental microorganisms at bay or reflect a rather stable situation with no harmful bacteria overgrowing the established flora (Ege et al. 2011).

Additionally, the inverse association of asthma prevalence with microbial diversity was well in line with the biodiversity hypothesis, which was proposed in the same year as the effect of microbial diversity (von Hertzen et al. 2011). The biodiversity hypothesis postulates that the increase of inflammatory disorders such as asthma and allergies is a consequence of the currently witnessed dramatic loss of biodiversity. The link between both phenomena is seen in the human microbiome, the entirety of microorganisms colonizing skin and mucosal surfaces of the human body. The microbiome complements the human organism with metabolic functions that are not encoded in the human genome, but in the human metagenome, the sum of all genes of the microbiome. The interplay between microorganisms and host affects the human immune system far beyond defence against pathogens. The microbiome, in turn, is susceptible to external stimuli by the environmental microbiota thereby transmitting and integrating signals of the environment to the human body.

In face of the enthusiasm the biodiversity hypothesis elicited, the effect of environmental microorganisms on asthma was often reduced to diversity, and more specific effects were overlooked. From a statistical point of view, microbial diversity yields higher power as it represents a compound signal encoded by a continuous variable. The evaluation of the biological relevance, however, requires a more detailed approach, since the innate immunity is redundantly organized with only few receptors, which cannot conduct the highly resolving signal of microbial diversity.

As microorganisms live in communities they might exert shared effects on the human immune system; hence the *composition* of the microbial communities should be in the focus of the analyses. For the assessment of the high-dimensional microbial data various types of linear and non-linear dimensionality reduction techniques have been explored such as non-metric multidimensional scaling (NMDS), principal coordinate analysis (PCoA), principal components analysis (PCA), and factor analysis. The common denominator of these statistical approaches is a focus on few dimensions that can be illustrated graphically and interpreted.

In the PARSIFAL study, e.g., a factor analysis of PCR-SSCP bands revealed clusters of asthma-protective bacteria, which again explained 57% of the diversity effect on asthma. These clusters contained sequences of *Listeria monocytogenes, Bacillus licheniformis, Staphylococcus sciuri,* and of the genera *Corynebacterium, Methylobacterium, Xanthomonas, Enterobacter, Pantoea, Staphylococcus, Salinicoccus, Macrococcus, Bacillus, Jeotgalicoccus,* and others (Ege et al. 2011). Some of these candidates were already known from murine asthma models such as *Listeria monocytogenes* (Hansen et al. 2000). Many of the listed genera of the family *Staphylococcaceae* were found in regions with a very low prevalence of asthma and atopy in the framework of the Karelia studies, which impressively illustrate the effects of disparate environmental exposures in geographically adjacent areas (Pakarinen et al. 2008).

6 The Role of Fungi

The first analyses on fungal exposure among the farm studies relied on generic markers such as extracellular polysaccharides (EPS) of *Penicillium* sp. and *Aspergillus* sp. and β $(1{\rightarrow}3)$ glucans (Schram et al. 2005). These analyses revealed, in contrast to their role as triggers for autoallergy (chapter "Microbial Triggers in Autoimmunity, Severe Allergy, and Autoallergy"), protective effects of fungal exposure on asthma and possibly also atopy (Table 2) (Ege et al. 2007). Both markers have been shown to be correlated and also related to the concentration of colony forming units of total culturable fungi, though EPS of *Penicillium* was not related to culturable *Penicillium* concentration (Chew et al. 2001). This may indicate that growth conditions *in vitro* differ from those in the habitats where fungi actually exert their effects on human health.

In analogy to the association of asthma and bacterial diversity (Fig. 3), a similar picture was found for fungal diversity as determined by classical culturing and microscopy in the GABRIELA study (Table 1), though the effect of fungal diversity (odds ratio 0.86 [0.75–0.99] per additional taxon) was fully consumed by the presence of the taxa *Penicillium* sp. and *Eurotium* sp., a sexual form of some *Aspergillus* sp. (Ege et al. 2011).

A more advanced analysis of the same study using PCR-SSCP targeting ITS essentially confirmed these candidate taxa and suggested *Pseudotaeniolina globosa* and species of the genera *Metschnikowia, Aureobasidium, Epicoccum,* and *Galactomyces* as novel candidates (Mueller-Rompa et al. 2017). Also in urban environments an inverse association of fungal diversity with atopic sensitization and wheeze was observed (Tischer et al. 2016).

The idea of moulds such as *Eurotium* and *Penicillium* being asthma-protective seems to be counter-intuitive as moulds are known risk factors for asthma, particularly in the presence of wall humidity and water damage (Reponen et al. 2012). In turn, the relation of moisture to fungal diversity has been reported to differ with visibility of mould spots, which may stand for secretion of antifungal metabolites by moulds (Dannemiller et al. 2014). So, the complexity of these interactions and the variability of the respective growth conditions may explain the seemingly paradoxical effects of mould exposure on asthma.

Furthermore, fungi produce metabolites with powerful effects on other microbiota; many antibiotics, e.g. penicillins and cephalosporins, are derived from fungal products. In addition, mycotoxins can modulate the human immune system as illustrated by immunosuppressive drugs such as ciclosporin. It is therefore conceivable that fungi may shape the composition of environmental bacteria, which themselves impact on human health. Alternatively, fungi may directly influence our immune system by their immunomodulatory properties, thereby inducing tolerance towards allergens.

7 The Special Case of Gram-Negative Bacteria and Endotoxin

In contrast to the clear association of asthma and microbial diversity, atopic sensitization and atopic disease in the form of hay fever were completely unrelated to microbial diversity (Ege et al. 2011). However, the presence of Gram-negative rods, the main source of endotoxins halved the risk of atopy. In search of candidate bacteria for atopy (Ege et al. 2012), *Acinetobacter lwoffii* was detected, a Gram-negative species of the family *Moraxellaceae* with proven atopy-protective properties in murine models of allergic airway inflammation (Debarry et al. 2007; Conrad et al. 2009).

The well-established effect of endotoxin on atopy (Braun-Fahrlander et al. 2002) has been explored further. A comparison between the Amish and the Hutterites, two populations in the U.S., who share a common genetic ancestry and a traditional lifestyle in agricultural communities, differed in their asthma and atopy prevalence (Stein et al. 2016). Dust from Amish homes prevented airway hyperresponsiveness in a murine model of allergic asthma whereas dust from Hutterite homes did not. The two populations mainly differ in their style of farming: Hutterites run modern industrialized farms, whereas the Amish keep to a traditional style of farming as their ancestors did 200 years ago. Though dust samples differed in the distribution of bacterial families, the most striking difference were the high endotoxin levels in Amish homes exceeding those of Hutterite homes by seven times. Knockout mice deficient for MyD88 and Trif did not respond to Amish dust thereby verifying that endotoxin operates through innate immune signalling.

In addition, the following modes of action are discussed for endotoxins: They suppress the activation of epithelial and dendritic cells of the airway mucosa by induction of the ubiquitin-modifying enzyme A20 (Schuijs et al. 2015). The expression of A20 is induced in intestinal epithelium by bacterial colonization and contributes to tolerance towards TLR ligands and bacteria (Wang et al. 2009). Another explanation of the endotoxin effect is related to the structure of lipopolysaccharides, which may downregulate TLR4 and NF-κB expression thereby inducing endotoxin tolerance in the gut (Vatanen et al. 2016). These explanations, however, are restricted to an effect of endotoxin on the intestinal mucosa; how this relates to airway mucosa remains still elusive. Endotoxin particles from animal sheds are usually too large to enter the lower airways (Berger et al. 2005), thus leaving only the nasal mucosa for interactions with the human immune system, presumably in a similar way as described for the intestinal mucosa. The united airways hypothesis postulates common pathomechanisms for allergic diseases of both the upper and lower airways, and allergic rhinitis often precedes the onset of asthma (Rochat et al. 2010); so a signal from environmental exposure received at the mucosa of the upper airways may affect the epithelium of the entire respiratory tract and the associated immune tissue.

8 Translation from Environmental Microbiota to Mucosal Microbiomes

What is true for small endotoxin particles is even more compelling for entire bacteria: How can bacteria, viable or not, impact on the airway mucosa? Bioactive metabolites produced by environmental microorganisms might be inhaled and penetrate also smaller airways. There they might exert their effects at the mucosal surfaces. Alternatively, environmental microorganisms may influence the colonization of the mucosa of the upper airways and the proximal sections of the lower airways thereby impacting the immune system or complementing the host with metabolic functions. Also horizontal gene-transfer from airborne to mucosal bacteria is conceivable.

The GABRIELA study (Table 1) provided the opportunity to directly compare nasal swabs to mattress dust samples with respect to 16s rRNA sequencing (Birzele et al. 2017). As a simple measure of α-diversity, i.e. the diversity within a sample, the number of individual species or any other taxa can be summed up. This sum is called 'richness' and can also be based on operational taxonomic units (OTUs). OTUs are clusters of DNA sequences usually defined by a sequence similarity of 97%, which approximately corresponds to the traditional definition of a species.

The relation of α-diversity in nasal swabs and mattress dust can be assessed by plotting the difference in richness against the mean of the richness values from both sources ('Bland-Altman plot'). Figure 4 demonstrates substantial agreement of richness in both sampling sources as the majority of data pairs range within two standard deviations of the mean difference as indicated by the dashed lines. The mean difference of about 200 (solid line) indicates that richness in mattress dust exceeds that in nasal swabs on average by 200 OTUs. In addition we see that farm children have higher difference and average values as compared to rural non-farm children, which means that their environment is richer in bacterial diversity and that this richness also translates to nasal swabs.

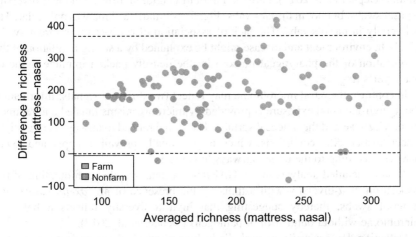

Fig. 4 Bland-Altman plot for bacterial diversity in environmental and mucosal samples. Reprinted from Allergy, Vol. 72, Birzele LT, Depner M, Ege MJ, Engel M, Kublik S, Bernau C, Loss GJ, Genuneit J, Horak E, Schloter M, Braun-Fahrlander C, Danielewicz H, Heederik D, von Mutius E, Legatzki A. Environmental and mucosal microbiota and their role in childhood asthma. Pages 109–119. Copyright 2017, with permission from John Wiley and Sons

Whether these environmental bacteria truly colonize the nasal mucosa or whether they are just caught by the vibrissae, the hairs protecting the nares from larger particles cannot be answered by these data. Nevertheless they demonstrate a relevant association between bacteria from environments and those coming into close contact with the airway mucosa.

Among the bacteria more prominently found in mattress dust samples of farm children were the genera *Bacteroides* and *Ruminococcus* (Birzele et al. 2017), which ferment plant fibres and produce propionate (Reichardt et al. 2014). Short chain fatty acids such as propionate have been shown to dampen Th2 cell–mediated allergic airway inflammation in a mouse model (Trompette et al. 2014). This may explain a share of the farm effect that is related to exposure to cows, whose stomachs are basically large-scale fermenters. Taken together, asthma- and atopy-protective effects by microbial metabolites are quite likely though they have not yet directly been assessed in human populations.

9 Microbial Virulence or Host Factors?

Whether colonization of the human mucosa is involved in the effects of environmental bacteria on asthma and allergies is not only a question of exposure but also of host factors.

In a different subsample of the Bavarian arm of the GABRIELA study (Table 1) nasal and pharyngeal microbiota were compared with respect to their effect on asthma (Depner et al. 2017). Farm and nonfarm children differed in the α-diversity in nasal swabs, but not in throat swabs (Fig. 5). Similarly asthma was only related to α-diversity in nasal swabs. The lack of association of α-diversity in throat swabs with both environment and disease might be explained by a strong regulation of the colonization of the pharyngeal mucosa by the densely packed immune tissue of Waldeyer's ring.

In contrast, the nasal microbiome may reflect two different signals: Information about environmental exposure is provided by microorganisms inhaled and retained by the vibrissae and the mucosal surface. The signal related to airway disease might reflect host-specific conditions, which might equally prevail at upper and lower airways according to the united airways hypothesis.

A more detailed analysis of the GABRIELA data revealed confounding of the association of α-diversity with asthma by the presence of *Moraxella catarrhalis* in nasal swabs, thereby suggesting that high α-diversity reflects a balanced microbiome without outgrowth of pathogens (Depner et al. 2017).

Detection of *Moraxella catarrhalis* in hypopharyngeal swabs in neonates has previously been described to precede development both of asthma and pneumonia by years (Bisgaard et al. 2007; Vissing et al. 2013). The ubiquitous occurrence of *Moraxella catarrhalis* in the nose, the hypopharynx, and the lung (Hilty et al. 2010) and its unspecific effect on asthma and pneumonia point towards host conditions such as an inadequate immune defence (Ege and von Mutius 2013). The virulence

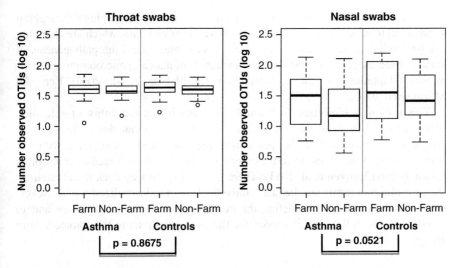

Fig. 5 Bacterial diversity in throat and nasal swabs. Reprinted with modifications from Journal of Allergy and Clinical Immunology, Vol 139, Depner M, Ege MJ, Cox MJ, Dwyer S, Walker AW, Birzele LT, Genuneit J, Horak E, Braun-Fahrlander C, Danielewicz H, Maier RM, Moffatt MF, Cookson WO, Heederik D, von Mutius E, Legatzki A. Bacterial microbiota of the upper respiratory tract and childhood asthma. Pages 826–834, Copyright 2017, with permission from Elsevier

of *Moraxella catarrhalis* might still be enhanced by exogenous factors such as horizontal gene transfer from community acquired microorganisms.

Finally, a different body region has also been implied in the development of asthma, the intestinal mucosa. The concept of the gut-lung axis postulates a link between the intestinal microbiome and effects on the lung-specific pathologies involving various types of immune cells (Legatzki et al. 2014). In the PARSIFAL study (Table 1), a reduced bacterial diversity was found in faecal samples from farm children, which was explained by a higher consumption of milk (Dicksved et al. 2007). These analyses, however, were based on terminal restriction fragment length polymorphism, which may not adequately cover the entire range of bacteria. Preliminary analyses of the PASTURE birth cohort (Table 1) suggest a higher bacterial diversity in faecal samples from farm children at the age of 12 months (Depner and Ege 2017, unpublished) though evidence for a role of the intestinal microbiome in the farm effect on asthma is still lacking.

10   The Right Microbiome

Though the gut microbiome has long been addressed as the 'forgotten organ' (O'Hara and Shanahan 2006), the importance of the microbiota on other mucosal surfaces, on the skin, and in our environment is currently only emerging. The 'microbiome revolution' has been catalysed by the availability of high-throughput sequencing techniques. These novel tools have generated data, which are changing our perception of the origins of disease; we now understand the pathogenesis of various diseases in the context of the co-evolution of man and microbiome and their mutual adaptation or just maladaptation (Dethlefsen et al. 2007; Lee and Mazmanian 2010).

This notion is important as it emphasizes the host's capability of selecting, within limits, beneficial microorganisms as colonizers (Van den Abbeele et al. 2011). The environmental changes during the last decades associated with our Western lifestyle have restricted the diversity of the biosphere including microbiota seriously (von Hertzen et al. 2011). There is probably no way back to the environments traditional farms are the last remnants of in an industrialized world. Nevertheless, it is up to us to redefine the conditions of microbial diversity and to re-establish the full range of choice for the right microbiome that protects from allergy.

References

Alfven T, Braun-Fahrlander C, Brunekreef B, von Mutius E, Riedler J, Scheynius A, van Hage M, Wickman M, Benz MR, Budde J, Michels KB, Schram D, Ublagger E, Waser M, Pershagen G, Group Ps (2006) Allergic diseases and atopic sensitization in children related to farming and anthroposophic lifestyle--the PARSIFAL study. Allergy 61(4):414–421. https://doi.org/10.1111/j.1398-9995.2005.00939.x

Berger I, Schierl R, Ochmann U, Egger U, Scharrer E, Nowak D (2005) Concentrations of dust, allergens and endotoxin in stables, living rooms and mattresses from cattle farmers in southern Bavaria. Ann Agric Environ Med 12(1):101–107

Birzele LT, Depner M, Ege MJ, Engel M, Kublik S, Bernau C, Loss GJ, Genuneit J, Horak E, Schloter M, Braun-Fahrlander C, Danielewicz H, Heederik D, von Mutius E, Legatzki A (2017) Environmental and mucosal microbiota and their role in childhood asthma. Allergy 72 (1):109–119. https://doi.org/10.1111/all.13002

Bisgaard H, Hermansen MN, Buchvald F, Loland L, Halkjaer LB, Bonnelykke K, Brasholt M, Heltberg A, Vissing NH, Thorsen SV, Stage M, Pipper CB (2007) Childhood asthma after bacterial colonization of the airway in neonates. N Engl J Med 357(15):1487–1495. https://doi.org/10.1056/NEJMoa052632

Braun-Fahrlander C, Gassner M, Grize L, Neu U, Sennhauser FH, Varonier HS, Vuille JC, Wuthrich B (1999) Prevalence of hay fever and allergic sensitization in farmer's children and their peers living in the same rural community. SCARPOL team. Swiss Study on Childhood Allergy and Respiratory Symptoms with Respect to Air Pollution. Clin Exp Allergy 29 (1):28–34

Braun-Fahrlander C, Riedler J, Herz U, Eder W, Waser M, Grize L, Maisch S, Carr D, Gerlach F, Bufe A, Lauener RP, Schierl R, Renz H, Nowak D, von Mutius E, Allergy, Endotoxin Study T (2002) Environmental exposure to endotoxin and its relation to asthma in school-age children. N Engl J Med 347(12):869–877. https://doi.org/10.1056/NEJMoa020057

Chew GL, Douwes J, Doekes G, Higgins KM, van Strien R, Spithoven J, Brunekreef B (2001) Fungal extracellular polysaccharides, beta (1-->3)-glucans and culturable fungi in repeated sampling of house dust. Indoor Air 11(3):171–178

Conrad ML, Ferstl R, Teich R, Brand S, Blumer N, Yildirim AO, Patrascan CC, Hanuszkiewicz A, Akira S, Wagner H, Holst O, von Mutius E, Pfefferle PI, Kirschning CJ, Garn H, Renz H

(2009) Maternal TLR signaling is required for prenatal asthma protection by the nonpathogenic microbe Acinetobacter lwoffii F78. J Exp Med 206(13):2869–2877. https://doi.org/10.1084/jem.20090845

Dannemiller KC, Mendell MJ, Macher JM, Kumagai K, Bradman A, Holland N, Harley K, Eskenazi B, Peccia J (2014) Next-generation DNA sequencing reveals that low fungal diversity in house dust is associated with childhood asthma development. Indoor Air 24(3):236–247

Debarry J, Garn H, Hanuszkiewicz A, Dickgreber N, Blumer N, von Mutius E, Bufe A, Gatermann S, Renz H, Holst O, Heine H (2007) Acinetobacter lwoffii and Lactococcus lactis strains isolated from farm cowsheds possess strong allergy-protective properties. J Allergy Clin Immunol 119(6):1514–1521. https://doi.org/10.1016/j.jaci.2007.03.023

Depner M, Ege MJ, Cox MJ, Dwyer S, Walker AW, Birzele LT, Genuneit J, Horak E, Braun-Fahrlander C, Danielewicz H, Maier RM, Moffatt MF, Cookson WO, Heederik D, von Mutius E, Legatzki A (2017) Bacterial microbiota of the upper respiratory tract and childhood asthma. J Allergy Clin Immunol 139(3):826–834. e813. https://doi.org/10.1016/j.jaci.2016.05.050

Dethlefsen L, McFall-Ngai M, Relman DA (2007) An ecological and evolutionary perspective on human-microbe mutualism and disease. Nature 449(7164):811–818

Dicksved J, Floistrup H, Bergstrom A, Rosenquist M, Pershagen G, Scheynius A, Roos S, Alm JS, Engstrand L, Braun-Fahrlander C, von Mutius E, Jansson JK (2007) Molecular fingerprinting of the fecal microbiota of children raised according to different lifestyles. Appl Environ Microbiol 73(7):2284–2289. https://doi.org/10.1128/AEM.02223-06

Ege M, Rompa S (2016) The hygiene hypothesis of allergy and asthma. In: Ratcliffe MJH (ed) Encyclopedia of immunobiology. Academic Press, Oxford, pp 328–335. https://doi.org/10.1016/B978-0-12-374279-7.16004-7

Ege M, von Mutius E (2013) Microbial airway colonization: a cause of asthma and pneumonia? Am J Respir Crit Care Med 188(10):1188–1189. https://doi.org/10.1164/rccm.201309-1680ED

Ege MJ, Bieli C, Frei R, van Strien RT, Riedler J, Ublagger E, Schram-Bijkerk D, Brunekreef B, van Hage M, Scheynius A, Pershagen G, Benz MR, Lauener R, von Mutius E, Braun-Fahrlander C, Study t P (2006) Prenatal farm exposure is related to the expression of receptors of the innate immunity and to atopic sensitization in school-age children. J Allergy Clin Immunol 117(4):817–823. https://doi.org/10.1016/j.jaci.2005.12.1307

Ege MJ, Frei R, Bieli C, Schram-Bijkerk D, Waser M, Benz MR, Weiss G, Nyberg F, van Hage M, Pershagen G, Brunekreef B, Riedler J, Lauener R, Braun-Fahrlander C, von Mutius E, team PS (2007) Not all farming environments protect against the development of asthma and wheeze in children. J Allergy Clin Immunol 119(5):1140–1147. https://doi.org/10.1016/j.jaci.2007.01.037

Ege MJ, Mayer M, Normand AC, Genuneit J, Cookson WO, Braun-Fahrlander C, Heederik D, Piarroux R, von Mutius E, Group GTS (2011) Exposure to environmental microorganisms and childhood asthma. N Engl J Med 364(8):701–709. https://doi.org/10.1056/NEJMoa1007302

Ege MJ, Mayer M, Schwaiger K, Mattes J, Pershagen G, van Hage M, Scheynius A, Bauer J, von Mutius E (2012) Environmental bacteria and childhood asthma. Allergy 67(12):1565–1571. https://doi.org/10.1111/all.12028

Gassner-Bachmann M (1989) Allergie und Umwelt. Allergologie 12(12):492–502

Gassner-Bachmann M, Wüthrich B (2000) Bauernkinder leiden selten an heuschnupfen und asthma (null). Dtsch Med Wochenschr 125(31/32):924–931. https://doi.org/10.1055/s-2000-6778

Gehring U, Spithoven J, Schmid S, Bitter S, Braun-Fahrlander C, Dalphin JC, Hyvarinen A, Pekkanen J, Riedler J, Weiland SK, Buchele G, von Mutius E, Vuitton DA, Brunekreef B, Group Ps (2008) Endotoxin levels in cow's milk samples from farming and non-farming families - the PASTURE study. Environ Int 34(8):1132–1136. https://doi.org/10.1016/j.envint.2008.04.003

Genuneit J, Buchele G, Waser M, Kovacs K, Debinska A, Boznanski A, Strunz-Lehner C, Horak E, Cullinan P, Heederik D, Braun-Fahrlander C, von Mutius E (2011) The GABRIEL advanced surveys: study design, participation and evaluation of bias. Paediatr Perinat Epidemiol 25(5):436–447

Hansen G, Yeung VP, Berry G, Umetsu DT, DeKruyff RH (2000) Vaccination with heat-killed *Listeria* as adjuvant reverses established allergen-induced airway hyperreactivity and inflammation: role of CD8+ T cells and IL-18. J Immunol 164(1):223–230

Hilty M, Burke C, Pedro H, Cardenas P, Bush A, Bossley C, Davies J, Ervine A, Poulter L, Pachter L, Moffatt MF, Cookson WO (2010) Disordered microbial communities in asthmatic airways. PloS One 5(1):e8578. https://doi.org/10.1371/journal.pone.0008578

Karvonen AM, Hyvarinen A, Rintala H, Korppi M, Taubel M, Doekes G, Gehring U, Renz H, Pfefferle PI, Genuneit J, Keski-Nisula L, Remes S, Lampi J, von Mutius E, Pekkanen J (2014) Quantity and diversity of environmental microbial exposure and development of asthma: a birth cohort study. Allergy 69(8):1092–1101. https://doi.org/10.1111/all.12439

Korthals M, Ege MJ, Tebbe CC, von Mutius E, Bauer J (2008) Application of PCR-SSCP for molecular epidemiological studies on the exposure of farm children to bacteria in environmental dust. J Microbiol Methods 73(1):49–56. https://doi.org/10.1016/j.mimet.2008.01.010

Langman LJ, Kapur BM (2006) Toxicology: then and now. Clin Biochem 39(5):498–510. https://doi.org/10.1016/j.clinbiochem.2006.03.004

Lauener RP, Birchler T, Adamski J, Braun-Fahrlander C, Bufe A, Herz U, von Mutius E, Nowak D, Riedler J, Waser M, Sennhauser FH (2002) Expression of CD14 and Toll-like receptor 2 in farmers' and non-farmers' children. Lancet 360(9331):465–466

Lee YK, Mazmanian SK (2010) Has the microbiota played a critical role in the evolution of the adaptive immune system? Science 330(6012):1768–1773

Legatzki A, Rosler B, von Mutius E (2014) Microbiome diversity and asthma and allergy risk. Curr Allergy Asthma Rep 14(10):466. https://doi.org/10.1007/s11882-014-0466-0

Loss G, Apprich S, Waser M, Kneifel W, Genuneit J, Buchele G, Weber J, Sozanska B, Danielewicz H, Horak E, van Neerven RJ, Heederik D, Lorenzen PC, von Mutius E, Braun-Fahrlander C, Group Gs (2011) The protective effect of farm milk consumption on childhood asthma and atopy: the GABRIELA study. J Allergy Clin Immunol 128(4):766–773. e764. https://doi.org/10.1016/j.jaci.2011.07.048

Mapp CE, Boschetto P, Maestrelli P, Fabbri LM (2005) Occupational asthma. Am J Respir Crit Care Med 172(3):280–305. https://doi.org/10.1164/rccm.200311-1575SO

Marescaux A, Degano B, Soumagne T, Thaon I, Laplante JJ, Dalphin JC (2016) Impact of farm modernity on the prevalence of chronic obstructive pulmonary disease in dairy farmers. Occup Environ Med 73(2):127–133. https://doi.org/10.1136/oemed-2014-102697

Matricardi PM, Rosmini F, Ferrigno L, Nisini R, Rapicetta M, Chionne P, Stroffolini T, Pasquini P, D'Amelio R (1997) Cross sectional retrospective study of prevalence of atopy among Italian military students with antibodies against hepatitis A virus. BMJ 314 (7086):999–1003

Matricardi PM, Rosmini F, Riondino S, Fortini M, Ferrigno L, Rapicetta M, Bonini S (2000) Exposure to foodborne and orofecal microbes versus airborne viruses in relation to atopy and allergic asthma: epidemiological study. BMJ 320(7232):412–417

Mueller-Rompa S, Janke T, Schwaiger K, Mayer M, Bauer J, Genuneit J, Braun-Fahrlaender C, Horak E, Boznanski A, von Mutius E, Ege MJ, Group Gs (2017) Identification of fungal candidates for asthma protection in a large population-based study. Pediatr Allergy Immunol 28(1):72–78. https://doi.org/10.1111/pai.12665

O'Hara AM, Shanahan F (2006) The gut flora as a forgotten organ. EMBO Rep 7(7):688–693. https://doi.org/10.1038/sj.embor.7400731

Pakarinen J, Hyvarinen A, Salkinoja-Salonen M, Laitinen S, Nevalainen A, Makela MJ, Haahtela T, von Hertzen L (2008) Predominance of gram-positive bacteria in house dust in the low-allergy risk Russian Karelia. Environ Microbiol 10(12):3317–3325. https://doi.org/10.1111/j.1462-2920.2008.01723.x

Reichardt N, Duncan SH, Young P, Belenguer A, McWilliam Leitch C, Scott KP, Flint HJ, Louis P (2014) Phylogenetic distribution of three pathways for propionate production within the human gut microbiota. ISME J 8(6):1323–1335. https://doi.org/10.1038/ismej.2014.14

Reponen T, Lockey J, Bernstein DI, Vesper SJ, Levin L, Khurana Hershey GK, Zheng S, Ryan P, Grinshpun SA, Villareal M, Lemasters G (2012) Infant origins of childhood asthma associated with specific molds. J Allergy Clin Immunol 130(3):639–644. e635. https://doi.org/10.1016/j.jaci.2012.05.030

Riedler J, Eder W, Oberfeld G, Schreuer M (2000) Austrian children living on a farm have less hay fever, asthma and allergic sensitization. Clin Exp Allergy 30(2):194–200

Riedler J, Braun-Fahrlander C, Eder W, Schreuer M, Waser M, Maisch S, Carr D, Schierl R, Nowak D, von Mutius E, Team AS (2001) Exposure to farming in early life and development of asthma and allergy: a cross-sectional survey. Lancet 358(9288):1129–1133. https://doi.org/10.1016/S0140-6736(01)06252-3

Rochat MK, Illi S, Ege MJ, Lau S, Keil T, Wahn U, von Mutius E, Multicentre Allergy Study Group (2010) Allergic rhinitis as a predictor for wheezing onset in school-aged children. J Allergy Clin Immunol 126(6):1170–1175. e1172. https://doi.org/10.1016/j.jaci.2010.09.008

Schram D, Doekes G, Boeve M, Douwes J, Riedler J, Ublagger E, von Mutius E, Budde J, Pershagen G, Nyberg F, Alm J, Braun-Fahrlander C, Waser M, Brunekreef B, Group PS (2005) Bacterial and fungal components in house dust of farm children, Rudolf Steiner school children and reference children--the PARSIFAL Study. Allergy 60(5):611–618. https://doi.org/10.1111/j.1398-9995.2005.00748.x

Schuijs MJ, Willart MA, Vergote K, Gras D, Deswarte K, Ege MJ, Madeira FB, Beyaert R, van Loo G, Bracher F, von Mutius E, Chanez P, Lambrecht BN, Hammad H (2015) Farm dust and endotoxin protect against allergy through A20 induction in lung epithelial cells. Science 349 (6252):1106–1110. https://doi.org/10.1126/science.aac6623

Shikhagaie MM, Andersson CK, Mori M, Kortekaas Krohn I, Bergqvist A, Dahl R, Ekblad E, Hoffmann HJ, Bjermer L, Erjefalt JS (2014) Mapping of TLR5 and TLR7 in central and distal human airways and identification of reduced TLR expression in severe asthma. Clin Exp Allergy 44(2):184–196. https://doi.org/10.1111/cea.12176

Stein MM, Hrusch CL, Gozdz J, Igartua C, Pivniouk V, Murray SE, Ledford JG, Marques dos Santos M, Anderson RL, Metwali N, Neilson JW, Maier RM, Gilbert JA, Holbreich M, Thorne PS, Martinez FD, von Mutius E, Vercelli D, Ober C, Sperling AI (2016) Innate immunity and asthma risk in amish and hutterite farm children. N Engl J Med 375(5):411–421. https://doi.org/10.1056/NEJMoa1508749

Strachan DP (1989) Hay fever, hygiene, and household size. BMJ 299(6710):1259–1260

Tischer C, Weikl F, Probst AJ, Standl M, Heinrich J, Pritsch K (2016) Urban dust microbiome: impact on later atopy and wheezing. Environ Health Perspect 124(12):1919–1923. https://doi.org/10.1289/EHP158

Trompette A, Gollwitzer ES, Yadava K, Sichelstiel AK, Sprenger N, Ngom-Bru C, Blanchard C, Junt T, Nicod LP, Harris NL, Marsland BJ (2014) Gut microbiota metabolism of dietary fiber influences allergic airway disease and hematopoiesis. Nat Med 20(2):159–166. https://doi.org/10.1038/nm.3444

Van den Abbeele P, Van de Wiele T, Verstraete W, Possemiers S (2011) The host selects mucosal and luminal associations of coevolved gut microorganisms: a novel concept. FEMS Microbiology Reviews 35(4):681–704

van Strien RT, Engel R, Holst O, Bufe A, Eder W, Waser M, Braun-Fahrlander C, Riedler J, Nowak D, von Mutius E (2004) Microbial exposure of rural school children, as assessed by levels of N-acetyl-muramic acid in mattress dust, and its association with respiratory health. J Allergy Clin Immunol 113(5):860–867

Vatanen T, Kostic AD, d'Hennezel E, Siljander H, Franzosa EA, Yassour M, Kolde R, Vlamakis H, Arthur TD, Hamalainen AM, Peet A, Tillmann V, Uibo R, Mokurov S, Dorshakova N, Ilonen J, Virtanen SM, Szabo SJ, Porter JA, Lahdesmaki H, Huttenhower C, Gevers D, Cullen TW, Knip M, Group DS, Xavier RJ (2016) Variation in microbiome LPS immunogenicity contributes to autoimmunity in humans. Cell 165(4):842–853. https://doi.org/10.1016/j.cell.2016.04.007

Vissing NH, Chawes BL, Bisgaard H (2013) Increased risk of pneumonia and bronchiolitis after bacterial colonization of the airways as neonates. Am J Respir Crit Care Med 188 (10):1246–1252. https://doi.org/10.1164/rccm.201302-0215OC

Von Ehrenstein OS, Von Mutius E, Illi S, Baumann L, Bohm O, von Kries R (2000) Reduced risk of hay fever and asthma among children of farmers. Clin Exp Allergy 30(2):187–193

von Hertzen L, Hanski I, Haahtela T (2011) Natural immunity. Biodiversity loss and inflammatory diseases are two global megatrends that might be related. EMBO Rep 12(11):1089–1093. https://doi.org/10.1038/embor.2011.195

von Mutius E, Illi S, Nicolai T, Martinez FD (1996) Relation of indoor heating with asthma, allergic sensitisation, and bronchial responsiveness: survey of children in south Bavaria. BMJ 312(7044):1448–1450

von Mutius E, Schmid S, Group PS (2006) The PASTURE project: EU support for the improvement of knowledge about risk factors and preventive factors for atopy in Europe. Allergy 61 (4):407–413. https://doi.org/10.1111/j.1398-9995.2006.01009.x

Wang J, Ouyang Y, Guner Y, Ford HR, Grishin AV (2009) Ubiquitin-editing enzyme A20 promotes tolerance to lipopolysaccharide in enterocytes. J Immunol 183(2):1384–1392. https://doi.org/10.4049/jimmunol.0803987

Weber J, Illi S, Nowak D, Schierl R, Holst O, von Mutius E, Ege MJ (2015) Asthma and the hygiene hypothesis. Does cleanliness matter? Am J Respir Crit Care Med 191(5):522–529. https://doi.org/10.1164/rccm.201410-1899OC

The Lost Friend: *H. pylori*

Raphaela P. Semper and Markus Gerhard

Abstract The Gram-negative bacterium *Helicobacter pylori* is predominantly known for its pathogenic role inducing peptic ulcer disease and gastric cancer. However, a considerable number of reports have linked *H. pylori* infection with the development of as well as protection from extra-gastrointestinal immunological disorders. Thus, recent epidemiological and experimental studies suggest that chronic infection with *H. pylori* might be beneficial for the host by conferring protection against allergic and chronic autoimmune diseases. In this chapter, we focus on the proposed beneficial role of *H. pylori* on atopic diseases, specifically asthma. We summarize the epidemiological data that support or refute a possible inverse correlation of *H. pylori* infection with atopic diseases. Experimental mouse studies show that asthma fails to develop in the presence of *H. pylori*. We examined and discussed involved bacterial factors and proposed mechanisms for this phenomenon. Mechanistically, the proposed mechanism involved in asthma protection is the particular property of *H. pylori* to induce regulatory T-cells (Tregs) with highly suppressive activity. These cells not only allow the bacterium to persist for decades in infected individuals but at the same time also work outside the stomach and confer protection against several diseases such as asthma and auto-inflammatory diseases.

1 Background

In the last decades prevalence of asthma and other atopic disorders have steadily increased in the western world (Austin et al. 1999; Mannino et al. 2002; Centers for Disease C, Prevention 2011; Akinbami et al. 2012). Although genetic predisposition contributes to asthma development (Castro-Giner et al. 2006), it cannot be solely responsible for this dramatic trend in the last decades. Concurrent to the

R.P. Semper • M. Gerhard (✉)
Institute for Medical Microbiology, Immunology and Hygiene, School of Medicine, Technical University Munich, Munich, Germany
e-mail: Markus.Gerhard@tum.de

© Springer International Publishing AG 2017   69
C.B. Schmidt-Weber (ed.), *Allergy Prevention and Exacerbation*, Birkhäuser
Advances in Infectious Diseases, https://doi.org/10.1007/978-3-319-69968-4_5

increasing incident of allergic diseases, infection with many pathogens has been decreasing. The hygiene hypothesis postulates that exposure to certain infection agents may protect against the development of allergic diseases (Bloomfield et al. 2006), in particular paediatric asthma and atrophic disorders. There are numerous studies showing that gut microbiota correlates with the development of atopic disorders (Ly et al. 2011). Until now it is not fully clear how the microbiota affects the immune system but believed that commensal microbiota may balance T cell responses. Immune response in neonates and also in germ-free mice is skewed toward a T_H2-like response (Prescott et al. 1998; Mazmanian et al. 2005). However, exposure to gut microbial antigens leads to a shift by stimulation of T_H1 cells (Mazmanian et al. 2005; Russell et al. 2013). This shift is crucial for the development of immune tolerance and protects the host from the development of atopic disease. Further, gut microbiota induce the production of retinoic acid, IL-10 and TGF-β from intestinal DCs and macrophages, which favours the differentiation of regulatory T cells, which in turn mediate anti-inflammatory reactions (McLoughlin and Mills 2011; Denning et al. 2007).

2   H. pylori

Helicobacter pylori (*H. pylori*) is a gram-negative ε-proteobacterium that colonizes the stomach of 50% of the world's population. *H. pylori* infection usually takes place during early childhood. The road of transmission is not completely clear. Person-to-person transmission of the bacterium through oral/oral or faecal/oral contact seems most likely. Thus, it is not surprising that low socio-economic status, density of housing, and lack of running water have been linked with higher infection rates (Webb et al. 1994). The incidence of *H. pylori* infection continuously declined over the last decades in western countries, and fewer children are now infected [see Fig. 1; numbers adopted from overview articles (Breckan et al. 2016; Ford and Axon 2010; Goh et al. 2011; Khedmat et al. 2013; Muhammad et al. 2012) or the original study (Ategbo et al. 2013; Fawcett et al. 2005; Gomez et al. 2004; Heuberger et al. 2003; Jaime et al. 2013; Miftahussurur et al. 2014; Mishra et al. 2008; Wewer et al. 1998)]. In developing countries, the infection in adults is still high with over 80% of the population being infected, but in many developed countries the incidence is now less than 20% [see Fig. 2 numbers adopted from overview articles (Breckan et al. 2016; Ford and Axon 2010; Goh et al. 2011; Khedmat et al. 2013; Muhammad et al. 2012) or the original study (Abdallah et al. 2014; Abebaw et al. 2014; Agreus et al. 2016; Aguemon et al. 2005; Ahmad et al. 1997; Al-Balushi et al. 2013; Andersen et al. 1996; Andoulo et al. 2013; Archampong et al. 2015; Asrat et al. 2004; Ayana et al. 2014; Bani-Hani et al. 2006; Begue et al. 1998; Benberin et al. 2013; Carrilho et al. 2009; Carter et al. 2011; Cataldo et al. 2004; Chong et al. 2008; Colmers-Gray et al. 2016; Darko et al. 2015; Destura et al. 2004; Diomande et al. 1991; Dorji et al. 2014; Edwards et al. 1997; Girdaladze et al. 2008; Goto et al. 2016; Gubina et al. 2006; Gunaid et al. 2003; Hoang et al. 2005; Hussein et al. 2008; Jiang et al. 2016; Jonaitis et al. 2013;

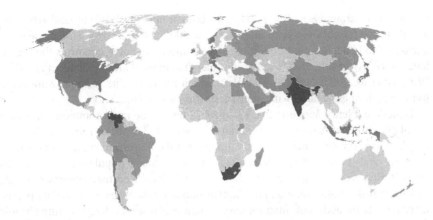

Fig. 1 Global prevalence of infection with *H. pylori* in children: blue ≤25%, green ≤50%, orange ≤75%, red >75% of children are infected with *H. pylori*. Blank world map adopted from Canuckguy (2006)

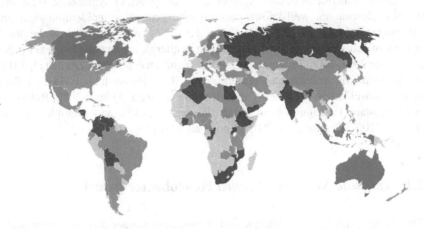

Fig. 2 Global prevalence of infection with *H. pylori* in adults: blue ≤25%, green ≤50%, orange ≤75%, red >75% of the population are infected with *H. pylori*. Blank world map adopted from Canuckguy (2006)

Kibru et al. 2014; Kim et al. 2001; Kimang'a et al. 2010; Kuzela et al. 2012; Li et al. 2010; Lindo et al. 1999; McLaughlin et al. 2003; Miftahussurur et al. 2014, 2015; Milman et al. 2003; Myint et al. 2015; Naja et al. 2007; Naous et al. 2007; Newton et al. 2006; Ntagirabiri et al. 2014; Olivares et al. 2006; Ontsira Ngoyi et al. 2015; Ortega et al. 2010; Sanchez Ceballos et al. 2007; Sasaki et al. 2009; Secka et al. 2011; Shiota et al. 2014; Sporea et al. 2003; Storskrubb et al. 2008; Tadesse et al. 2014; Telaranta-Keerie et al. 2010; Thjodleifsson et al. 2007; Uchida et al. 2015; Walker et al. 2014; Werme et al. 2015)] and is expected to decline further (Grad et al. 2012), e.g. in Japan the prevalence of *H. pylori* infection has dramatically declined in the last decades. Older generations born before 1950 show a high prevalence of around 80–90%, individuals born around the 1990s around 10%, and less than 2% for children born after the year 2000 (Inoue 2016). This decrease is attributed to better sanitary conditions. Interestingly, in some countries the prevalence of *H. pylori* infection is lower in urban than in rural areas (Cheng et al. 2009; Windsor et al. 2005; Nagy et al. 2016), which might be due to improvements in living standard associated with industrialization and urbanization.

Genetic evidence suggests that *H. pylori* is one of the most common and oldest members of the human gastro-intestinal microbiome. This co-existence of humans and *H. pylori* for thousands of years implies that the bacterium confers a benefit to its host and is therefore maintained in the human population. During its co-evolution with the human host, *H. pylori* has developed mechanisms to evade and subvert the host's innate and adaptive immune responses to ensure its persistence despite induction of inflammatory responses. However, longstanding chronic colonization and constant inflammation result in some individuals in the development of gastric pathologies. Thus, infection with *H. pylori* has been linked to gastric and duodenal ulcers (in 1–10% of infected patients), gastric carcinoma (0.1–3%) and gastric mucosa-associated lymphoid tissue (MALT) lymphoma (less than 0.01%). Despite its well-known carcinogenic properties, epidemiological and experimental data point to a protective effect of *H. pylori* infection on the development of many extra-gastric diseases, such as allergic asthma and hay fever (Chen and Blaser 2007, 2008; Reibman et al. 2008; Imamura et al. 2010; Zevit et al. 2012), atopic dermatitis and eczema (Amberbir et al. 2011; Herbarth et al. 2007), Inflammatory Bowel Disease (Luther et al. 2010; Sonnenberg and Genta 2012; Roka et al. 2014; Castano-Rodriguez et al. 2015; Wu et al. 2015), and multiple sclerosis (Li et al. 2007; Pedrini et al. 2015; Yao et al. 2016).

2.1 Immune Response Against Helicobacter pylori

This ability to protect from allergic and chronic autoimmune diseases might be due to *H. pylori*'s ability to modulate immune responses. In asymptomatic carriers and specifically in children, *H. pylori* suppresses immune responses by interfering with effector T cell activation, proliferation und function by inducing highly suppressive regulatory T cells. Thus, the development of gastric malignancies might be due to a misbalance in effector/regulatory responses to the infection due to genetic factors of the bacterium and host, such as bacterial virulence factors or host's polymorphisms. Studies in mice as well as in humans indicated that one predictor and driver of disease is the polarization of gastric *H. pylori*-specific T cells. Upon *H. pylori* infection, regulatory T cells home and accumulate in the gastric mucosa where they can suppress *H. pylori* specific effector T cell responses (Lundgren et al. 2003, 2005a, b). Individuals with peptic ulcer disease showed much higher T_H1 and T_H17 responses but lower regulatory T cell responses than asymptomatic carriers. These studies imply that in asymptomatic *H. pylori*-infected individuals a predominate regulatory T cell response is induced, resulting in persistent infection without immunopathology, while in symptomatic carriers an effector T cell response dominates. Specifically, IL-10-producing regulatory T cells were found in the gastric mucosa of asymptomatic carriers. Interestingly, IL-10 levels correlated directly with bacterial density, with high IL-10 amounts and high colonization scores in asymptomatic individuals, and low IL-10 levels accompanied with lower bacterial density in ulcer patients (Robinson et al. 2008). This implies that IL-10 production from regulatory T cells enables persistent infection but at the same time prevents severe immunopathology. *H. pylori* is normally acquired in early childhood. At that time, the immune system is immature and prone to develop immune tolerance rather than immunity towards foreign antigens and the microbiome. Thus, *H. pylori*-infected children show high amounts of gastric regulatory T cells and associated cytokines IL-10 and TGF-β (Harris et al. 2008). Animal studies which try to consider the time of infection revealed that neonatal infected mice showed higher regulatory T cell numbers than mice infected with *H. pylori* as adults. Whereas the latter displayed high numbers of T_H1 response, gastric inflammation and lower bacterial load compared to neonatal infected animals (Arnold et al. 2011a; Oertli and Muller 2012). These regulatory T cells tightly control effector T cell response (Arnold et al. 2011b). The depletion of regulatory T cells resulted in improved clearing of the infection by induction of strong T_H1 and T_H17 responses but at the same time the development of gastritis, atrophy and intestinal metaplasia was induced (Zhang et al. 2010; Arnold et al. 2011b). Mice whose $CD4^+$ cells were deficient in TGF-β receptor II also showed enhanced gastric inflammation and reduced bacterial burden (Arnold et al. 2011b), indicating that TGF-β-dependent inducible regulatory T cells and not TGF-β-independent natural regulatory T cells mediate this *H. pylori*-specific immune tolerance.

The induction and differentiation of T cells depend on dendritic cells (DCs). DCs are crucial in the development of immune tolerance to allergens, auto-antigens and harmless antigens of commensal species (Yogev et al. 2012; Coombes and Maloy 2007; Maldonado and von Andrian 2010; Diehl et al. 2013). Thus, it is supposed that inducible regulatory T cells are generated in the periphery by tolerogenic DCs which convert in absence of co-stimulatory signals and cytokines, either alone or in combination with membrane-bound tolerogenic factors such as IL-10, TGF-β or retinoic acid, naïve T cells into $FoxP3^+$ regulatory T cells upon antigen presentation (Kretschmer et al. 2005; Maldonado and von Andrian 2010). Several studies showed that *H. pylori* targets DCs to promote tolerance. Thus, stimulation of human and murine DCs with *H. pylori* and co-culture with T cells resulted in the inability to prime T_H1 and T_H17 responses and instead regulatory T cells were induced (Kao et al. 2010; Zhang et al. 2010; Oertli and Muller 2012; Kaebisch et al. 2014). This inability to prime effector T cells was due the fact that *H. pylori* failed to induce maturation of DCs, reflected by upregulation of maturation markers as CD80, CD86 or CD40 and secretion of T cell-differentiating cytokines such as TNF, IL-6, IL-12 and IL-23, while instead these DCs secreted high amounts of IL-10 (Oertli and Muller 2012; Kaebisch et al. 2014). Moreover, *H. pylori*-exposed DCs failed to upregulate maturation markers and secrete inflammatory cytokines in presence of strong TLR stimuli such as *E. coli*, LPS, Pam3Cys or CpG (Oertli and Muller 2012; Kim et al. 2011; Kaebisch et al. 2014). These semi-mature or tolerogenic DCs were strong inducers of regulatory T cells in vitro (Kaebisch et al. 2014; Oertli and Muller 2012) but also in vivo (Kao et al. 2010). Also in *H. pylori*-infected humans such semi-mature DCs could be found in the gastric mucosa (Oertli and Muller 2012). Thus, even though DCs were recruited upon *H. pylori* infection to the gastric mucosa, these DCs fail to mature and thus activate effector T cells responses. For the induction of these semi-mature DCs several virulence factors might be involved: thus, in murine cells, *H. pylori* deficient for the two secreted proteins γ-glutamyltranspeptidase (gGT) (Oertli et al. 2013) or the vacuolating cytotoxin A (VacA) could not induce such tolerogenic DCs and regulatory T cells (Oertli et al. 2013). Concurringly, stimulation with recombinant VacA resulted in inhibited maturation of DCs and inhibited cytokine production (Kim et al. 2011). In contrast, one other study did not observe differences in the induction of effector or regulatory T cells with a VacA-deficient bacterium compared to the wild type *H. pylori* (Kao et al. 2010). A functional CagPAI or CagA seemed to be dispensable for the induction of these tolerogenic DCs followed by induction of regulatory T cells in mice (Oertli et al. 2013; Rizzuti et al. 2015). In contrast, in human DCs, CagA was crucial for the semi-maturation of DCs and followed induction of regulatory T cells, whereas after stimulation with VacA- or gGT-deficient bacteria, DCs did not differ in maturation or pro-inflammatory cytokine secretion (Kaebisch et al. 2014). However, gGT activity could induce regulatory T cells by a maturation-independent mechanism. Instead, through the gGT activity glutamate is produced, which activates glutamate receptors on DCs, inhibiting cAMP signalling and the release of the proinflammatory cytokine IL-6 (Kabisch et al. 2016). IL-6 is a key cytokine in the induction of T_H17 responses and suppression of regulatory T cell (Kimura and Kishimoto 2010). Thus, the expansion of regulatory T cells was favoured over the development of T_H17 cells in presence of gGT activity.

Beside IL-10 production from DCs, secretion of IL-18 was also shown to be crucial for conversion of regulatory T cells. Thus, *H. pylori* exposed IL-18-deficient DCs were not able to convert naïve wild type T cells into regulatory T cells in vitro and ex vivo (Oertli et al. 2012). Likewise, T cells lacking the IL-18R were incapable to express FoxP3$^+$ upon co-culture of *H. pylori* exposed wild type DCs. Both, IL-18 KO and IL-18R KO mice exhibited lower overall regulatory T cell numbers in the mesenteric lymph nodes (MLNs). Thus, DC-derived IL-18 acts via IL-18R signalling directly on T cells to promote regulatory T cell conversion and *H. pylori*-induced tolerance in vivo and in vitro. Interestingly, IL-18 KO mice failed to induce regulatory T cells but showed a stronger T_H17 polarization than wild type mice (Hitzler et al. 2012; Oertli and Muller 2012). Regulatory T cells and T_H17 cells represent developmentally related lineages and retain a high plasticity even when fully differentiated. While both subtypes require TGF-β, in presence of IL-6 and supported by IL-1β, naïve T cells as well as naturally occurring regulatory $CD4^+CD25^+$ T cells may differentiate into T_H17 cells (Veldhoen and Stockinger 2006; Yang et al. 2008; Leveque et al. 2009; Mailer et al. 2015). This might be the reason why IL-18-deficient mice clear *H. pylori* better than wild type mice but showed a stronger immunopathology (Oertli et al. 2012). Thus, the balance of IL-1β (pro-inflammatory) and IL-18 (regulatory) might determine the outcome of the infection. Interestingly, polymorphisms in *IL-1B*, which result in high IL-1β production (Rad et al. 2004), or transgenic gastric expression of IL-1β are associated with strong inflammation and the development of gastric pathologies (Tu 2008), whereas IL-1β- and IL-1R-deficient animals were better protected from gastric pathologies (Shigematsu et al. 2013; Serizawa et al. 2015). The production of active IL-1β in immune cells was dependent on the CagPAI (Semper et al. 2014), whose presence is associated with higher risk for gastric cancer development (Cover 2016).

Dendritic cells are important for immune regulation during *H. pylori* infection. Transfer of *H. pylori*-exposed DCs resulted in induction of regulatory T cells in vivo (Kao et al. 2010) and the depletion of $CD11c^+$ DCs accelerated $T_{H1}/T_{H}17$ responses and improved bacterial clearance (Hitzler et al. 2011; Oertli and Muller 2012). Even in the tolerogenic neonatal model gastritis was observed in DCs-depleted animals (Oertli and Muller 2012). Thus, DCs seem not to be important for immunity but rather for tolerance as similar observations were also made in CNS autoimmunity (Yogev et al. 2012).

The induction of tolerogenic DCs by *H. pylori*, which favours the differentiation of regulatory T cells over induction of effector T cells, might be one factor by which this bacterium prevents its clearing and most of the carriers stay asymptomatic. The primary induction of regulatory T cell occurs by at least two different mechanisms: *H. pylori* induces—in mice via gGT and VacA and in humans via CagA—semimature or tolerogenic DCs which favour regulatory T cell induction. Second, urease induces IL-18 production via TLR2/Nlrp3 signalling, which is crucial for induction of regulatory T cells. In contrast, some virulence factors such as HP-NAP (in mice and human) induce cytokines which favour development of T_H1 or T_H17 responses.

3 Allergy Prevalence Caused by Environmental Impact?

In 2002–2003 the World Health Survey (WHS) of the WHO analysed the global prevalence of asthma in adults. Based on these data, Fig. 3 [adapted from To et al. (2012)] shows a world map of the estimated prevalence of clinical asthma in adults. The prevalence of clinical asthma differed among the 70 participating countries. Most European countries and Australia showed high prevalence of asthma. In contrast, in most African and Southeast Asian countries the prevalence was <5%. The incidence of paediatric asthma showed a similar pattern (see Fig. 4). Globally, 13% of the adolescence (13–14-year-old) suffered from asthma. High prevalence of asthma was found in Australia, New Zealand and UK (≥20%), followed by North America (17%), Western Europe (15%), Latin America (14%), Asia-Pacific (11%) and Africa (11%). The lowest rates were found in Eastern Mediterranean (7%), the Indian Subcontinent (6%), and Northern and Eastern Europe (5%). This study also

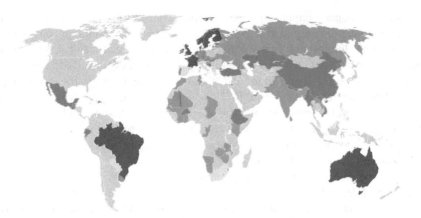

Fig. 3 Incidence of clinical asthma in adults: red ≥10%, orange >7.5%, yellow >5%, green >2.5%, blue ≤2.5% suffer from clinical asthma; the 70 countries which participate on the WHO study are shown. Blank world map adopted from Canuckguy (2006)

Fig. 4 Incidence of clinical asthma in children (13–14 years): red ≥10%, orange >7.5%, yellow >5%, green >2.5%, blue ≤2.5% suffer from clinical asthma; the 98 countries which participate on the ISAAC phase III study are shown. We acknowledge the ISAAC Steering Committee for the data provided. Blank world map adopted from Canuckguy (2006)

analysed younger children (6–7 years). In these groups the total incidence was lower (9%) but the pattern in the different world regions was similar (Lai et al. 2009). Until now three ISAAC studies were conducted (phase I in 1994 and 1995 and phase III 2001–2003) to examine time trends in the prevalence of asthma. The changes in asthma prevalence showed different regional patterns. Although the global prevalence of asthma increased in total during this time, some regions that had previously shown the highest asthma rates witnessed a decrease in asthmatic symptoms in children. These regions were mainly in Oceania and in the United Kingdom. In contrast, regions with low asthma incidence in the first study, such as in Africa and parts of Asia, witnessed an increase in asthmatic symptoms in children (Pearce et al. 2007; Asher et al. 2006). Interestingly, some countries witnessed no increase in the prevalence of asthma and allergy during the past few years, suggesting an end of the asthma epidemic in some countries (Asher et al. 2006). This might be because environmental influences associated with modern life may have reached the maximum in inducing asthma in genetically susceptible individuals, but also due to earlier detection and improved treatment of asthmatics (von Hertzen and Haahtela 2005; Ronchetti et al. 2001; Akinbami and Schoendorf 2002).

4 *Helicobacter pylori* and Allergy

4.1 *Human Studies*

There are several studies analysing the correlation between *H. pylori* infection and atopic/allergic diseases. Some studies found an inverse correlation (Kosunen et al. 2002; Linneberg et al. 2003; McCune et al. 2003; Seiskari et al. 2007; Janson et al. 2007; Shiotani et al. 2008; Chen and Blaser 2008; Holster et al. 2012) but there are also studies which did not find a correlation at all (Figura et al. 1999b; Matricardi et al. 2000; Bodner et al. 2000; Tsang et al. 2000; Linneberg et al. 2003; Radon et al. 2004; Jarvis et al. 2004; Jun et al. 2005; Reibman et al. 2008; Baccioglu et al. 2008; Cam et al. 2009; Holster et al. 2012). In 2014 Lionetti et al. conducted a meta-analysis of studies assessing a possible relationship between *H. pylori* infection and atopy/allergic diseases. In case-control studies they found an inverse correlation between development of atopy/allergic diseases and *H. pylori* infection, whereas in cross-sectional studies such correlation could not be found (Lionetti et al. 2014).

Since allergy can be triggered by different ways such as ingestion or inhalation of the allergic antigen, it might be possible that *H. pylori* infection would differentially influence the various allergic diseases. For food allergy, there are only a few studies analysing the influence of an *H. pylori* infection (Table 1). These studies show either no correlation or an increased risk for food allergy by *H. pylori*. It is difficult to interpret these contradictory results, but all of these studies were limited by small sample sizes and heterogeneous methodologies, e.g. group of enrolled individuals (age, symptoms, food allergen).

In contrast, there are several asthma-related studies showing an inverse association between *H. pylori* infection and asthma. However, there are also studies which did not find any correlation or even observed a positive correlation. Until now, many cross-sectional and case-control studies have been performed. A summary of these are depicted in Tables 2 and 3. Recent meta-analyses have been performed to elucidate these contradictory findings. Analysing 19 studies (nine cross-sectional studies, seven case–control studies, and three prospective cohort studies), Wang et al. found an inverse association between asthma and *H. pylori* infection both in children (OR 0.81, 95% CI: 0.72–0.91) and in adults (OR 0.88,

Table 1 Association between *Helicobacter pylori* and food allergy (OR rations from the paper or calculated with MedCalc)

Age (years)	Population	n	Main findings OR (95%CI) in relation to *H. pylori*[+]	References
17–74	Turkey	90	No association 2.08 (0.24–17.67)	Baccioglu et al. (2008)
5–12	Italy	60	Positive correlation 35.97 (2.00–645.97) vs. IBD group Positive correlation 5.21 (1.28–21.24) vs. asthma group Positive correlation 11.00 (2.77–43.64) vs. IBD and asthma group	Corrado et al. (1998)
4–12	Italy	90	Positive correlation vs. asthma group	Corrado et al. (2000)
6–60	Italy	91	No association 0.81 (0.35–1.89) Positive association with CagA[+] 4.29 (1.13–16.31)	Figura et al. (1999a)
25–65	Germany	62	No correlation 0.75 (0.25–2.26)	Konturek et al. (2008)
5–15	Finland	74	No correlation 1.42 (0.47–4.25)	Kolho et al. (2005)

Table 2 Association between *H pylori* and asthma in cross-sectional studies (OR rations from the paper or calculated with MedCalc)

Age (years)	Population	n	Main findings OR (95%CI) in relation to *H. pylori*[+]	References
18–20	Germany, Spain	1368	No association 0.99 (0.57–1.64)	Uter et al. (2003)
Adults	UK	3244	Inverse association 0.78 (0.59–1.05)	McCune et al. (2003)
Adults Children	USA	7663	CagA[+] inverse association 0.79 (0.63–0.99) Subgroup ≤15 years 0.63 (0.43–0.93) >15 years 0.97 (0.72–1.32)	Chen and Blaser (2007)
3–19	USA	7412	Inverse association 0.69 (0.45–1.06) Subgroup <5 years: 0.58 (0.38–0.88) 3–13 years 0.14 (0.24–0.69)	Chen and Blaser (2008)
17–74	Turkey	90	Inverse association (0.27 (0.07–0.96)	Baccioglu et al. (2008)
Adults	UK	2437	No association 1.09 (0.77–1.54)	Fullerton et al. (2009)
5–18	Israel	6959	Inverse association 0.82 (0.69–0.08)	Zevit et al. (2012)

95% CI: 0.71–1.08) (Wang et al. 2013). Similar results were found reviewing 14 studies with over 28,200 individuals by Zhou et al. (OR 0.84, 95% CI: 0.73–0.96) (Zhou et al. 2013).

Table 3 Association between *H. pylori* and asthma in case-control studies (OR rations from the paper or calculated with MedCalc)

Age (years)	Population	N (case)	N (control)	Main findings OR (95%CI) in relation to *H. pylori*[+]	References
Adults	China	90	97	No association 1.55 (0.87–2.78)	Tsang et al. (2000)
Adults	Finland	245	405	Inverse relationship 0.86 (0.63–1.19)	Pessi et al. (2005)
Adult	Japan	46	48	No association 1.20 (0.53–2.72)	Jun et al. (2005)
1–10	Saudi Arabia	220	543	No association 0.84 (0.56–1.25)	Jaber (2006)
5–15	Turkey	79	36	No association 1.69 (0.62–4.67)	Annagur et al. (2007)
Adults	USA	318	208	CagA[+] inverse association 0.57 (0.36–0.89)	Reibman et al. (2008)
7–9	The Netherlands	98	447	No association 0.74 (0.32–1.70)	Holster et al. (2012)
6–12	Iran	98	98	No association 0.73 (0.37–1.47)	Karimi et al. (2013)
Adults	Korea	359	14,673	No association 1.02 (0.82–1.27) <40 years Inverse association 0.52 (0.30–0.92) ≥40 years No association 1.22 (0.94–1.59)	Lim et al. (2016)
6	The Netherlands	242	2820	Positive association 1.59 (1.01–2.50) Positive association CagA[−] 1.86 (1.13–3.08) No association CagA[+] 0.93 (0.34–2.58)	den Hollander et al. (2016)

Although epidemiological data hint to an inverse association between *H. pylori* and asthma, demonstration of a possible causal relationship is extremely complicated to show. In a recent study, Hussain et al. investigated the mechanism to explain the inverse correlation between *H. pylori* infection and allergies in humans. They found an inverse correlation between IL-10 producing blood regulatory T cell numbers and IgE levels in PBMCs from *H. pylori* infected individuals but not in uninfected persons. Blockage of IL-10 resulted in higher IgE production in PBMCs of *H. pylori* infected individuals. Thus, these IL-10-secreting leukocytes could control IgE production (Hussain et al. 2016). *H. pylori*-induced regulatory T cell response and increased IL-10 production might protect against allergy. Further, individuals infected with CagA[+] positive strains showed much lower IgE levels which could explain the stronger inverse association between asthma and infection with CagA$^+$ *H. pylori* strains (Reibman et al. 2008; Chen and Blaser 2007), since CagA$^+$ strains induce higher IL-10 levels and stronger regulatory T cell response (Kaebisch et al. 2014). Also, plasma IgE concentration of individuals infected with *vacA i1* strains were lower than the ones from individuals infected with *vacA i2* strains (Hussain et al. 2016). This might be in part since *vacA i1* strains are usually CagA$^+$ (Memon et al. 2014), but also mediated by VacA activity itself (Gonzalez-Rivera et al. 2012).

One other protein of *H. pylori* which might contribute to lower asthma incidence in *H. pylori*-infected individuals is *Helicobacter* neutrophil activating protein (HP-NAP). Amedei et al. could show that stimulation of allergen-induced T-cell lines from asthmatic individuals with HP-NAP resulted in a strong increase of IFNγ-secreting T cells and a decrease of IL-4-secreting ones, thus a redirection of the immune response from T_H2 to T_H1 (Amedei et al. 2006).

4.2 Animal Studies

Following up on various observational studies in human populations, mechanistic studies in experimental animal models have been conducted. There are no experimental animal studies analysing the role of *H. pylori* on food allergy until now, but there are a few groups that studied the role of *H. pylori* on asthma development. Direct proof that *H. pylori* infection can protect from asthma comes from studies from Arnold al. They could show that experimental *H. pylori* infection prevents allergen-induced asthma in mice. Thus, airway hyper-responsiveness, lung inflammation, and goblet cell metaplasia were lower in *H. pylori* -infected mice upon ovalbumin or house dust mite allergen-induced asthma. Most indicators of asthma were preferentially reduced in neonatal-infected animals (Arnold et al. 2011a), thus at an age when humans typically get infected with *H. pylori*. Consistently, also in humans, the inverse correlation was stronger in children and was more pronounced in early-onset asthma (Chen and Blaser 2007, 2008). Presence of the bacterium was necessary for *H pylori* -induced asthma protection, since antibiotic treatment prior to allergen challenge abolished asthma protection (Arnold et al. 2011a). Several studies in humans and mice proposed an important role of regulatory T cells in controlling asthma (Ryanna et al. 2009). *H. pylori* infection in children preferentially induces regulatory T cell responses in children (Harris et al. 2008), which might explain the particular benefit of *H. pylori* at this age. Also in experimental infection of mice, protection was most evident in neonatally infected animals, which developed *H. pylori*-induced immunological tolerance with higher numbers of inducible CD4$^+$CD25$^+$FoxP3$^+$ regulatory T cells (Arnold et al. 2011b; Oertli et al. 2012). During the neonatal period of life the immune system is highly plastic and innately biased towards tolerance. Regulatory T cells generated during the earliest stages of life are supposed to differ qualitatively in their suppressive capacity compared to regulatory T cells induced in adults. Thus, more than 70% of CD4$^+$ T cells from neonate mice differentiated under T cell receptor stimulation into $CD4^+FoxP3^+$ T cells, whereas less than 10% of $CD4^+$ T cells from adults did so under the same conditions. These regulatory T cells from neonates stably express FoxP3 and display suppressive functions. Similar to the observations in vitro, injection of anti-CD3 in vivo resulted in higher regulatory T cell amounts in neonates than in adults. Interestingly, this ability to generate regulatory responses vanished gradually within two weeks of birth (Wang et al. 2010). These regulatory T cells are crucial in *H. pylori*-mediated asthma protection, since adoptive transfer of *H. pylori*-induced regulatory T cells isolated from MLNs was sufficient to transfer protection to naïve mice, whereas the depletion of regulatory T cells abrogates the protection (Arnold et al. 2011a). Later, the same group showed that *H. pylori*-induced tolerogenic DCs, which preferentially induce regulatory T cells, were not only generated in the stomach, but also provided evidence that the transfer of *H. pylori*-experienced DCs from MLNs could protect mice from OVA-induced asthma (Oertli and Muller 2012). *H. pylori*-challenged mice not only showed a tolerogenic DCs phenotype in the stomach; also in the lungs, semi-mature DCs with low to intermediate MHCII expression were present, which are believed to induce peripheral T cell tolerance (Dhodapkar et al. 2001; Arnold et al. 2011a). Although there were similar numbers of DCs recruited to the lung of sensitized mice, the proportion of different DCs types differed. In *H. pylori*-infected and thus asthma-protected mice, there were more plasmacytoid DCs (pDCs, $CD11c^{lo}$, $MHCII^{low}$, $B220^+$), whereas non- infected asthmatic animals showed more conventional DCs (cDCs, $CD11c^+$, $MHCII^{high}$, $B220^-$). In *H. pylori*- infected mice these pDCs were recruited upon OVA-challenge. Further, DCs of asthmatic and protected mice differed in $CD11b^+$ and $CD103^+$ cDCs subtypes: asthmatic mice showed high numbers of $CD11b^+DCs$ whereas the protected mice had more $CD103^+DCs$ (Engler et al. 2014). In the gut it was shown that $CD11b^+CD8\alpha^-$ DCs induce preferential T_H2 responses (Iwasaki and Kelsall 2001), whereas $CD103^+$ DCs were shown to induce regulatory T cells (Coombes et al. 2007). The proof that tolerogenic DCs protect from asthma is hard to show, since depletion of $CD11c^+$ cells abrogates asthma completely (van Rijt et al. 2005; see chapter "The Role of the Gut in Type 2 Immunity" for more information on gut resident Tregs).

Not only a live *H. pylori* infection protected mice against OVA-induced asthma, but also oral administration of *H. pylori* extracts (Engler et al. 2014). This lysate—similar as live bacteria—altered the infiltration of DC subsets, and protected against OVA-induced asthma (Engler et al. 2014). Some *H. pylori* factors, which were important for this protection, could be pinpointed. Allergic asthma is driven by T_H2 responses, and since *H. pylori* mainly stimulates T_H1 responses, it was postulated that this response might inhibit asthma development. Indeed, administration of the *H. pylori* protein HP-NAP, which is a strong inducer of T_H1 responses, by systemic administration and even more pronounced through the intranasal route, reduced the infiltration of basophiles in to the lung and decreased IgE, IL-4, IL-5 and GM-CSF production in OVA-sensitized and challenged mice (Codolo et al. 2008; D'Elios et al. 2009). This was recently confirmed by Zhou et al, who further observed higher levels of IFNγ and IL-10 in HP-NAP-pre-treated animals than in untreated mice after OVA-induced asthma (Zhou et al. 2017). Recently, Liu et al showed that a plasmid encoding HP-NAP together with the soluble IL-4 receptor, which is the decoy receptor for IL-4, was more potent to inhibit airway inflammation in a murine asthma model than a plasmid encoding the soluble IL-4 receptor alone, which has been reported to be effective in treating asthma in a phase I/II clinical trial in humans. The lower airway inflammation and serum OVA-specific IgE levels were due to changes in the T_H1/T_H2 balance towards the T_H1 side (Liu et al. 2016). Also, systemic application of recombinant VacA and gGT could protect mice from allergen-induced asthma (Engler et al. 2014). These two virulence factors were shown before to be important for reprogramming DCs and induction of regulatory T cells in mice (Oertli et al. 2013). Thus, administration of *H. pylori* proteins might be a strategy to prevent asthma in high-risk individuals. This protection was dependent on IL-10 production of basic leucine zipper AFT-like 3 (BATF-3)-dependent DCs (Engler et al. 2014). BATF-3 is the key transcription factor for the development of CD8α lymphoid tissue DCs and CD103[+] DCs (Edelson et al. 2010).

One other cytokine, which was shown to be crucial for *H. pylori*-mediated asthma prevention is IL-18 (Oertli and Muller 2012). As mentioned before, DC-derived IL-18 acts via IL-18R signalling directly on T cells to promote regulatory T cell conversion and *H. pylori*-induced tolerance in vivo and in vitro. Thus, DCs from IL-18 KO mice could not convert naïve T cells into FoxP3[+]/CD25[+] cells, and T cells lacking IL-18R signalling were not able to develop into functional regulatory T cells upon co-culture with *H. pylori*-experienced wild type DCs. Therefore, transfer of regulatory T cells from IL-18- or IL-18R-deficient mice could not protect OVA-sensitized and challenged wild type recipients against asthma (Oertli and Muller 2012). Equally, administration of a blocking anti-IL-18 antibody abolished *H. pylori*-mediated asthma protection (Koch and Muller 2015).

Production of IL-18 and asthma-protection were highly dependent of one other virulence factor of *H. pylori*, the enzyme urease. Mice infected with urease-deficient *H. pylori* were not protected against asthma to a similar extent as non-infected animals. Equally, adoptive transfer of regulatory T cells from mice challenged with urease-deficient strain, in contrast to regulatory T cells from mice infected with wild type bacterium, could not protect against allergen-induced asthma (Koch and Muller 2015). The loss of protection might be because of lower IL-18 levels and thus lower numbers of regulatory T induced upon infection with the urease-deficient bacterium, since IL-18 production from DCs is crucial for induction of regulatory T cells and thus asthma protection (Oertli et al. 2012). In the studies of Koch et al. urease-induced IL-18 production was highly dependent on TLR2- and Nlrp3-dependent inflammasome activation. Activation of this multiprotein complex is crucial for functional IL-18 (Dinarello 1999). Thus, adoptive transfer of regulatory T cells from *H. pylori*-infected Nlrp3- or TLR2-deficient mice into OVA-sensitized wild type animals failed to protect them from asthma (Koch and Muller 2015).

In human studies CagA[+] strains correlated inversely with asthma; however, in mice, asthma protection was not linked with the expression of a functional type-IV secretion system, since infection with a CagE mutant strain also prevents allergic airway inflammation (Arnold et al. 2011a). In contrast to humans, in vitro CagPAI in mice was not important for inducing semi-mature DCs and thus regulatory T cells (Kaebisch et al. 2014; Oertli et al. 2013; Rizzuti et al. 2015).

Another characteristic feature of mice infected with *H. pylori* as neonates, and subjected to the asthma-inducing protocol, was their pulmonary infiltration by semi-mature DCs with low to intermediate MHCII expression. Since DCs are not known to migrate between stomach and lung directly, it might be that *H. pylori*-exposed DCs induce regulatory T cells in the draining lymph nodes such as mesenteric lymph nodes or Payer's Patches. These regulatory T cells not only inhibit gastric inflammatory T_H1/T_H17 responses but also might be transported via the blood stream to the lung where they induce tolerogenic lung DCs, and to directly reduce asthmatic T_H2 and T_H17 responses. This is also supported by in vitro studies, were regulatory T cells and DCs regulate each other in a negative feedback loop, promoting a tolerogenic environment. Thus, regulatory T cells interact with DCs and block their maturation. These immature or tolerogenic DCs in turn were not able to induce naïve T cell even in presence of strong stimuli (Onishi et al. 2008).

5 Concluding Remarks

The epidemic increase of allergic, chronic inflammatory and auto-inflammatory disorders is one of the great challenges for the public health in this century. The better sanitary conditions, the "sterile" life style, and use of antibiotics during childhood were considered to contribute to the rise in incidence of allergic and autoimmune disorders (Bach 2002) due to disappearance of some members of the "ancestral" microbiota. Epidemiological and experimental data indicate that *H. pylori* might be one of these vanished members. Substantial epidemiological evidence supports the impression that *H. pylori* is not just a pathogen but that this bacterium might have also beneficial functions. Probably, this bacterium was once a normal, ancient member of the human gastric microbiome that protected mankind from immunological disorders. If it is possible to separate *H. pylori*'s pathogenic property from its beneficial ones, *H. pylori*'s immunomodulating properties could be one day used as a therapeutic agent against these kinds of diseases (summarized in Fig. 5).

In this context, it is tempting to reflect upon how one could use this knowledge in the clinic. Is an eradication of *H. pylori* e.g. by antibiotic treatment of all infected individuals or a prophylactic immunization of children a good way to go? While obviously this will result in lowering gastric cancer incidence, this severe disease occurs only in 1% of infected individuals, whereas most people stay asymptomatic throughout their life. Based on the data discussed above, the eradication of *H. pylori* will likely lead to a further increase of the global incidence of atopic diseases (as well as other inflammatory diseases). Researchers are more and more starting to appreciate *H. pylori* as a commensal, and in this context, and in the light of our exponentially growing understanding of the beneficial effects of the human

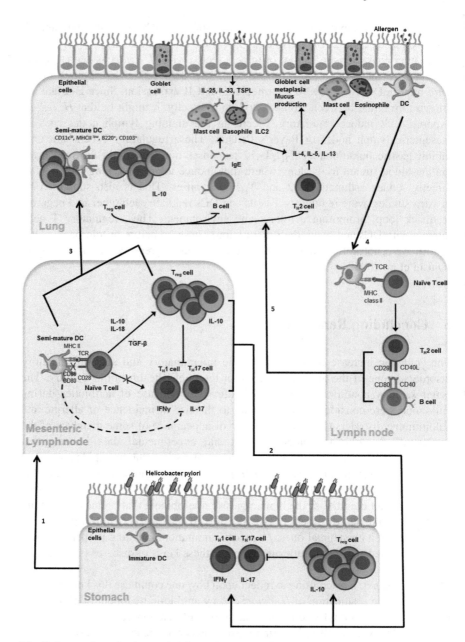

Fig. 5 Proposed mechanism underlying asthma protection by *H. pylori*: (1) *H. pylori* activates semi-mature DCs. (2) These semi-mature DCs move to the gut-draining mesenteric lymphnodes (MLN), where they activate mainly regulatory T cells. (3) These T$_{reg}$s not only return to the stomach, where they inhibit inflammatory responses and avoid clearing of the bacterium (3) but also migrate to the lung, where they induce tolerogenic lung DCs and inhibit allergen-induced T$_H$2 responses. (4) Asthma is induced upon allergen-loaded DCs migrate to local lymph nodes, where they interact with naive T cells and differentiate into T$_H$2. Also IgE-B cells are induced. (5) These cells migrate back into the lung, where their cytokines togeter with epithlium-derived cytokines and chemokines attrack and activate mast cells, eosinophiles and basophiles. The inflammatory responses lead to changes in the airway mucosa that predispose to asthma development

microbiome, one could consider a prophylactic infection of toddlers with an "apathogenic" strain as a possibility to protect them from the epidemic allergy. However, it might be quite challenging to define a "safe strain" in regulatory terms. The dilemma is even more complicated by the fact that epidemiologic studies and mouse studies suggest that CagA and VacA involved in asthma protection, but at the same time these are the main virulence factors which are linked to development of severe gastric diseases. Thus, it is possible to shut down the detrimental properties but maintain the good ones? There is still a long way to go to understand the mechanism behind the protective role of *H. pylori* on atopic diseases before artificial infection with "protective" *H. pylori* might be an option.

Alternatively, it is debated whether a prophylactic immunisation of children with *H. pylori* lysate or the "beneficial" *H. pylori* proteins could represent an approach to protect humans from both, allergies and—at the same time—from the infection itself and thus from diseases such as ulcers and cancer.

References

Abdallah TM, Mohammed HB, Mohammed MH, Ali AAA (2014 Apr) Sero-prevalence and factors associated with *Helicobacter pylori* infection in Eastern Sudan. Asian Pac J Trop Dis 4(2):115–119. https://doi.org/10.1016/S2222-1808(14)60326-1

Abebaw W, Kibret M, Abera B (2014) Prevalence and risk factors of *H. pylori* from dyspeptic patients in northwest Ethiopia: a hospital based cross-sectional study. Asian Pac J Cancer Prev 15(11):4459–4463

Agreus L, Hellstrom PM, Talley NJ, Wallner B, Forsberg A, Vieth M, Veits L, Bjorkegren K, Engstrand L, Andreasson A (2016) Towards a healthy stomach? *Helicobacter pylori* prevalence has dramatically decreased over 23 years in adults in a Swedish community. United Eur Gastroenterol J 4(5):686–696. https://doi.org/10.1177/2050640615623369

Aguemon BD, Struelens MJ, Massougbodji A, Ouendo EM (2005) Prevalence and risk-factors for *Helicobacter pylori* infection in urban and rural Beninese populations. Clin Microbiol Infect 11 (8):611–617. https://doi.org/10.1111/j.1469-0691.2005.01189.x

Ahmad MM, Rahman M, Rumi AK, Islam S, Huq F, Chowdhury MF, Jinnah F, Morshed MG, Hassan MS, Khan AK, Hasan M (1997) Prevalence of *Helicobacter pylori* in asymptomatic population—a pilot serological study in Bangladesh. J Epidemiol 7(4):251–254

Akinbami LJ, Schoendorf KC (2002) Trends in childhood asthma: prevalence, health care utilization, and mortality. Pediatrics 110(2 Pt 1):315–322

Akinbami LJ, Moorman JE, Bailey C, Zahran HS, King M, Johnson CA, Liu X (2012) Trends in asthma prevalence, health care use, and mortality in the United States, 2001–2010. NCHS Data Brief 94:1–8

Al-Balushi MS, Al-Busaidi JZ, Al-Daihani MS, Shafeeq MO, Hasson SS (2013) Sero-prevalence of *Helicobacter pylori* infection among asymptomatic healthy Omani blood donors. Asian Pac J Trop Dis 3(2):146–149. https://doi.org/10.1016/S2222-1808(13)60059-6

Amberbir A, Medhin G, Erku W, Alem A, Simms R, Robinson K, Fogarty A, Britton J, Venn A, Davey G (2011) Effects of *Helicobacter pylori*, geohelminth infection and selected commensal bacteria on the risk of allergic disease and sensitization in 3-year-old Ethiopian children. Clin Exp Allergy 41(10):1422–1430. https://doi.org/10.1111/j.1365-2222.2011.03831.x

Amedei A, Cappon A, Codolo G, Cabrelle A, Polenghi A, Benagiano M, Tasca E, Azzurri A, D'Elios MM, Del Prete G, de Bernard M (2006) The neutrophil-activating protein of *Helicobacter pylori* promotes Th1 immune responses. J Clin Investig 116(4):1092–1101. https://doi.org/10.1172/JCI27177

Andersen LP, Rosenstock SJ, Bonnevie O, Jorgensen T (1996) Seroprevalence of immunoglobulin G, M, and A antibodies to *Helicobacter pylori* in an unselected Danish population. Am J Epidemiol 143(11):1157–1164

Andoulo FA, Noah DN, Tagni-Sartre M, Ndam EC, Blackett KN (2013) Epidemiology of infection *Helicobacter pylori* in Yaounde: specificity of the African enigma. Pan Afr Med J 16:115. https://doi.org/10.11604/pamj.2013.16.115.3007

Annagur A, Kendirli SG, Yilmaz M, Altintas DU, Inal A (2007) Is there any relationship between asthma and asthma attack in children and atypical bacterial infections; *Chlamydia pneumoniae*, *Mycoplasma pneumoniae* and *Helicobacter pylori*. J Trop Pediatr 53(5):313–318. https://doi. org/10.1093/tropej/fmm040

Archampong TN, Asmah RH, Wiredu EK, Gyasi RK, Nkrumah KN, Rajakumar K (2015) Epidemiology of *Helicobacter pylori* infection in dyspeptic Ghanaian patients. Pan Afr Med J 20:178. 10.11604/pamj.2015.20.178.5024

Arnold IC, Dehzad N, Reuter S, Martin H, Becher B, Taube C, Muller A (2011a) *Helicobacter pylori* infection prevents allergic asthma in mouse models through the induction of regulatory T cells. J Clin Investig 121(8):3088–3093. https://doi.org/10.1172/JCI45041

Arnold IC, Lee JY, Amieva MR, Roers A, Flavell RA, Sparwasser T, Muller A (2011b) Tolerance rather than immunity protects from *Helicobacter pylori*-induced gastric preneoplasia. Gastroenterology 140(1):199–209. https://doi.org/10.1053/j.gastro.2010.06.047

Asher MI, Montefort S, Bjorksten B, Lai CK, Strachan DP, Weiland SK, Williams H, Group IPTS (2006) Worldwide time trends in the prevalence of symptoms of asthma, allergic rhinoconjunctivitis, and eczema in childhood: ISAAC phases one and three repeat multicountry cross-sectional surveys. Lancet 368(9537):733–743. https://doi.org/10.1016/S0140-6736(06)69283-0

Asrat D, Nilsson I, Mengistu Y, Ashenafi S, Ayenew K, Al-Soud WA, Wadstrom T, Kassa E (2004) Prevalence of *Helicobacter pylori* infection among adult dyspeptic patients in Ethiopia. Ann Trop Med Parasitol 98(2):181–189. https://doi.org/10.1179/000349804225003190

Ategbo S, Rogombe SMO, Ngoungou E, Midili T, Moussavou A (2013) Épidémiologie de l'infection à *Helicobacter pylori* chez l'enfant de 6 mois à 7 ans à Libreville, Gabon. Clin Mother Child Health 10

Austin JB, Kaur B, Anderson HR, Burr M, Harkins LS, Strachan DP, Warner JO (1999) Hay fever, eczema, and wheeze: a nationwide UK study (ISAAC, international study of asthma and allergies in childhood). Arch Dis Child 81(3):225–230

Ayana SM, Swai B, Maro VP, Kibiki GS (2014) Upper gastrointestinal endoscopic findings and prevalence of *Helicobacter pylori* infection among adult patients with dyspepsia in northern Tanzania. Tanzan J Health Res 16(1):16–22

Baccioglu A, Kalpaklioglu F, Guliter S, Yakaryilmaz F (2008) *Helicobacter pylori* in allergic inflammation—fact or fiction? Allergol Immunopathol 36(2):85–89

Bach JF (2002) The effect of infections on susceptibility to autoimmune and allergic diseases. N Engl J Med 347(12):911–920. https://doi.org/10.1056/NEJMra020100

Bani-Hani KE, Shatnawi NJ, El Qaderi S, Khader YS, Bani-Hani BK (2006) Prevalence and risk factors of *Helicobacter pylori* infection in healthy schoolchildren. Chin J Dig Dis 7(1):55–60. https://doi.org/10.1111/j.1443-9573.2006.00245.x

Begue RE, Gonzales JL, Correa-Gracian H, Tang SC (1998) Dietary risk factors associated with the transmission of *Helicobacter pylori* in Lima, Peru. Am J Trop Med Hyg 59(4):637–640

Benberin V, Bektayeva R, Karabayeva R, Lebedev A, Akemeyeva K, Paloheimo L, Syrjanen K (2013) Prevalence of *H. pylori* infection and atrophic gastritis among symptomatic and dyspeptic adults in Kazakhstan. A hospital-based screening study using a panel of serum biomarkers. Anticancer Res 33(10):4595–4602

Bloomfield SF, Stanwell-Smith R, Crevel RW, Pickup J (2006) Too clean, or not too clean: the hygiene hypothesis and home hygiene. Clin Exp Allergy 36(4):402–425. https://doi.org/10.1111/j.1365-2222.2006.02463.x

Bodner C, Anderson WJ, Reid TS, Godden DJ (2000) Childhood exposure to infection and risk of adult onset wheeze and atopy. Thorax 55(5):383–387

Breckan RK, Paulssen EJ, Asfeldt AM, Kvamme JM, Straume B, Florholmen J (2016) The all-age prevalence of *Helicobacter pylori* Infection and potential transmission routes. A population-based study. Helicobacter 21(6):586–595. https://doi.org/10.1111/hel.12316

Cam S, Ertem D, Bahceciler N, Akkoc T, Barlan I, Pehlivanoglu E (2009) The interaction between *Helicobacter pylori* and atopy: does inverse association really exist? Helicobacter 14(1):1–8. https://doi.org/10.1111/j.1523-5378.2009.00660.x

Canuckguy (2006) BlankMap-World6.svg. https://de.wikipedia.org/wiki/Datei:BlankMap-World6.svg#file

Carrilho C, Modcoicar P, Cunha L, Ismail M, Guisseve A, Lorenzoni C, Fernandes F, Peleteiro B, Almeida R, Figueiredo C, David L, Lunet N (2009) Prevalence of *Helicobacter pylori* infection, chronic gastritis, and intestinal metaplasia in Mozambican dyspeptic patients. Virchows Arch 454(2):153–160. https://doi.org/10.1007/s00428-008-0713-7

Carter FP, Frankson T, Pintard J, Edgecombe B (2011) Seroprevalence of *Helicobacter pylori* infection in adults in the Bahamas. West Indian Med J 60(6):662–665

Castano-Rodriguez N, Kaakoush NO, Lee WS, Mitchell HM (2015) Dual role of *Helicobacter* and *Campylobacter* species in IBD: a systematic review and meta-analysis. Gut. https://doi.org/10.1136/gutjnl-2015-310545

Castro-Giner F, Kauffmann F, de Cid R, Kogevinas M (2006) Gene-environment interactions in asthma. Occup Environ Med 63(11):776–786, 761. https://doi.org/10.1136/oem.2004.019216

Cataldo F, Simpore J, Greco P, Ilboudo D, Musumeci S (2004) *Helicobacter pylori* infection in Burkina Faso: an enigma within an enigma. Dig Liver Dis 36(9):589–593. https://doi.org/10.1016/j.dld.2004.05.005

Centers for Disease C, Prevention (2011) Vital signs: asthma prevalence, disease characteristics, and self-management education: United States, 2001–2009. MMWR Morb Mortal Wkly Rep 60(17):547–552

Chen Y, Blaser MJ (2007) Inverse associations of *Helicobacter pylori* with asthma and allergy. Arch Intern Med 167(8):821–827. https://doi.org/10.1001/archinte.167.8.821

Chen Y, Blaser MJ (2008) *Helicobacter pylori* colonization is inversely associated with childhood asthma. J Infect Dis 198(4):553–560. https://doi.org/10.1086/590158

Cheng H, Hu F, Zhang L, Yang G, Ma J, Hu J, Wang W, Gao W, Dong X (2009) Prevalence of *Helicobacter pylori* infection and identification of risk factors in rural and urban Beijing, China. Helicobacter 14(2):128–133. https://doi.org/10.1111/j.1523-5378.2009.00668.x

Chong VH, Lim KC, Rajendran N (2008) Prevalence of active *Helicobacter pylori* infection among patients referred for endoscopy in Brunei Darussalam. Singapore Med J 49(1):42–46

Codolo G, Mazzi P, Amedei A, Del Prete G, Berton G, D'Elios MM, de Bernard M (2008) The neutrophil-activating protein of *Helicobacter pylori* down-modulates Th2 inflammation in ovalbumin-induced allergic asthma. Cell Microbiol 10(11):2355–2363. https://doi.org/10.1111/j.1462-5822.2008.01217.x

Colmers-Gray IN, Vandermeer B, Greidanus RI, Kolber MR (2016) *Helicobacter pylori* status among patients undergoing gastroscopy in rural northern Alberta. Can Fam Physician 62(9): e547–e554

Coombes JL, Maloy KJ (2007) Control of intestinal homeostasis by regulatory T cells and dendritic cells. Semin Immunol 19(2):116–126. https://doi.org/10.1016/j.smim.2007.01.001

Coombes JL, Siddiqui KR, Arancibia-Carcamo CV, Hall J, Sun CM, Belkaid Y, Powrie F (2007) A functionally specialized population of mucosal CD103+ DCs induces Foxp3+ regulatory T cells via a TGF-beta and retinoic acid-dependent mechanism. J Exp Med 204(8):1757–1764. https://doi.org/10.1084/jem.20070590

Corrado G, Luzzi I, Lucarelli S, Frediani T, Pacchiarotti C, Cavaliere M, Rea P, Cardi E (1998) Positive association between *Helicobacter pylori* infection and food allergy in children. Scand J Gastroenterol 33(11):1135–1139

Corrado G, Luzzi I, Pacchiarotti C, Lucarelli S, Frediani T, Cavaliere M, Rea P, Cardi E (2000) *Helicobacter pylori* seropositivity in children with atopic dermatitis as sole manifestation of food allergy. Pediatr Allergy Immunol 11(2):101–105

Cover TL (2016) *Helicobacter pylori* diversity and gastric cancer risk. mBio 7(1):e01869–e01815. https://doi.org/10.1128/mBio.01869-15

D'Elios MM, Codolo G, Amedei A, Mazzi P, Berton G, Zanotti G, Del Prete G, de Bernard M (2009) *Helicobacter pylori*, asthma and allergy. FEMS Immunol Med Microbiol 56(1):1–8. https://doi.org/10.1111/j.1574-695X.2009.00537.x

Darko R, Yawson AE, Osei V, Owusu-Ansah J, Aluze-Ele S (2015) Changing patterns of the prevalence of *Helicobacter pylori* among patients at a Corporate Hospital in Ghana. Ghana Med J 49(3):147–153

den Hollander WJ, Sonnenschein-van der Voort AM, Holster IL, de Jongste JC, Jaddoe VW, Hofman A, Perez-Perez GI, Moll HA, Blaser MJ, Duijts L, Kuipers EJ (2016) *Helicobacter pylori* in children with asthmatic conditions at school age, and their mothers. Aliment Pharmacol Ther. https://doi.org/10.1111/apt.13572

Denning TL, Wang YC, Patel SR, Williams IR, Pulendran B (2007) Lamina propria macrophages and dendritic cells differentially induce regulatory and interleukin 17-producing T cell responses. Nat Immunol 8(10):1086–1094. https://doi.org/10.1038/ni1511

Destura RV, Labio ED, Barrett LJ, Alcantara CS, Gloria VI, Daez ML, Guerrant RL (2004) Laboratory diagnosis and susceptibility profile of *Helicobacter pylori* infection in the Philippines. Ann Clin Microbiol Antimicrob 3:25. https://doi.org/10.1186/1476-0711-3-25

Dhodapkar MV, Steinman RM, Krasovsky J, Munz C, Bhardwaj N (2001) Antigen-specific inhibition of effector T cell function in humans after injection of immature dendritic cells. J Exp Med 193(2):233–238

Diehl GE, Longman RS, Zhang JX, Breart B, Galan C, Cuesta A, Schwab SR, Littman DR (2013) Microbiota restricts trafficking of bacteria to mesenteric lymph nodes by CX(3)CR1(hi) cells. Nature 494(7435):116–120. https://doi.org/10.1038/nature11809

Dinarello CA (1999) Interleukin-18. Methods 19(1):121–132. https://doi.org/10.1006/meth.1999.0837

Diomande MI, Flejou JF, Potet F, Dago-Akribi A, Ouattara D, Kadjo K, Niamkey E, Beaumel A, Gbe K, Beda BY (1991) Chronic gastritis and *Helicobacter pylori* infection on the Ivory Coast. A series of 277 symptomatic patients. Gastroenterol Clin Biol 15(10):711–716

Dorji D, Dendup T, Malaty HM, Wangchuk K, Yangzom D, Richter JM (2014) Epidemiology of *Helicobacter pylori* in Bhutan: the role of environment and Geographic location. Helicobacter 19(1):69–73. https://doi.org/10.1111/hel.12088

Edelson BT, Kc W, Juang R, Kohyama M, Benoit LA, Klekotka PA, Moon C, Albring JC, Ise W, Michael DG, Bhattacharya D, Stappenbeck TS, Holtzman MJ, Sung SS, Murphy TL, Hildner K, Murphy KM (2010) Peripheral CD103+ dendritic cells form a unified subset developmentally related to CD8alpha+ conventional dendritic cells. J Exp Med 207 (4):823–836. https://doi.org/10.1084/jem.20091627

Edwards CN, Douglin CP, Prussia PR, Garriques SA, Levett PN (1997) Epidemiology of *Helicobacter pylori* infection in Barbados. West Indian Med J 46(1):3–7

Engler DB, Reuter S, van Wijck Y, Urban S, Kyburz A, Maxeiner J, Martin H, Yogev N, Waisman A, Gerhard M, Cover TL, Taube C, Muller A (2014) Effective treatment of allergic airway inflammation with *Helicobacter pylori* immunomodulators requires BATF3-dependent dendritic cells and IL-10. Proc Natl Acad Sci USA 111(32):11810–11815. https://doi.org/10.1073/pnas.1410579111

Fawcett JP, Barbezat GO, Poulton R, Milne BJ, Xia HH, Talley NJ (2005) *Helicobacter pylori* serology in a birth cohort of New Zealanders from age 11 to 26. World J Gastroenterol 11 (21):3273–3276

Figura N, Perrone A, Gennari C, Orlandini G, Bianciardi L, Giannace R, Vaira D, Vagliasinti M, Rottoli P (1999a) Food allergy and *Helicobacter pylori* infection. Ital J Gastroenterol Hepatol 31(3):186–191

Figura N, Perrone A, Gennari C, Orlandini G, Giannace R, Lenzi C, Vagliasindi M, Bianciardi L, Rottoli P (1999b) CagA-positive *Helicobacter pylori* infection may increase the risk of food allergy development. J Physiol Pharmacol 50(5):827–831

Ford AC, Axon AT (2010) Epidemiology of *Helicobacter pylori* infection and public health implications. Helicobacter 15(Suppl 1):1–6. https://doi.org/10.1111/j.1523-5378.2010.00779.x

Fullerton D, Britton JR, Lewis SA, Pavord ID, McKeever TM, Fogarty AW (2009) *Helicobacter pylori* and lung function, asthma, atopy and allergic disease—a population-based cross-sectional study in adults. Int J Epidemiol 38(2):419–426. https://doi.org/10.1093/ije/dyn348

Girdaladze AM, Mosidze BA, Tsertsvadze TN, Shartava Ts K, Girdaladze SA (2008) The prevalence of *Helicobacter pylori* infection among the Georgian population. Georgian Med News 157:34–39

Goh KL, Chan WK, Shiota S, Yamaoka Y (2011) Epidemiology of *Helicobacter pylori* infection and public health implications. Helicobacter 16(Suppl 1):1–9. https://doi.org/10.1111/j.1523-5378.2011.00874.x

Gomez NA, Salvador A, Vargas PE, Zapatier JA, Alvarez J (2004) Seroprevalence of *Helicobacter pylori* among the child population of ecuador. Rev Gastroenterol Peru 24 (3):230–233

Gonzalez-Rivera C, Algood HM, Radin JN, McClain MS, Cover TL (2012) The intermediate region of *Helicobacter pylori* VacA is a determinant of toxin potency in a Jurkat T cell assay. Infect Immun 80(8):2578–2588. https://doi.org/10.1128/IAI.00052-12

Goto Y, Syam AF, Darnindro N, Puspita Hapsari FC (2016) Risk factors for and prevalence of *Helicobacter Pylori* Infection among healthy inhabitants in Northern Jakarta, Indonesia. Asian Pac J Cancer Prev 17(9):4469–4475

Grad YH, Lipsitch M, Aiello AE (2012) Secular trends in *Helicobacter pylori* seroprevalence in adults in the United States: evidence for sustained race/ethnic disparities. Am J Epidemiol 175 (1):54–59. https://doi.org/10.1093/aje/kwr288

Gubina AV, Tsaregorodtseva TM, Sokolova GN, Serova TI (2006) Chronic pancreatitis and anti-*Helicobacter pylori* antibodies. Exp Clin Gastroenterol 1:69–71

Gunaid AA, Hassan NA, Murray-Lyon I (2003) Prevalence and risk factors for *Helicobacter pylori* infection among Yemeni dyspeptic patients. Saudi Med J 24(5):512–517

Harris PR, Wright SW, Serrano C, Riera F, Duarte I, Torres J, Pena A, Rollan A, Viviani P, Guiraldes E, Schmitz JM, Lorenz RG, Novak L, Smythies LE, Smith PD (2008) *Helicobacter pylori* gastritis in children is associated with a regulatory T-cell response. Gastroenterology 134(2):491–499. https://doi.org/10.1053/j.gastro.2007.11.006

Herbarth O, Bauer M, Fritz GJ, Herbarth P, Rolle-Kampczyk U, Krumbiegel P, Richter M, Richter T (2007) *Helicobacter pylori* colonisation and eczema. J Epidemiol Commun Health 61 (7):638–640. https://doi.org/10.1136/jech.2006.046706

Heuberger F, Pantoflickova D, Gassner M, Oneta C, Grehn M, Blum AL, Dorta G (2003) *Helicobacter pylori* infection in Swiss adolescents: prevalence and risk factors. Eur J Gastroenterol Hepatol 15(2):179–183

Hitzler I, Oertli M, Becher B, Agger EM, Muller A (2011) Dendritic cells prevent rather than promote immunity conferred by a helicobacter vaccine using a mycobacterial adjuvant. Gastroenterology 141(1):186–196, 196.e181. https://doi.org/10.1053/j.gastro.2011.04.009

Hitzler I, Sayi A, Kohler E, Engler DB, Koch KN, Hardt WD, Muller A (2012) Caspase-1 has both proinflammatory and regulatory properties in Helicobacter infections, which are differentially mediated by its substrates IL-1beta and IL-18. J Immunol 188(8):3594–3602. https://doi.org/10.4049/jimmunol.1103212

Hoang TT, Bengtsson C, Phung DC, Sorberg M, Granstrom M (2005) Seroprevalence of *Helicobacter pylori* infection in urban and rural Vietnam. Clin Diagn Lab Immunol 12 (1):81–85. https://doi.org/10.1128/CDLI.12.1.81-85.2005

Holster IL, Vila AM, Caudri D, den Hoed CM, Perez-Perez GI, Blaser MJ, de Jongste JC, Kuipers EJ (2012) The impact of *Helicobacter pylori* on atopic disorders in childhood. Helicobacter 17 (3):232–237. https://doi.org/10.1111/j.1523-5378.2012.00934.x

Hussain K, Letley DP, Greenaway AB, Kenefeck R, Winter JA, Tomlinson W, Rhead J, Staples E, Kaneko K, Atherton JC, Robinson K (2016) *Helicobacter pylori*-mediated protection from allergy is associated with IL-10-secreting peripheral blood regulatory T cells. Front Immunol 7:71. https://doi.org/10.3389/fimmu.2016.00071

Hussein NR, Robinson K, Atherton JC (2008) A study of age-specific *Helicobacter pylori* seropositivity rates in Iraq. Helicobacter 13(4):306–307. https://doi.org/10.1111/j.1523-5378. 2008.00618.x

Imamura S, Sugimoto M, Kanemasa K, Sumida Y, Okanoue T, Yoshikawa T, Yamaoka Y (2010) Inverse association between *Helicobacter pylori* infection and allergic rhinitis in young Japanese. J Gastroenterol Hepatol 25(7):1244–1249. https://doi.org/10.1111/j.1440-1746. 2010.06307.x

Inoue M (2016) Changing epidemiology of *Helicobacter pylori* in Japan. Gastric Cancer. https://doi.org/10.1007/s10120-016-0658-5

Iwasaki A, Kelsall BL (2001) Unique functions of CD11b+, CD8 alpha+, and double-negative Peyer's patch dendritic cells. J Immunol 166(8):4884–4890

Jaber SM (2006) *Helicobacter pylori* seropositivity in children with chronic disease in Jeddah, Saudi Arabia. Saudi J Gastroenterol 12(1):21–26

Jaime F, Villagran A, Serrano C, Cerda J, Harris PR (2013) Frequency of *Helicobacter pylori* infection in 144 school age Chilean children. Rev Med Chil 141(10):1249–1254. https://doi. org/10.4067/s0034-98872013001000003

Janson C, Asbjornsdottir H, Birgisdottir A, Sigurjonsdottir RB, Gunnbjornsdottir M, Gislason D, Olafsson I, Cook E, Jogi R, Gislason T, Thjodleifsson B (2007) The effect of infectious burden on the prevalence of atopy and respiratory allergies in Iceland, Estonia, and Sweden. J Allergy Clin Immunol 120(3):673–679. https://doi.org/10.1016/j.jaci.2007.05.003

Jarvis D, Luczynska C, Chinn S, Burney P (2004) The association of hepatitis A and *Helicobacter pylori* with sensitization to common allergens, asthma and hay fever in a population of young British adults. Allergy 59(10):1063–1067. https://doi.org/10.1111/j.1398-9995.2004.00539.x

Jiang JX, Liu Q, Mao XY, Zhang HH, Zhang GX, Xu SF (2016) Downward trend in the prevalence of *Helicobacter pylori* infections and corresponding frequent upper gastrointestinal diseases profile changes in Southeastern China between 2003 and 2012. SpringerPlus 5(1):1601. https://doi.org/10.1186/s40064-016-3185-2

Jonaitis L, Kiudelis G, Kupcinskas L, Kupcinskas J (2013) Prevalence of helicobacter pylori among outpatient middle-aged patients in Lithuania and its relation to dyspeptic symptoms. Helicobacter 18:104

Jun ZJ, Lei Y, Shimizu Y, Dobashi K, Mori M (2005) *Helicobacter pylori* seroprevalence in patients with mild asthma. Tohoku J Exp Med 207(4):287–291

Kabisch R, Semper RP, Wustner S, Gerhard M, Mejias-Luque R (2016) *Helicobacter pylori* gamma-glutamyltranspeptidase induces tolerogenic human dendritic cells by activation of glutamate receptors. J Immunol 196(10):4246–4252. https://doi.org/10.4049/jimmunol. 1501062

Kaebisch R, Mejias-Luque R, Prinz C, Gerhard M (2014) *Helicobacter pylori* cytotoxin-associated gene A impairs human dendritic cell maturation and function through IL-10-mediated activation of STAT3. J Immunol 192(1):316–323. https://doi.org/10.4049/jimmunol.1302476

Kao JY, Zhang M, Miller MJ, Mills JC, Wang B, Liu M, Eaton KA, Zou W, Berndt BE, Cole TS, Takeuchi T, Owyang SY, Luther J (2010) *Helicobacter pylori* immune escape is mediated by dendritic cell-induced Treg skewing and Th17 suppression in mice. Gastroenterology 138 (3):1046–1054. https://doi.org/10.1053/j.gastro.2009.11.043

Karimi A, Fakhimi-Derakhshan K, Imanzadeh F, Rezaei M, Cavoshzadeh Z, Maham S (2013) *Helicobacter pylori* infection and pediatric asthma. Iran J Microbiol 5(2):132–135

Khedmat H, Karbasi-Afshar R, Agah S, Taheri S (2013) *Helicobacter pylori* Infection in the general population: a middle Eastern perspective. Caspian J Intern Med 4(4):745–753

Kibru D, Gelaw B, Alemu A, Addis Z (2014) *Helicobacter pylori* infection and its association with anemia among adult dyspeptic patients attending Butajira Hospital, Ethiopia. BMC Infect Dis 14:656. https://doi.org/10.1186/s12879-014-0656-3

Kim JH, Kim HY, Kim NY, Kim SW, Kim JG, Kim JJ, Roe IH, Seo JK, Sim JG, Ahn H, Yoon BC, Lee SW, Lee YC, Chung IS, Jung HY, Hong WS, Choi KW (2001) Seroepidemiological study of *Helicobacter pylori* infection in asymptomatic people in South Korea. J Gastroenterol Hepatol 16(9):969–975

Kim JM, Kim JS, Yoo DY, Ko SH, Kim N, Kim H, Kim YJ (2011) Stimulation of dendritic cells with *Helicobacter pylori* vacuolating cytotoxin negatively regulates their maturation via the restoration of E2F1. Clin Exp Immunol 166(1):34–45. https://doi.org/10.1111/j.1365-2249.2011.04447.x

Kimang'a AN, Revathi G, Kariuki S, Sayed S, Devani S (2010) *Helicobacter pylori*: prevalence and antibiotic susceptibility among Kenyans. SAMJ 100:53–57

Kimura A, Kishimoto T (2010) IL-6: regulator of Treg/Th17 balance. Eur J Immunol 40 (7):1830–1835. https://doi.org/10.1002/eji.201040391

Koch KN, Muller A (2015) *Helicobacter pylori* activates the TLR2/NLRP3/caspase-1/IL-18 axis to induce regulatory T-cells, establish persistent infection and promote tolerance to allergens. Gut Microbes 6(6):382–387. https://doi.org/10.1080/19490976.2015.1105427

Kolho KL, Haapaniemi A, Haahtela T, Rautelin H (2005) *Helicobacter pylori* and specific immunoglobulin E antibodies to food allergens in children. J Pediatr Gastroenterol Nutr 40 (2):180–183

Konturek PC, Rienecker H, Hahn EG, Raithel M (2008) *Helicobacter pylori* as a protective factor against food allergy. Med Sci Monit 14(9):Cr452–Cr458

Kosunen TU, Hook-Nikanne J, Salomaa A, Sarna S, Aromaa A, Haahtela T (2002) Increase of allergen-specific immunoglobulin E antibodies from 1973 to 1994 in a Finnish population and a possible relationship to *Helicobacter pylori* infections. Clin Exp Allergy 32(3):373–378

Kretschmer K, Apostolou I, Hawiger D, Khazaie K, Nussenzweig MC, von Boehmer H (2005) Inducing and expanding regulatory T cell populations by foreign antigen. Nat Immunol 6 (12):1219–1227. https://doi.org/10.1038/ni1265

Kuzela L, Oltman M, Sutka J, Zacharova B, Nagy M (2012) Epidemiology of *Helicobacter pylori* infection in the Slovak Republic. Hepato-Gastroenterol 59(115):754–756. https://doi.org/10.5754/hge11457

Lai CK, Beasley R, Crane J, Foliaki S, Shah J, Weiland S, International Study of A, Allergies in Childhood Phase Three Study G (2009) Global variation in the prevalence and severity of asthma symptoms: phase three of the International Study of Asthma and Allergies in Childhood (ISAAC). Thorax 64(6):476–483. https://doi.org/10.1136/thx.2008.106609

Leveque L, Deknuydt F, Bioley G, Old LJ, Matsuzaki J, Odunsi K, Ayyoub M, Valmori D (2009) Interleukin 2-mediated conversion of ovarian cancer-associated CD4+ regulatory T cells into proinflammatory interleukin 17-producing helper T cells. J Immunother 32(2):101–108. https://doi.org/10.1097/CJI.0b013e318195b59e

Li W, Minohara M, Su JJ, Matsuoka T, Osoegawa M, Ishizu T, Kira J (2007) *Helicobacter pylori* infection is a potential protective factor against conventional multiple sclerosis in the Japanese population. J Neuroimmunol 184(1-2):227–231. https://doi.org/10.1016/j.jneuroim.2006.12.010

Li Z, Zou D, Ma X, Chen J, Shi X, Gong Y, Man X, Gao L, Zhao Y, Wang R, Yan X, Dent J, Sung JJ, Wernersson B, Johansson S, Liu W, He J (2010) Epidemiology of peptic ulcer disease: endoscopic results of the systematic investigation of gastrointestinal disease in China. Am J Gastroenterol 105(12):2570–2577. https://doi.org/10.1038/ajg.2010.324

Lim JH, Kim N, Lim SH, Kwon JW, Shin CM, Chang YS, Kim JS, Jung HC, Cho SH (2016) Inverse relationship between *Helicobacter pylori* infection and asthma among adults younger than 40 years: a cross-sectional study. Medicine 95(8):e2609. https://doi.org/10.1097/MD.0000000000002609

Lindo JF, Lyn-Sue AE, Palmer CJ, Lee MG, Vogel P, Robinson RD (1999) Seroepidemiology of *Helicobacter pylori* infection in a Jamaican community. Trop Med Int Health 4(12):862–866

Linneberg A, Ostergaard C, Tvede M, Andersen LP, Nielsen NH, Madsen F, Frolund L, Dirksen A, Jorgensen T (2003) IgG antibodies against microorganisms and atopic disease in Danish adults: the Copenhagen Allergy Study. J Allergy Clin Immunol 111(4):847–853

Lionetti E, Leonardi S, Lanzafame A, Garozzo MT, Filippelli M, Tomarchio S, Ferrara V, Salpietro C, Pulvirenti A, Francavilla R, Catassi C (2014) *Helicobacter pylori* infection and atopic diseases: is there a relationship? A systematic review and meta-analysis. World J Gastroenterol 20(46):17635–17647. https://doi.org/10.3748/wjg.v20.i46.17635

Liu X, Fu G, Ji Z, Huang X, Ding C, Jiang H, Wang X, Du M, Wang T, Kang Q (2016) A recombinant DNA plasmid encoding the sIL-4R-NAP fusion protein suppress airway inflammation in an OVA-induced mouse model of asthma. Inflammation 39(4):1434–1440. https://doi.org/10.1007/s10753-016-0375-6

Lundgren A, Suri-Payer E, Enarsson K, Svennerholm AM, Lundin BS (2003) *Helicobacter pylori*-specific CD4+ CD25high regulatory T cells suppress memory T-cell responses to *H. pylori* in infected individuals. Infect Immun 71(4):1755–1762

Lundgren A, Stromberg E, Sjoling A, Lindholm C, Enarsson K, Edebo A, Johnsson E, Suri-Payer E, Larsson P, Rudin A, Svennerholm AM, Lundin BS (2005a) Mucosal FOXP3-expressing CD4+ CD25high regulatory T cells in *Helicobacter pylori*-infected patients. Infect Immun 73(1):523–531. https://doi.org/10.1128/IAI.73.1.523-531.2005

Lundgren A, Trollmo C, Edebo A, Svennerholm AM, Lundin BS (2005b) *Helicobacter pylori*-specific CD4+ T cells home to and accumulate in the human *Helicobacter pylori*-infected gastric mucosa. Infect Immun 73(9):5612–5619. https://doi.org/10.1128/IAI.73.9.5612-5619.2005

Luther J, Dave M, Higgins PD, Kao JY (2010) Association between *Helicobacter pylori* infection and inflammatory bowel disease: a meta-analysis and systematic review of the literature. Inflamm Bowel Dis 16(6):1077–1084. https://doi.org/10.1002/ibd.21116

Ly NP, Litonjua A, Gold DR, Celedon JC (2011) Gut microbiota, probiotics, and vitamin D: interrelated exposures influencing allergy, asthma, and obesity? J Allergy Clin Immunol 127 (5):1087–1094; quiz 1095–1086. https://doi.org/10.1016/j.jaci.2011.02.015

Mailer RK, Joly AL, Liu S, Elias S, Tegner J, Andersson J (2015) IL-1beta promotes Th17 differentiation by inducing alternative splicing of FOXP3. Sci Rep 5:14674. https://doi.org/10.1038/srep14674

Maldonado RA, von Andrian UH (2010) How tolerogenic dendritic cells induce regulatory T cells. Adv Immunol 108:111–165. https://doi.org/10.1016/B978-0-12-380995-7.00004-5

Mannino DM, Homa DM, Akinbami LJ, Moorman JE, Gwynn C, Redd SC (2002) Surveillance for asthma—United States, 1980–1999. Morb Mortal Wkly Rep Surveill Summ 51(1):1–13

Matricardi PM, Rosmini F, Riondino S, Fortini M, Ferrigno L, Rapicetta M, Bonini S (2000) Exposure to foodborne and orofecal microbes versus airborne viruses in relation to atopy and allergic asthma: epidemiological study. BMJ 320(7232):412–417

Mazmanian SK, Liu CH, Tzianabos AO, Kasper DL (2005) An immunomodulatory molecule of symbiotic bacteria directs maturation of the host immune system. Cell 122(1):107–118. https://doi.org/10.1016/j.cell.2005.05.007

McCune A, Lane A, Murray L, Harvey I, Nair P, Donovan J, Harvey R (2003) Reduced risk of atopic disorders in adults with *Helicobacter pylori* infection. Eur J Gastroenterol Hepatol 15 (6):637–640. https://doi.org/10.1097/01.meg.0000059127.68845.57

McLaughlin NJ, McLaughlin DI, Lefcort H (2003) The influence of socio-economic factors on *Helicobacter pylori* infection rates of students in rural Zambia. Centr Afr J Med 49(3-4):38–41

McLoughlin RM, Mills KH (2011) Influence of gastrointestinal commensal bacteria on the immune responses that mediate allergy and asthma. J Allergy Clin Immunol 127 (5):1097–1107; quiz 1108–1099. https://doi.org/10.1016/j.jaci.2011.02.012

Memon AA, Hussein NR, Miendje Deyi VY, Burette A, Atherton JC (2014) Vacuolating cytotoxin genotypes are strong markers of gastric cancer and duodenal ulcer-associated *Helicobacter pylori* strains: a matched case-control study. J Clin Microbiol 52(8):2984–2989. https://doi.org/ 10.1128/JCM.00551-14

Miftahussurur M, Tuda J, Suzuki R, Kido Y, Kawamoto F, Matsuda M, Tantular IS, Pusarawati S, Nasronudin, Harijanto PN, Yamaoka Y (2014) Extremely low *Helicobacter pylori* prevalence in North Sulawesi, Indonesia and identification of a Maori-tribe type strain: a cross sectional study. Gut Pathogens 6(1):42. https://doi.org/10.1186/s13099-014-0042-0

Miftahussurur M, Sharma RP, Shrestha PK, Maharjan RK, Shiota S, Uchida T, Sato H, Yamaoka Y (2015) *Helicobacter pylori* infection and gastric mucosal atrophy in two ethnic groups in Nepal. Asian Pac J Cancer Prev 16(17):7911–7916

Milman N, Byg KE, Andersen LP, Mulvad G, Pedersen HS, Bjerregaard P (2003) Indigenous Greenlanders have a higher sero-prevalence of IgG antibodies to *Helicobacter pylori* than Danes. Int J Circumpolar Health 62(1):54–60

Mishra S, Singh V, Rao GR, Dixit VK, Gulati AK, Nath G (2008) Prevalence of *Helicobacter pylori* in asymptomatic subjects—a nested PCR based study. Infect Genet Evol 8(6):815–819. https://doi.org/10.1016/j.meegid.2008.08.001

Muhammad JS, Zaidi SF, Sugiyama T (2012) Epidemiological ins and outs of *helicobacter pylori*: a review. JPMA J Pak Med Assoc 62(9):955–959

Myint T, Shiota S, Vilaichone RK, Ni N, Aye TT, Matsuda M, Tran TT, Uchida T, Mahachai V, Yamaoka Y (2015) Prevalence of *Helicobacter pylori* infection and atrophic gastritis in patients with dyspeptic symptoms in Myanmar. World J Gastroenterol 21(2):629–636. https://doi.org/10.3748/wjg.v21.i2.629

Nagy P, Johansson S, Molloy-Bland M (2016) Systematic review of time trends in the prevalence of *Helicobacter pylori* infection in China and the USA. Gut Pathogens 8:8. https://doi.org/10. 1186/s13099-016-0091-7

Naja F, Kreiger N, Sullivan T (2007) *Helicobacter pylori* infection in Ontario: prevalence and risk factors. Can J Gastroenterol 21(8):501–506

Naous A, Al-Tannir M, Naja Z, Ziade F, El-Rajab M (2007) Fecoprevalence and determinants of *Helicobacter pylori* infection among asymptomatic children in Lebanon. Leban Med J 55 (3):138–144

Newton R, Ziegler JL, Casabonne D, Carpenter L, Gold BD, Owens M, Beral V, Mbidde E, Parkin DM, Wabinga H, Mbulaiteye S, Jaffe H (2006) *Helicobacter pylori* and cancer among adults in Uganda. Infect Agents Cancer 1:5. https://doi.org/10.1186/1750-9378-1-5

Ntagirabiri R, Harerimana S, Makuraza F, Ndirahisha E, Kaze H, Moibeni A (2014) *Helicobacter pylori* au Burundi: première évaluation de la prévalence en endoscopie et de l'éradication. J Afr d'Hépato-Gastroentérol 8(4):217–222. https://doi.org/10.1007/s12157-014-0567-3

Oertli M, Muller A (2012) *Helicobacter pylori* targets dendritic cells to induce immune tolerance, promote persistence and confer protection against allergic asthma. Gut Microbes 3 (6):566–571. https://doi.org/10.4161/gmic.21750

Oertli M, Sundquist M, Hitzler I, Engler DB, Arnold IC, Reuter S, Maxeiner J, Hansson M, Taube C, Quiding-Jarbrink M, Muller A (2012) DC-derived IL-18 drives Treg differentiation, murine *Helicobacter pylori*-specific immune tolerance, and asthma protection. J Clin Investig 122(3):1082–1096. https://doi.org/10.1172/JCI61029

Oertli M, Noben M, Engler DB, Semper RP, Reuter S, Maxeiner J, Gerhard M, Taube C, Muller A (2013) *Helicobacter pylori* gamma-glutamyl transpeptidase and vacuolating cytotoxin promote gastric persistence and immune tolerance. Proc Natl Acad Sci USA 110(8):3047–3052. https://doi.org/10.1073/pnas.1211248110

Olivares A, Buadze M, Kutubidze T, Lobjanidze M, Labauri L, Kutubidze R, Chikviladze D, Zhvania M, Kharzeishvili O, Lomidze N, Perez-Perez GI (2006) Prevalence of *Helicobacter pylori* in Georgian patients with dyspepsia. Helicobacter 11(2):81–85. https://doi.org/10.1111/ j.1523-5378.2006.00367.x

Onishi Y, Fehervari Z, Yamaguchi T, Sakaguchi S (2008) Foxp3+ natural regulatory T cells preferentially form aggregates on dendritic cells in vitro and actively inhibit their maturation. Proc Natl Acad Sci USA 105(29):10113–10118. https://doi.org/10.1073/pnas.0711106105

Ontsira Ngoyi EN, Atipo Ibara BI, Moyen R, Ahoui Apendi PC, Ibara JR, Obengui O, Ossibi Ibara RB, Nguimbi E, Niama RF, Ouamba JM, Yala F, Abena AA, Vadivelu J, Goh KL, Menard A, Benejat L, Sifre E, Lehours P, Megraud F (2015) Molecular detection of *Helicobacter pylori* and its antimicrobial resistance in Brazzaville, Congo. Helicobacter 20(4):316–320. https://doi.org/10.1111/hel.12204

Ortega JP, Espino A, Calvo BA, Verdugo P, Pruyas M, Nilsen E, Villarroel L, Padilla O, Riquelme A, Rollan A (2010) *Helicobacter pylori* infection in symptomatic patients with benign gastroduodenal diseases: analysis of 5.664 cases. Rev Med Chil 138(5):529–535. https://doi.org/10.4067/S0034-98872010000500001

Pearce N, Ait-Khaled N, Beasley R, Mallol J, Keil U, Mitchell E, Robertson C, Group IPTS (2007) Worldwide trends in the prevalence of asthma symptoms: phase III of the International Study of Asthma and Allergies in Childhood (ISAAC). Thorax 62(9):758–766. https://doi.org/10.1136/thx.2006.070169

Pedrini MJ, Seewann A, Bennett KA, Wood AJ, James I, Burton J, Marshall BJ, Carroll WM, Kermode AG (2015) *Helicobacter pylori* infection as a protective factor against multiple sclerosis risk in females. J Neurol Neurosurg Psychiatry 86(6):603–607. https://doi.org/10.1136/jnnp-2014-309495

Pessi T, Virta M, Adjers K, Karjalainen J, Rautelin H, Kosunen TU, Hurme M (2005) Genetic and environmental factors in the immunopathogenesis of atopy: interaction of *Helicobacter pylori* infection and IL4 genetics. Int Arch Allergy Immunol 137(4):282–288. https://doi.org/10.1159/000086421

Prescott SL, Macaubas C, Holt BJ, Smallacombe TB, Loh R, Sly PD, Holt PG (1998) Transplacental priming of the human immune system to environmental allergens: universal skewing of initial T cell responses toward the Th2 cytokine profile. J Immunol 160(10):4730–4737

Rad R, Dossumbekova A, Neu B, Lang R, Bauer S, Saur D, Gerhard M, Prinz C (2004) Cytokine gene polymorphisms influence mucosal cytokine expression, gastric inflammation, and host specific colonisation during *Helicobacter pylori* infection. Gut 53(8):1082–1089. https://doi.org/10.1136/gut.2003.029736

Radon K, Windstetter D, Eckart J, Dressel H, Leitritz L, Reichert J, Schmid M, Praml G, Schosser M, von Mutius E, Nowak D (2004) Farming exposure in childhood, exposure to markers of infections and the development of atopy in rural subjects. Clin Exp Allergy 34 (8):1178–1183. https://doi.org/10.1111/j.1365-2222.2004.02005.x

Reibman J, Marmor M, Filner J, Fernandez-Beros ME, Rogers L, Perez-Perez GI, Blaser MJ (2008) Asthma is inversely associated with *Helicobacter pylori* status in an urban population. PloS One 3(12):e4060. https://doi.org/10.1371/journal.pone.0004060

Rizzuti D, Ang M, Sokollik C, Wu T, Abdullah M, Greenfield L, Fattouh R, Reardon C, Tang M, Diao J, Schindler C, Cattral M, Jones NL (2015) *Helicobacter pylori* inhibits dendritic cell maturation via interleukin-10-mediated activation of the signal transducer and activator of transcription 3 pathway. J Innate Immun 7(2):199–211. https://doi.org/10.1159/000368232

Robinson K, Kenefeck R, Pidgeon EL, Shakib S, Patel S, Polson RJ, Zaitoun AM, Atherton JC (2008) *Helicobacter pylori*-induced peptic ulcer disease is associated with inadequate regulatory T cell responses. Gut 57(10):1375–1385. https://doi.org/10.1136/gut.2007.137539

Roka K, Roubani A, Stefanaki K, Panayotou I, Roma E, Chouliaras G (2014) The prevalence of *Helicobacter pylori* gastritis in newly diagnosed children with inflammatory bowel disease. Helicobacter 19(5):400–405. https://doi.org/10.1111/hel.12141

Ronchetti R, Villa MP, Barreto M, Rota R, Pagani J, Martella S, Falasca C, Paggi B, Guglielmi F, Ciofetta G (2001) Is the increase in childhood asthma coming to an end? Findings from three surveys of schoolchildren in Rome, Italy. Eur Respir J 17(5):881–886

Russell SL, Gold MJ, Willing BP, Thorson L, McNagny KM, Finlay BB (2013) Perinatal antibiotic treatment affects murine microbiota, immune responses and allergic asthma. Gut Microbes 4(2):158–164. https://doi.org/10.4161/gmic.23567

Ryanna K, Stratigou V, Safinia N, Hawrylowicz C (2009) Regulatory T cells in bronchial asthma. Allergy 64(3):335–347. https://doi.org/10.1111/j.1398-9995.2009.01972.x

Sanchez Ceballos F, Taxonera Samso C, Garcia Alonso C, Alba Lopez C, Sainz de Los Terreros Soler L, Diaz-Rubio M (2007) Prevalence of *Helicobacter pylori* infection in the healthy population of Madrid (Spain). Rev Esp Enferm Dig 99(9):497–501

Sasaki T, Hirai I, Yamamoto Y (2009) Analysis of *Helicobacter pylori* infection in a healthy Panamanian population. Kansenshogaku Zasshi 83(2):127–132

Secka O, Antonio M, Berg DE, Tapgun M, Bottomley C, Thomas V, Walton R, Corrah T, Thomas JE, Adegbola RA (2011) Mixed infection with cagA positive and cagA negative strains of *Helicobacter pylori* lowers disease burden in the Gambia. PloS One 6(11):e27954. https://doi. org/10.1371/journal.pone.0027954

Seiskari T, Kondrashova A, Viskari H, Kaila M, Haapala AM, Aittoniemi J, Virta M, Hurme M, Uibo R, Knip M, Hyoty H (2007) Allergic sensitization and microbial load—a comparison between Finland and Russian Karelia. Clin Exp Immunol 148(1):47–52. https://doi.org/10. 1111/j.1365-2249.2007.03333.x

Semper RP, Mejias-Luque R, Gross C, Anderl F, Muller A, Vieth M, Busch DH, Prazeres da Costa C, Ruland J, Gross O, Gerhard M (2014) *Helicobacter pylori*-induced IL-1beta secretion in innate immune cells is regulated by the NLRP3 inflammasome and requires the cag pathogenicity island. J Immunol 193(7):3566–3576. https://doi.org/10.4049/jimmunol. 1400362

Serizawa T, Hirata Y, Hayakawa Y, Suzuki N, Sakitani K, Hikiba Y, Ihara S, Kinoshita H, Nakagawa H, Tateishi K, Koike K (2015) Gastric metaplasia induced by *Helicobacter pylori* is associated with enhanced SOX9 expression via Interleukin-1 signaling. Infect Immun 84 (2):562–572. https://doi.org/10.1128/iai.01437-15

Shigematsu Y, Niwa T, Rehnberg E, Toyoda T, Yoshida S, Mori A, Wakabayashi M, Iwakura Y, Ichinose M, Kim YJ, Ushijima T (2013) Interleukin-1beta induced by *Helicobacter pylori* infection enhances mouse gastric carcinogenesis. Cancer Lett 340(1):141–147. https://doi.org/ 10.1016/j.canlet.2013.07.034

Shiota S, Cruz M, Abreu JA, Mitsui T, Terao H, Disla M, Iwatani S, Nagashima H, Matsuda M, Uchida T, Tronilo L, Rodriguez E, Yamaoka Y (2014) Virulence genes of *Helicobacter pylori* in the Dominican Republic. J Med Microbiol 63(Pt 9):1189–1196. https://doi.org/10.1099/ jmm.0.075275-0

Shiotani A, Miyanishi T, Kamada T, Haruma K (2008) *Helicobacter pylori* infection and allergic diseases: epidemiological study in Japanese university students. J Gastroenterol Hepatol 23 (7 Pt 2):e29–e33. https://doi.org/10.1111/j.1440-1746.2007.05107.x

Sonnenberg A, Genta RM (2012) Low prevalence of *Helicobacter pylori* infection among patients with inflammatory bowel disease. Aliment Pharmacol Ther 35(4):469–476. https://doi.org/10. 1111/j.1365-2036.2011.04969.x

Sporea I, Popescu A, van Blankenstein M, Sirli R, Focsea M, Danila M (2003) The prevalence of *Helicobacter pylori* infection in western Romania. Rom J Gastroenterol 12(1):15–18

Storskrubb T, Aro P, Ronkainen J, Sipponen P, Nyhlin H, Talley NJ, Engstrand L, Stolte M, Vieth M, Walker M, Agreus L (2008) Serum biomarkers provide an accurate method for diagnosis of atrophic gastritis in a general population: the Kalixanda study. Scand J Gastroenterol 43(12):1448–1455. https://doi.org/10.1080/00365520802273025

Tadesse E, Daka D, Yemane D, Shimelis T (2014) Seroprevalence of *Helicobacter pylori* infection and its related risk factors in symptomatic patients in southern Ethiopia. BMC Res Notes 7:834. https://doi.org/10.1186/1756-0500-7-834

Telaranta-Keerie A, Kara R, Paloheimo L, Harkonen M, Sipponen P (2010) Prevalence of undiagnosed advanced atrophic corpus gastritis in Finland: an observational study among 4,256 volunteers without specific complaints. Scand J Gastroenterol 45(9):1036–1041. https://doi.org/10.3109/00365521.2010.487918

Thjodleifsson B, Asbjornsdottir H, Sigurjonsdottir RB, Gislason D, Olafsson I, Cook E, Gislason T, Jogi R, Janson C (2007) Seroprevalence of *Helicobacter pylori* and cagA anti-bodies in Iceland, Estonia and Sweden. Scand J Infect Dis 39(8):683–689. https://doi.org/10. 1080/00365540701225736

To T, Stanojevic S, Moores G, Gershon AS, Bateman ED, Cruz AA, Boulet LP (2012) Global asthma prevalence in adults: findings from the cross-sectional world health survey. BMC Public Health 12:204. https://doi.org/10.1186/1471-2458-12-204

Tsang KW, Lam WK, Chan KN, Hu W, Wu A, Kwok E, Zheng L, Wong BC, Lam SK (2000) *Helicobacter pylori* sero-prevalence in asthma. Respir Med 94(8):756–759. https://doi.org/10.1053/rmed.2000.0817

Tu S (2008) Overexpression of interleukin-1beta induces gastric inflammation and cancer and mobilizes myeloid-derived suppressor cells in mice. Cancer Cell 14(5):408–419. https://doi.org/10.1016/j.ccr.2008.10.011

Uchida T, Miftahussurur M, Pittayanon R, Vilaichone RK, Wisedopas N, Ratanachu-Ek T, Kishida T, Moriyama M, Yamaoka Y, Mahachai V (2015) *Helicobacter pylori* Infection in Thailand: a nationwide study of the CagA phenotype. PloS One 10(9):e0136775. https://doi.org/10.1371/journal.pone.0136775

Uter W, Stock C, Pfahlberg A, Guillen-Grima F, Aguinaga-Ontoso I, Brun-Sandiumenge C, Kramer A (2003) Association between infections and signs and symptoms of 'atopic' hypersensitivity—results of a cross-sectional survey among first-year university students in Germany and Spain. Allergy 58(7):580–584

van Rijt LS, Jung S, Kleinjan A, Vos N, Willart M, Duez C, Hoogsteden HC, Lambrecht BN (2005) In vivo depletion of lung CD11c+ dendritic cells during allergen challenge abrogates the characteristic features of asthma. J Exp Med 201(6):981–991. https://doi.org/10.1084/jem.20042311

Veldhoen M, Stockinger B (2006) TGFbeta1, a "Jack of all trades": the link with pro-inflammatory IL-17-producing T cells. Trends Immunol 27(8):358–361. https://doi.org/10.1016/j.it.2006.06.001

von Hertzen L, Haahtela T (2005) Signs of reversing trends in prevalence of asthma. Allergy 60 (3):283–292. https://doi.org/10.1111/j.1398-9995.2005.00769.x

Walker TD, Karemera M, Ngabonziza F, Kyamanywa P (2014) *Helicobacter pylori* status and associated gastroscopic diagnoses in a tertiary hospital endoscopy population in Rwanda. Trans R Soc Trop Med Hyg 108(5):305–307. https://doi.org/10.1093/trstmh/tru029

Wang G, Miyahara Y, Guo Z, Khattar M, Stepkowski SM, Chen W (2010) "Default" generation of neonatal regulatory T cells. J Immunol 185(1):71–78. https://doi.org/10.4049/jimmunol.0903806

Wang Q, Yu C, Sun Y (2013) The association between asthma and *Helicobacter pylori*: a meta-analysis. Helicobacter 18(1):41–53. https://doi.org/10.1111/hel.12012

Webb PM, Knight T, Greaves S, Wilson A, Newell DG, Elder J, Forman D (1994) Relation between infection with *Helicobacter pylori* and living conditions in childhood: evidence for person to person transmission in early life. BMJ 308(6931):750–753

Werme K, Bisseye C, Ouedraogo I, Yonli AT, Ouermi D, Djigma F, Moret R, Gnoula C, Nikiema JB, Simpore J (2015) Molecular diagnostics of *Helicobacter pylori* by PCR in patients in gastroenterology consultation at Saint Camille Medical Centre in Ouagadougou. Pan Afr Med J 21:123. 10.11604/pamj.2015.21.123.6001

Wewer V, Andersen LP, Paerregaard A, Gernow AB, Hart Hansen JP, Matzen P, Krasilnikoff PA (1998) The prevalence and related symptomatology of *Helicobacter pylori* in children with recurrent abdominal pain. Acta Paediatr 87(8):830–835

Windsor HM, Abioye-Kuteyi EA, Leber JM, Morrow SD, Bulsara MK, Marshall BJ (2005) Prevalence of *Helicobacter pylori* in Indigenous Western Australians: comparison between urban and remote rural populations. Med J Aust 182(5):210–213

Wu XW, Ji HZ, Yang MF, Wu L, Wang FY (2015) *Helicobacter pylori* infection and inflammatory bowel disease in Asians: a meta-analysis. World J Gastroenterol 21(15):4750–4756. https://doi.org/10.3748/wjg.v21.i15.4750

Yang XO, Nurieva R, Martinez GJ, Kang HS, Chung Y, Pappu BP, Shah B, Chang SH, Schluns KS, Watowich SS, Feng XH, Jetten AM, Dong C (2008) Molecular antagonism and plasticity of regulatory and inflammatory T cell programs. Immunity 29(1):44–56. https://doi.org/10. 1016/j.immuni.2008.05.007

Yao G, Wang P, Luo XD, Yu TM, Harris RA, Zhang XM (2016) Meta-analysis of association between *Helicobacter pylori* infection and multiple sclerosis. Neurosci Lett 620:1–7. https://doi.org/10.1016/j.neulet.2016.03.037

Yogev N, Frommer F, Lukas D, Kautz-Neu K, Karram K, Ielo D, von Stebut E, Probst HC, van den Broek M, Riethmacher D, Birnberg T, Blank T, Reizis B, Korn T, Wiendl H, Jung S, Prinz M, Kurschus FC, Waisman A (2012) Dendritic cells ameliorate autoimmunity in the CNS by controlling the homeostasis of PD-1 receptor(+) regulatory T cells. Immunity 37(2):264–275. https://doi.org/10.1016/j.immuni.2012.05.025

Zevit N, Balicer RD, Cohen HA, Karsh D, Niv Y, Shamir R (2012) Inverse association between *Helicobacter pylori* and pediatric asthma in a high-prevalence population. Helicobacter 17 (1):30–35. https://doi.org/10.1111/j.1523-5378.2011.00895.x

Zhang M, Liu M, Luther J, Kao JY (2010) *Helicobacter pylori* directs tolerogenic programming of dendritic cells. Gut Microbes 1(5):325–329. https://doi.org/10.4161/gmic.1.5.13052

Zhou S, Huang Y, Liang B, Dong H, Yao S, Chen Y, Xie Y, Long Y, Gong S, Zhou Z (2017) Systemic and mucosal pre-administration of recombinant *Helicobacter pylori* neutrophil-activating protein prevents ovalbumin-induced allergic asthma in mice. FEMS Microbiol Lett. https://doi.org/10.1093/femsle/fnw288

Zhou X, Wu J, Zhang G (2013) Association between *Helicobacter pylori* and asthma: a meta-analysis. Eur J Gastroenterol Hepatol 25(4):460–468. https://doi.org/10.1097/MEG. 0b013e32835c280a

Parasite Mediated Protection Against Allergy

Julia Esser-von Bieren

Abstract Throughout evolution, mammals have been exposed to large multicellular worm parasites (helminths). This co-evolution has resulted in a well-balanced network of host-parasite interactions that shape the host's immune system. In particular, helminth parasites have acquired an impressive repertoire of immunomodulatory strategies to ensure their own survival and reproduction whilst avoiding excessive harm to their host. Moreover, work over the recent decades has shown that the immunomodulatory potential of helminths may not only entail an intriguing capacity for immune evasion, but also provide a means to protect against chronic inflammatory diseases. Allergy is driven by pathological type 2 immune responses, including type 2 cytokine production, IgE class switching, mast cell activation as well as eosinophil recruitment and activation, which are also at play during helminth infection. The overlap between pathological mechanisms of allergy and protective immune mechanisms targeted at parasite killing or expulsion may provide an (evolutionary) explanation for the potential of helminths to suppress allergic inflammation. This chapter will describe the epidemiological and mechanistic basis of parasite-mediated protection against allergy with a focus on the cellular mechanisms employed by helminth parasites to suppress type 2 inflammation.

1 Introduction

Parasitic helminths (*hélmins* (greek) = worm) represent a diverse group of macroparasites, ranging from millimeters to meters in size. Helminths have complex lifecycles often involving multiple larval and adult stages and several hosts (e.g. insects and humans). The lifecycle usually begins with eggs being shed by the host (e.g. through the feces). In settings of poor sanitation, eggs or hatched free-

J. Esser-von Bieren
Center of Allergy and Environment (ZAUM), Technical University of Munich and Helmholtz Center Munich, Munich, Germany
e-mail: julia.esser@tum.de

© Springer International Publishing AG 2017
C.B. Schmidt-Weber (ed.), *Allergy Prevention and Exacerbation*, Birkhäuser
Advances in Infectious Diseases, https://doi.org/10.1007/978-3-319-69968-4_6

living larvae can be transmitted via contaminated soil or water. Infection often occurs via oral ingestion or via the skin. Indeed, most helminths are well-equipped to migrate through host tissues such as the skin, lung, liver and/or intestine. During this migration, the mammalian host typically develops a type 2 immune response, characterized by eosinophilia, M2 macrophage polarization, activation of T_H2 cells and high levels of IgE. How exactly this type 2 response is initiated remains only partially understood, but alarmin secretion by damaged epithelial cells as well as direct activation of the innate immune system by parasite molecules have been suggested as key mechanisms of helminth-driven type 2 immunity (Humphreys et al. 2008; Massacand et al. 2009).

Considering that helminths induce type 2 immune responses, which are pathogenic in allergy, one might suspect that helminth infection would exacerbate allergic inflammation. However, instead epidemiological evidence over the last decades has put helminth parasites into the spotlight as guards against allergic and other chronic inflammatory diseases (Maizels 2016; Obeng et al. 2014; Stein et al. 2016; Yazdanbakhsh et al. 2002). A recent study, involving more than 1000 patients, has also shown that deworming can attenuate immunosuppression in helminth infected individuals (Wammes et al. 2016). Thus, helminth infection may represent a major cause of the lower incidence of chronic inflammatory diseases in developing countries. From an evolutionary perspective, the anti-allergic and anti-inflammatory effects of helminths are well explainable. Indeed, many of the pathological mechanisms in allergy represent protective effector mechanisms targeted at killing or expelling helminth parasites. Through their long co-evolution with their mammalian hosts, helminth parasites have thus evolved immune evasion strategies to suppress mechanisms of type 2 inflammation to their own benefit (Allen and Maizels 2011).

However, it is important to note that helminths may also induce allergy like symptoms and promote pathogenic inflammation at least acutely (Enobe et al. 2006; Nogueira et al. 2016). Allergy promoting effects are particularly evident for the human and pig parasite *Ascaris* and the fish parasite *Anisakis* (Enobe et al. 2006; Nogueira et al. 2016). Recent studies in mice and humans show that sensitization to *Ascaris* antigens is associated with allergic airway inflammation and that *Ascaris* infection may promote house dust mite (HDM) allergy, potentially due to the induction if crossreactive IgE (Ahumada et al. 2015; Suzuki et al. 2016). Yet, despite their pro-allergic potential also *Ascaris* and *Anisakis* have been reported to produce protective molecules, which can prevent allergic inflammation in rodent models (Cho et al. 2015; Itami et al. 2005; McConchie et al. 2006). Due to the potential of helminths to elicit both exacerbation and prevention of allergic inflammation, it is imperative to dissect the exact context, mechanisms and active molecules responsible for the immunoregulatory effects of individual helminth species. The major focus of this chapter will thus be on already identified immunemodulatory helminth products and the cellular and molecular mechanisms underlying their allergy protective effects (for an overview see Table 1).

Table 1 Examples of helminth-derived modulators of allergy and their cellular mechanisms of action

Helminth product	Mechanism of action	References
Acanthocheilonema viteae/Onchocerca volvulus Av17/Ov17 Cystatins	Induction of IL-10 producing regulatory T-cells and macrophages Interference with Mast cell activation	Daniłowicz-Luebert et al. (2013), Klotz et al. (2011), Schnoeller et al. (2008), Ziegler et al. (2015)
Acanthocheilonema viteae ES-62	Interference with IgE-mediated mast cell activation	Melendez et al. (2007)
Ancylostoma caninum AIP-2	Induction of regulatory T-cells, modulation of DC activation	Navarro et al. (2016)
Ascaris suum PAS-1	Induction of regulatory T-cells, induction of IL-10	de Araújo et al. (2010), Itami et al. (2005)
Ascaris suum PCF[a]	Induction of IL-10, suppression of DC function	McConchie et al. (2006)
Heligmosomoides polygyrus HES[a]	Induction of regulatory T-cells, TGF-β-like activity, suppression of IL-33 production, suppression of ILC2 activation	McSorley et al. (2012, 2014, 2015), Wilson et al. (2005)
Nippostrongylus brasiliensis NES[a] (nonprotein fraction)	Induction of regulatory T-cells[b]	Trujillo-Vargas et al. (2007)
Nippostrongylus brasiliensis Cystatin	Interference with antigen presentation	Dainichi et al. (2001)
Schistosoma mansoni eggs[a], egg antigens[a]	Induction of regulatory T-cells	Layland et al. (2013), Yang et al. (2007)
Schistosoma mansoni LysoPS	Modulation of DC function, induction of IL-10 producing regulatory T-cells	van der Kleij et al. (2002)
Schistosoma mansoni	IL-10 producing regulatory B-cells	Smits et al. (2007)
Schistosoma mansoni LNFP-III	Induction of IL-10 producing B-cells, prostaglandin E_2 synthesis	Velupillai and Harn (1994)

[a]Crude mixtures of secreted products or extracts
[b]Putative mechanism of action

2 Epidemiology of Parasite Mediated Protection Against Allergy

Epidemiological evidence for the allergy preventive effects of helminth parasites in humans has been obtained in a variety of studies mainly conducted in developing countries. Studies performed in Africa found an inverse correlation between atopy and helminth infection (e.g. with *Schistosoma haematobium* or hookworm) (Nyan et al. 2001; Scrivener et al. 2001; van den Biggelaar et al. 2000). In particular, high parasite burdens were associated with reduced asthma symptoms (Scrivener et al. 2001). Schistosome infection was also found to be negatively associated with mite atopy, but not wheeze or asthma (Obeng et al. 2014). Further work suggested a role for increased IL-10 production and the expansion of regulatory T- and B-cells in the allergy-protective effect of Schistosome infection (Schmiedel et al. 2015; van den Biggelaar et al. 2000; van der Vlugt et al. 2012).

In keeping with the helminth-mediated protection against allergy in African populations, studies in South America showed that *S. mansoni* infection was associated with a reduction in asthma severity (Medeiros et al. 2003). In regions, where helminths are endemic, chronically infected individuals, who showed high levels of total IgE and anti-*Ascaris lumbricoides* IgG4, developed reduced skin prick test reactivity (Cooper et al. 2003).

However, in contrast to Schistosome infection, which has repeatedly been shown to protect against allergy, *Ascaris* infection has also been reported to increase respiratory allergy symptoms. A study performed in South Africa reported that children with elevated Ascaris-sIgE showed an increased risk of positive skin prick test to aeroallergens (e.g. house dust mite) as well as atopic asthma, atopic rhinitis and increased airway hyper-responsiveness (Obihara et al. 2006). Different outcomes for particular helminth species (e.g. *Ascaris*) may be explained by additional environmental factors such as co-infections or the composition of the microbiota, which may fundamentally differ between communities and which are known to impact on the development of allergies (see chapters "Bacterial Allergens", "Good and Bad Farming: The Right Microbiome Protects from Allergy" and "The Lost Friend: H. pylori").

The maternal immune system is known to strongly influence the immune state of the neonate and evidence from rodent models (see Sect. 3) supported a role for maternal helminth infection in the protection against allergy in infants (Straubinger et al. 2014). Yet, a recent study in humans found no protective effect of maternal helminth infection against the development of eczema or wheeze in young children despite apparent reductions in skin prick test reactivity to house dust mite and perennial allergens (Cooper et al. 2016).

However, two recent studies provided further support for a direct allergy protective effect of helminth infection. The first study showed that Ethiopian immigrants, who moved to Israel and who were infected with helminths, had a low risk of allergy. However, already after one year, immigrants presented with lower helminth burdens and a higher risk of allergy (Stein et al. 2016). Although other environmental factors likely contributed to immunological changes in these individuals, this study suggested a link between the loss of immunemodulatory helminth infection and the development of allergic disease. The second study, performed in Indonesia, found that community deworming abolished the immuno-compromised state in helminth infected individuals (Wammes et al. 2016). This suggested a direct immunomodulatory effect of helminth infection, particularly with respect to allergy prevention.

Thus, helminth parasites appear to be powerful immuneregulators with preventive or even therapeutic potential in allergy. To test the therapeutic effects of helminth infection, several clinical trials were initiated using life infection with hookworms or whipworms. Unfortunately, clinical trials using (such) life infection with helminths have so far resulted in negative outcomes despite overall safety and feasibility [for a more comprehensive review see (Evans and Mitre 2015)]. As an example, experimental infection of asthmatic patients with hookworm larvae had no impact on asthma symptoms including airway hyperresponsiveness (Feary et al. 2010). In addition, infection with the whipworm *Trichuris suis* failed to improve allergic rhinitis (Bager et al. 2010). These disappointing results in therapeutic settings may suggest that helminth infection is more efficient in preventing rather than curing allergic diseases. In this context it is important to note that the most dramatic allergy-protective effects in rodent models have also been observed in preventive settings (see Sect. 3). However, as we will see, there is also evidence for a therapeutic effect of helminth infection or helminth-derived molecules in allergy models. Yet, to fully exploit the allergy-modulatory potential of helminths, future identification of the mode(s) of action and (bio)chemical characterization of active helminth molecules is warranted. Such studies should not remain limited to rodent models, but have to be translated to relevant human models of allergy.

3 Parasite Mediated Protection Against Allergy in Rodent Models

Rodent models combining helminth infection and allergic sensitization represent a useful tool for studying the direct impact of infection on allergy symptoms. These models offer the opportunity to study key aspects of helminth-based prevention or therapy (e.g. dose, timing, route of administration) in a highly controlled fashion, Furthermore, rodent models allow for the use of conditional knock-outs or adoptive transfer studies, which provide detailed mechanistic insights into the allergy-protective actions of helminths. In general, two types of rodent models can be discriminated: first, "natural" infection with a parasite that would usually infect the respective host in the wild [e.g. *Heligmosomoides polygyrus bakeri* (*H. polygyrus*) in mice] and second "artificial/xeno" infection with a parasite that would usually infect another species, but not the model species (e.g. *S. mansoni* or *A. suum* infection in mice). The latter models have several disadvantages, including the lack of a complete lifecycle as well as of co-evolution of the parasite and the host. Thus, findings obtained from rodent models of infection with human parasites should be interpreted with care and immunomodulatory mechanisms have to be confirmed in human settings.

Schistosoma infection is one of the best-studied helminth infections with regards to allergy prevention in both humans (see Sect. 2) and mice. Although mice do not represent a natural host of *S. mansoni*, most studies have shown that experimental *S. mansoni* infection can prevent allergic airway inflammation in murine allergy models (mostly OVA-alum sensitization). However, the allergy preventive effects were strongly dependent on the duration of the infection as chronic but not acute infection protected against allergy (Smits et al. 2007). In addition, the developmental stage of the parasite (i.e. eggs vs. adult worms) was suggested to differentiate between prevention versus exacerbation of allergic inflammation (Layland et al. 2013; Mangan et al. 2006). However, most studies reported a similar mechanistic basis for allergy protective effects of Schistosomes, which were also consistent between murine and human epidemiological studies: The parasite induces regulatory T-cells and high levels of IL-10 (Schmiedel et al. 2015; van den Biggelaar et al. 2000; van der Vlugt et al. 2012), which were found to be responsible for suppressing allergic airway inflammation in experimental infection (Layland et al. 2013; Smits et al. 2007). Interestingly, the protective effect of *S. mansoni* infection could even be transferred to the offspring of infected mothers, although this required mating in the T_H1 or regulatory phase (rather than the T_H2 phase) of the infection (Straubinger et al. 2014). Moreover, the maternally transmitted protection against allergy was independent on parasite antigens, but dependent on Ifnγ, particularly in the acute (T_H1) phase of *S. mansoni* infection.

In contrast to the trematode *S. mansoni*, the nematode *H. polygyrus* is a natural parasite of mice with a purely enteric lifecycle. When mice were infected with *H. polygyrus*, airway inflammation (particularly eosinophilia) was attenuated in mice sensitized to ovalbumin (OVA) or house dust mite allergen (Der p1) (Wilson et al. 2005). This effect was found to be dependent on regulatory T-cells, but independent of IL-10. Whilst confirming the allergy-protective effect of *H. polygyrus* and the central role for Tregs, another study suggested that the prevention of allergic airway inflammation was IL-10 dependent (Kitagaki et al. 2006). This discrepancy is likely explained by the fact that the latter study investigated infection of IL-10 deficient mice, whilst the earlier study used transfer of lymph node cells from IL-10 deficient animals into wildtype recipients. Thus, cells other than Tregs or Bregs likely represent the major source of regulatory IL-10 during *H. polygyrus* infection. Consistent with the central role of IL-10 in parasite-mediated prevention of allergy, the rat parasite *Nippostrongylus brasiliensis* (*N. brasiliensis*) could prevent OVA-alum induced airway inflammation in an IL-10-dependent fashion (Wohlleben et al. 2004).

Allergic diseases are not limited to the respiratory tract, but can also affect the intestine or skin. However, only few studies have investigated the effect of helminth infection on food allergy or atopic dermatitis. When comparing the effects of *H. polygyrus* infection on experimental allergic airway inflammation or atopic dermatitis (AD), one study found that the protection against allergy occurred in the lung, but not in the skin (Hartmann et al. 2009). This was associated with a failure to induce regulatory T-cells in skin-draining lymph nodes and with an increased recruitment of mast cells to the skin of allergic mice. These findings suggest a particular role for the gut-lung axis in the protection against allergy by intestinal parasites. A recent study provided new insights into the events underlying this immunomodulatory gut-lung crosstalk by identifying the intestinal microbiota as a key player in helminth-mediated prevention of allergy (Zaiss et al. 2015).

Filarial parasites such as the rodent parasite *Litomosoides sigmodontis* (*L. sigmodontis*) display a more complex lifecycle than soil-transmitted nematodes such as *H. polygyrus* as they are transmitted via an intermediate host (e.g. mites). However, filaria can employ similar strategies to modulate the immune system of their definitive host. Thus, infection of mice with *L. sigmodontis* suppressed sensitization, airway inflammation, and bronchial hyperreactivity in an OVA allergy model in a Treg dependent fashion (Dittrich et al. 2008). In a clinically relevant model of grass pollen allergy, filarial cystatin was shown to suppress airway eosinophilia, type 2 cytokine production and allergen-specific IgE production (Daniłowicz-Luebert et al. 2013). Importantly, filarial cystatin also had the capacity to modulate cytokine production in PBMCs from grass pollen allergic patients, shifting their profile towards T_H1 cytokines.

In contrast to *H. polygyrus* or *L. sigmodontis*, the parasitic nematodes *Ascaris* (host: human or pig) and *Anisakis* (host: fish) have shown positive associations with allergy (Levin et al. 2012; Nieuwenhuizen and Lopata 2014). However, even these potentially allergy-promoting parasites produce molecules, which can prevent allergic inflammation in rodent models (Cho et al. 2015; Itami et al. 2005; McConchie et al. 2006).

Taken together, mouse models of allergic airway inflammation have provided a bulk of evidence for the allergy-protective impact of helminth infection. Key mechanisms that were highlighted by multiple studies were the induction of IL-10 and the expansion of regulatory T-cells. However, helminths can manipulate the full spectrum of immune cells to create an optimal environment for their survival, maturation and reproduction (egg production). Current research aims at identifying the active molecules that are produced by helminth parasites to achieve this immunomodulation. The following chapter will give an overview of the helminthic immunomodulators that have already been identified and characterized as well as their cellular and (where known) molecular targets.

4 Mechanisms of Parasite Mediated Protection Against Allergy

Helminths can modulate all layers of their hosts' immune response, including both the innate and the adaptive arm as well as leukocytes and structural cells (Fig. 1). Whilst the tissue damage that is caused by migratory helminth larvae induces type 2 inflammation, helminths can secrete or shed a variety of molecules that control the pro-inflammatory tissue type 2 response (Grainger et al. 2010; McSorley et al. 2012; Schnoeller et al. 2008). Moreover, helminths can shape their host's immune response to ensure rapid and efficient repair of the damaged tissue (Chen et al. 2012; Esser-von Bieren et al. 2013, 2015).

4.1 Modulation of the Innate Immune System

Upon the first encounter between a helminth parasite and its host, the parasite is recognized by the innate immune system mainly via two principal mechanisms:

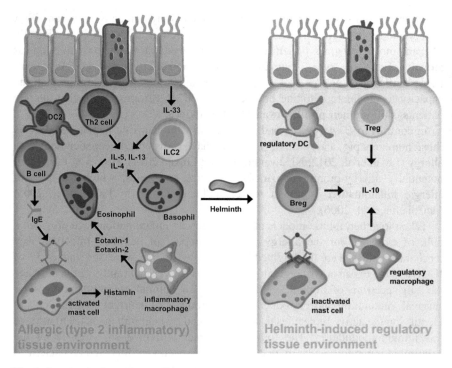

Fig. 1 Parasites and parasite products with preventive and/or therapeutic effects on allergy

First, the relatively large invading larvae damage the host's epithelial barrier (e.g. of the skin, intestine or lung) causing the release of danger associated molecular patterns (DAMPs) or alarmins, which can activate leukocytes such as monocytes/macrophages, basophils and innate lymphoid cells. Second, the parasite releases immunomodulatory molecules, which can directly bind to pattern recognition receptors (PRRs) to modulate innate immune functions.

4.1.1 Epithelial Cells

The epithelia of the skin and mucosal tissues provide a physical and immunological barrier, which represents the first line of defense against invading helminth parasites. However, epithelial cells are not only passive responders to parasite-driven tissue damage. Instead, epithelial cells e.g. in the respiratory tract express PPRs such as toll like receptors (TLRs), which allow them to directly respond to helminth derived molecules. In a mouse model of OVA/alum-induced allergic airway inflammation, an immunomodulatory molecule from the fish parasite *Anisakis simplex* (As-MIF) increased the expression of TLR2 on bronchial epithelial cells (Park et al. 2009). TLR2 signaling was responsible for suppressing several hallmarks of allergic inflammation including eosinophilia, type 2 cytokine production and airway hyper-responsiveness (Cho et al. 2015; Park et al. 2009).

4.1.2 Macrophages

Macrophages represent a highly heterogeneous and versatile cell type, which can be of embryonic origin or differentiate from bone marrow derived monocytes (for review see Perdiguero and Geissmann 2016). Historically and for reasons of simplicity, macrophage polarization during infection or inflammation is usually divided into the classical (M1) and the alternative (M2) activation states. However, more recently it has become clear that these states represent two rather artificial extremes of the macrophage polarization spectrum (Murray 2016; Xue et al. 2014). During type 2 immune responses, macrophages are usually polarized towards the M2 state, which is involved in helminth trapping, immunoregulation and wound healing or tissue remodeling. Owing to their high plasticity macrophages represent an optimal target for immunomodulation by helminth parasites. Thus, helminths have evolved strategies to modulate macrophage function during type 2 immune responses.

As an example, the filarial nematode *Acanthocheilonema viteae* (*A. viteae*) secretes an immunomodulatory protease inhibitor (cystatin, Av17), which could suppress allergic airway inflammation by inducing IL-10 producing macrophages (Schnoeller et al. 2008). Of note, the preventive and therapeutic effects of Av17 in allergy were independent of regulatory T-cells. Instead, Av cystatin-elicited regulatory macrophages were sufficient to suppress several hallmarks of allergic airway inflammation when transferred to mice before allergen challenge (Ziegler et al. 2015). This was associated with and expansion of IL-10 producing CD4$^+$ T-cells, both locally and systemically.

4.1.3 Neutrophils and Eosinophils

Infiltration of eosinophils and neutrophils represent two of the most commonly used read-outs of allergic airway inflammation. However, granulocyte infiltration is mostly regarded as a downstream consequence of cytokine or chemokine secretion by other cells (e.g. T-cells, macrophages or epithelial cells). Thus, most studies addressing the cellular basis of parasite-mediated allergy prevention have neglected granulocytes as direct targets of immunomodulatory helminth molecules.

Indeed, few studies have investigated direct effects of helminths and their molecules on the activation, recruitment or survival of eosinophils or neutrophils. Of note, both cell types express a broad variety of PRRs, allowing them to respond to microbial ligands, including those produced by helminths (Kvarnhammar and Cardell 2012; Thomas and Schroder 2013). Given that granulocytes accumulate rapidly in the proximity of invading parasites, they are likely to respond directly to the immunomodulatory factors that are released at the site of infection. However, today most studies suggest that helminth molecules such as schistosomal lysophosphatidylcholine or nematode products recruit and/ or activate eosinophils (Magalhães et al. 2010; Patnode et al. 2014). Furthermore, the filarial parasite

Onchocerca volvulus, which causes river blindness, harbors an endosymbiont that can directly activate human neutrophil migration and cytokine production in vitro (Tamarozzi et al. 2014). However, as granulocytes have a short lifespan, particularly when cultured under suboptimal in vitro conditions, eosinophil and neutrophil activation by helminths has mainly been studied in acute settings. Considering that a pro-inflammatory cytokine environment can largely extend the survival of granulocytes and that they play important roles not only during the early phase of inflammation, but also during its resolution, more long-term experiments are warranted (Campbell et al. 2014; Geering et al. 2013; Mantovani et al. 2011). In the context of allergy, it would be important to investigate how helminth molecules may affect the survival and mediator profile of eosinophils and neutrophils during allergic airway inflammation.

4.1.4 Mast Cells and Basophils

Mast cells and basophils are central players in allergy as they carry the high affinity receptor for IgE (Fc epsilon RI) and are thus activated by allergen-specific IgE, causing the release of pro-inflammatory mediators. Due to their potential to release a variety of type 2 cytokines and toxic molecules, mast cells and basophils are also implicated in the protective immune response against helminth parasites. Helminths on the other hand can directly interfere with the activation of these cells, thus contributing to their ability to suppress allergic inflammation. Historically, the suppression of mast cell and basophil activation during helminth infection was believed to depend on the induction of polyclonal IgE by helminth parasites, which would then compete for Fc epsilon RI binding with allergen specific IgE. However, competition experiments using high ratios of polyclonal to house dust mite specific IgE showed that polyclonal IgE levels in helminth infected individuals rarely reached the level that would be required to prevent allergen-specific IgE binding to FcepsilonRI (Mitre et al. 2005). Moreover, polyclonal IgE levels were not sufficiently high to saturate IgE receptors on basophils or mast cells.

Instead, a subsequent study found that a helminthic immunomodulator (ES-62) could directly interfere with FcepsilonRI signaling via TLR4-mediated degradation of protein kinase (Melendez et al. 2007). This resulted in deficient coupling of FcepsilonRI to phospholipase D, thus preventing mast cell activation by allergen specific IgE. Importantly, the same study showed that ES-62 could not only prevent mast cell-dependent allergic reactions in the lung, but also in the skin.

Evidence from mouse models of chronic infection with the helminths *S. mansoni* or *L. sigmodontis* further suggested that helminths can suppress basophil function. The reduced responsiveness of basophils from infected mice was dependent on IL-10, which downregulated components of IgE signaling (Larson et al. 2012). Thus, IL-10 derived from other helminth-responsive cell types (e.g. macrophages or regulatory T-cells) appeared to be responsible for the de-activation of basophils by chronic helminth infection. Based on these promising findings in helminth-infected rodents, a potential suppression of basophil-mediated allergic responses was investigated in helminth-infected children. Hookworm infection was associated with reduced basophil responsiveness, assessed as IgE-mediated histamine release, and reduced skin prick test reactivity to house dust mite allergens (Pinot de Moira et al. 2014). In contrast, no basophil-suppressive effects were observed for *S. mansoni* infected children, thus exemplifying that immunomodulatory effects observed in mouse models do not necessarily translate to human settings.

4.1.5 Innate Lymphoid Cells

Innate lymphoid cell (ILCs) represent a relatively recently discovered cell type of the innate immune system that participate in type 2 immune responses during helminth infection and allergy (Kim et al. 2012; Neill et al. 2010). In particular the ILC2 subtype, which is activated by the epithelial-derived cytokines IL-33 and IL-25 contributes to allergic inflammation both in the airways and the skin (Klein Wolterink et al. 2012; Salimi et al. 2013). When studying allergic airway inflammation in mice sensitized to ovalbumin or the mold *Alternaria*, McSorley et al. found that secreted products of the helminth *H. polygyrus* could suppress ILC2 activation (McSorley et al. 2014, 2015). However, the helminth-mediated suppression of ILC2 cytokine production likely represented an indirect rather than a direct effect on ILC2s. Indeed, *H. polygyrus* products abrogated early allergen-induced IL-33 release, possibly by acting on innate sources of this cytokine such as airway epithelial cells or macrophages.

4.1.6 Dendritic Cells

The potential of Schistosomes and their products to modulate dendritic cell (DC) responses is one of the best-studied immunomodulatory strategies employed by helminths. However, components of *S. mansoni*, particularly the schistosomal egg antigen (SEA) constituent omega-1, drive Th2 polarization by modulating DC activation and DC-Th cell interactions (Everts et al. 2009; Steinfelder et al. 2009). This would suggest that *S. mansoni*-exposed DCs display a Th2-inducing phenotype with the potential to exacerbate allergic inflammation. Yet, adult stages of the same parasite produce glycolipids that induce a Th1 skewing cytokine profile in DCs (van Stijn et al. 2010). In addition, an immunomodulatory peptide from a closely related parasite (*S. japonicum*) induced tolerogenic DCs, which supported the polarization of naïve into regulatory T-cells (Wang et al. 2009). Together, these studies illustrate that (particularly) for helminths with a complex lifecycle, the duration of the infection and thus the developmental stage of the parasite crucially determines the DC activation state and thus the nature of the Th response.

When DCs isolated from the spleens of *S. japonicum* infected mice 8 weeks post infection were transferred to mice prior to OVA allergen sensitization, this resulted in an efficient prevention of allergic airway inflammation (Liu et al. 2010). The protective effect was specifically mediated by CD8α⁻ DCs and was associated with a marked reduction in type 2 cytokine levels, whilst regulatory cytokines (IL-10 and TGF-β) were increased (Liu et al. 2014, 2010).

As Schistosoma, *Clonorchis sinensis* is a trematode parasite, which is transmitted via an intermediate host (fish) and which can infect the human liver. When administered in an OVA/alum model of allergic airway inflammation, total protein extract from *C. sinensis* cercaria, reduced airway eosinophilia and type 2 inflammation (Jeong et al. 2011). In vitro, *C. sinensis* proteins could directly modulate DC activation, thus suggesting that the allergy protective effects depended on modified T-cell priming by antigen presenting cells (Jeong et al. 2011).

As described under Sect. 4.1.2, *cystatin (Av17)* from *A. viteae* has the potential to induce regulatory IL-10-producing macrophages. In general, cystatins are protease inhibitors, which are conserved amongst helminths. However, immuneregulatory actions appear to be specific for cystatins from parasitic helminth species such as the filarial nematodes *A. viteae* and *O. volvulus* or the rat hookworm *N. brasiliensis* (Dainichi et al. 2001; Schierack et al. 2003). As cystatins from several nematodes (*N. brasiliensis, H. polygyrus, B. malayi*) interfered with antigen processing by DCs, this likely represents a conserved immunoregulatory mechanism with potential importance for allergy prevention (Dainichi et al. 2001). Of note, cystatins from *Angiostrongylus cantonensis* (*A. cantonensis*) and *A. viteae* could prevent respiratory allergies in mouse models of OVA/alum or grass pollen sensitization (Daniłowicz-Luebert et al. 2013; Ji et al. 2015; Schnoeller et al. 2008). However, most of these studies were performed in preventive rather than in therapeutic settings, thus suggesting that helminth-mediated modulation of DC function may be more relevant to allergy prevention than therapy.

A recent study identifying an immunomodulatory hookworm protein (anti-inflammatory protein-2, AIP-2), which targeted both human and murine DCs, confirmed the preventive potential of helminth-modulated DCs in allergy (Navarro et al. 2016). Of note, AIP-2 was equally efficacious in reducing the production of type 2 cytokines and allergen-specific IgE, when administered during the last allergen challenges as when administered preventively (during sensitization).

Thus, helminth-mediated protection against allergy at the level of the DC seems to be a particularly powerful strategy for immunomodulation. This can likely be explained by the central role of DCs as instructors of the adaptive arm of the immune system.

4.2 Modulation of the Adaptive Immune System

Depending on the signals they obtain from DCs, T-cells are polarized towards Th1, Th2, Th17 or regulatory T-cells. Whilst Th2 cells are central for type 2 immunity in parasite infection and allergy, regulatory T-cells can suppress Th2 cell proliferation and type 2 cytokine production. Thus, the induction of regulatory T-cells by helminth parasites provides a common and efficient strategy for parasitic immune evasion.

4.2.1 Regulatory T-Cells

A large number of studies have identified the induction of regulatory T-cells (Tregs) as a major mechanism for helminth-mediated prevention of allergy. A direct link between helminth-elicited Tregs and reduced allergic airway inflammation was first shown by a study using depletion and transfer of regulatory T-cells from helminth (*H. polygyrus*) infected mice (Wilson et al. 2005). Whilst Treg depletion by CD25 neutralizing antibodies abrogated protection in an OVA/alum allergy model, the protection could be transferred via adoptive transfer of cells from lymph nodes of infected mice (Wilson et al. 2005). Furthermore, this study showed that IL-10 was not the major mediator of Treg-mediated suppression of allergy. A later study demonstrated that the helminth itself produced a TGFb-like molecule, which could directly induce Treg differentiation (Grainger et al. 2010). When helminth product-induced Tregs were transferred to OVA/alum sensitized mice, they could suppress allergic airway inflammation to a similar extend as Tregs elicited by host TGFb1 (Grainger et al. 2010). However, co-administration of *H. polygyrus* excretory secretory products (HES) either during OVA-alum sensitization or at challenge resulted in different immunological outcomes. Thus, the effects of HES on T cell polarization and pro-inflammatory cytokine production were limited to the preventive setting, whilst innate type 2 markers and eosinophilia were efficiently reduced by HES administration even at challenge. Of note, these different immunemodulatory effects were likely mediated by distinct components in HES as heat treatment abrogated allergy-protection at challenge but not at sensitization (McSorley et al. 2012).

Similar to products of the murine nematode *H. polygyrus*, the pig nematode *A. suum* produces immunomodulatory proteins that induce regulatory T-cells and that modulate allergic airway inflammation. In particular the "Protein from *A. suum* 1" (PAS-1) could prevent airway eosinophilia and type 2 cytokine production, when administered during the sensitization of mice to OVA/alum (Araújo et al. 2008, p. 1). This allergy-preventive potential of PAS-1 was lost in mice deficient for either IL-10 or Ifng, suggesting a major role for these cytokines in PAS-1-mediated immunoregulation. A subsequent study showed that PAS-1 immunisation increased CD4$^+$ CD25$^+$ FoxP3$^+$ and CD8$^+$ γδTCR$^+$ T-cell populations and that the adoptive transfer of these PAS-1 primed T-cell subsets could reproduce the suppressive potential of PAS-1 immunization on allergic airway inflammation (de Araújo et al. 2010, p. 1).

Studies using proteins (Sm22.6, Sm29, PIII) of the trematode *S. mansoni* demonstrated that the Treg inducing potential is not limited to nematode proteins. Sm22.6 and Sm29, which belong to a group of tegumental antigens, as well as PIII, a fraction of adult worm extract were originally identified as potential vaccine candidates (Cardoso et al. 2008; Hirsch et al. 1997). Based on their IL-10 inducing potential in PBMCs from human *Schistosome* infected asthmatics (Cardoso et al. 2006) potential immunemodulatory effects of Sm22.6, Sm29 and PIII were evaluated in mice during OVA-alum induced allergic airway inflammation. When performed prior to sensitization, immunization of mice with all three antigens resulted in reduced lung pathology, reduced airway eosinophilia, lower levels of OVA-specific IgE and an expansion of Tregs (Cardoso et al. 2010).

More recently, the protein AIP-2 isolated from the hookworm *Ancylostoma caninum* (*A. caninum*) was shown to reduce OVA/alum induced allergic airway inflammation in mice in a DC and Treg dependent fashion (Navarro et al. 2016). In particular, AIP-2 treatment elicited the accumulation of allergy-protective CCR9 expressing Tregs in the trachea and small intestinal lamia propria in a highly tissue specific manner (Navarro et al. 2016). This study thus shed new light on the central role of the gut-lung axis in the helminth-mediated protection against allergic airway inflammation.

4.2.2 Regulatory B-Cells

Whilst the role of regulatory T-cells in helminth-mediated prevention of allergy has been studied extensively, regulatory B-cells have received less attention. However, the transfer of both Tregs and Bregs from mice chronically infected with *S. mansoni* to OVA-sensitized recipients could reduce allergic airway inflammation in an IL-10 dependent fashion (Smits et al. 2007). A later study confirmed the central role for Bregs in *S. mansoni*-mediated protection against allergy and identified CD1d, IL-10 production and the induction of Tregs as central immunomodulatory mechanisms underlying this protection. (Amu et al. 2010). Indeed, Bregs were also found to play a important role in the modulation of allergic airway inflammation by *H. polygyrus*, thus demonstrating that the induction of allergy-protective Bregs is not limited to *S. mansoni* infection (Wilson et al. 2010). However, in contrast to *S. mansoni*-elicited Bregs, *H. polygyrus*-induced Bregs suppressed pathogenic type 2 inflammation in an IL-10 independent manner. Thus, Breg phenotypes and functions may differ depending on the parasite and its specific life cycle as well as the respective immune evasion mechanisms.

5   Concluding Remarks and Outlook

Albeit partially heterogeneous, the immuneregulatory strategies of different helminth parasites are remarkably conserved and they efficiently target all arms of the immune system. Due to their broad and powerful immunomodulatory capacities, helminth parasites can modify a variety of chronic inflammatory diseases beyond allergy. However, as allergic diseases are driven by the very same mechanisms that have evolved to kill or expel helminths, parasitic immune evasion mechanisms may be particularly well adapted to suppress type 2 immunopathology in allergy.

Harnessing the immunomodulatory potential of helminth parasites for the prevention and therapy of allergic diseases has been proposed decades ago, but promising findings from rodent and epidemiological studies have not been successfully translated into current clinical practice. The potential reasons for this lack of translation have been comprehensively reviewed elsewhere (Evans and Mitre 2015). In particular, infecting allergy patients with life parasites raises considerable concerns of safety and acceptance and is therefore unlikely to enter clinical practice on a large scale. Thus, investing into research aimed at identifying and producing parasite derived immunomodulatory compounds may represent the more promising approach to translating allergy-modifying capacities of parasites into new therapies. This may also include the use of helminth products that indirectly suppress allergies by modifying the microbial composition in target organs, which may result in an increased production of protective metabolites such as short chain fatty acids (SCFA) (Hayes et al. 2010; Zaiss et al. 2015).

Given the early window of opportunity for tolerance induction, the treatment of relatively young children likely represents the most relevant use case for any new allergy prevention strategy (Young 2015). To allow for the application of parasitic immunemodulators for allergy prevention in high-risk children, high standards of safety and efficacy have to be met, which remain to be evaluated in suitable human model settings. The use of life infections or systemic administration of parasite molecules that has been used in most murine models of allergy prevention is unlikely to be translatable into human and especially into pediatric settings.

Furthermore, if parasitic immunemodulators are to enter clinical practice on a broader scale, it will be important to assess their efficacy in therapeutic rather than merely preventive settings and to translate them to human systems. Thus, future studies should evaluate promising parasitic immunemodulators in therapeutic rodent models using topical administration (e.g. McSorley et al. 2014) and in human tissue culture systems (e.g. precision cut lung slices). This is highly relevant as negative outcomes of human trials using *Trichuris suis* ova (TSO) for the treatment of allergy or IBD show that promising results from rodent studies may prove difficult to translate to human disease (Bager et al. 2010; Fleming and Weinstock 2015).

References

Ahumada V, García E, Dennis R, Rojas MX, Rondón MA, Pérez A, Peñaranda A, Barragán AM, Jimenez S, Kennedy MW, Caraballo L (2015) IgE responses to Ascaris and mite tropomyosins are risk factors for asthma. Clin Exp Allergy 45:1189–1200. https://doi.org/10.1111/cea.12513

Allen JE, Maizels RM (2011) Diversity and dialogue in immunity to helminths. Nat Rev Immunol 11:375–388. https://doi.org/10.1038/nri2992

Amu S, Saunders SP, Kronenberg M, Mangan NE, Atzberger A, Fallon PG (2010) Regulatory B cells prevent and reverse allergic airway inflammation via FoxP3-positive T regulatory cells in a murine model. J Allergy Clin Immunol 125:1114–1124.e8. https://doi.org/10.1016/j.jaci.2010.01.018

Araújo CA, Perini A, Martins MA, Macedo MS, Macedo-Soares MF (2008) PAS-1, a protein from *Ascaris suum*, modulates allergic inflammation via IL-10 and IFN-gamma, but not IL-12. Cytokine 44:335–341. https://doi.org/10.1016/j.cyto.2008.09.005

Bager P, Arnved J, Rønborg S, Wohlfahrt J, Poulsen LK, Westergaard T, Petersen HW, Kristensen B, Thamsborg S, Roepstorff A, Kapel C, Melbye M (2010) Trichuris suis ova therapy for allergic rhinitis: a randomized, double-blind, placebo-controlled clinical trial. J Allergy Clin Immunol 125:123–130. https://doi.org/10.1016/j.jaci.2009.08.006

Campbell EL, Bruyninckx WJ, Kelly CJ, Glover LE, McNamee EN, Bowers BE, Bayless AJ, Scully M, Saeedi BJ, Golden-Mason L, Ehrentraut SF, Curtis VF, Burgess A, Garvey JF, Sorensen A, Nemenoff R, Jedlicka P, Taylor CT, Kominsky DJ, Colgan SP (2014) Transmigrating neutrophils shape the mucosal microenvironment through localized oxygen depletion to influence resolution of inflammation. Immunity 40:66–77. https://doi.org/10.1016/j.immuni.2013.11.020

Cardoso LS, Oliveira SC, Pacífico LGG, Góes AM, Oliveira RR, Fonseca CT, Carvalho EM d, Araújo MI (2006) *Schistosoma mansoni* antigen-driven interleukin-10 production in infected asthmatic individuals. Mem Inst Oswaldo Cruz 101(Suppl 1):339–343

Cardoso FC, Macedo GC, Gava E, Kitten GT, Mati VL, de Melo AL, Caliari MV, Almeida GT, Venancio TM, Verjovski-Almeida S, Oliveira SC (2008) *Schistosoma mansoni* tegument protein Sm29 is able to induce a Th1-type of immune response and protection against parasite infection. PLoS Negl Trop Dis 2:e308. https://doi.org/10.1371/journal.pntd.0000308

Cardoso LS, Oliveira SC, Góes AM, Oliveira RR, Pacífico LG, Marinho FV, Fonseca CT, Cardoso FC, Carvalho EM, Araujo MI (2010) *Schistosoma mansoni* antigens modulate the allergic response in a murine model of ovalbumin-induced airway inflammation. Clin Exp Immunol 160:266–274. https://doi.org/10.1111/j.1365-2249.2009.04084.x

Chen F, Liu Z, Wu W, Rozo C, Bowdridge S, Millman A, Van Rooijen N, Urban JF Jr, Wynn TA, Gause WC (2012) An essential role for T(H)2-type responses in limiting acute tissue damage during experimental helminth infection. Nat Med. https://doi.org/10.1038/nm.2628

Cho MK, Park MK, Kang SA, Park SK, Lyu JH, Kim D-H, Park H-K, Yu HS (2015) TLR2-dependent amelioration of allergic airway inflammation by parasitic nematode type II MIF in mice. Parasite Immunol 37:180–191. https://doi.org/10.1111/pim.12172

Cooper PJ, Chico ME, Rodrigues LC, Ordonez M, Strachan D, Griffin GE, Nutman TB (2003) Reduced risk of atopy among school-age children infected with geohelminth parasites in a rural area of the tropics. J Allergy Clin Immunol 111:995–1000

Cooper PJ, Chico ME, Amorim LD, Sandoval C, Vaca M, Strina A, Campos AC, Rodrigues LC, Barreto ML, Strachan DP (2016) Effects of maternal geohelminth infections on allergy in early childhood. J Allergy Clin Immunol 137:899–906.e2. https://doi.org/10.1016/j.jaci.2015.07.044

Dainichi T, Maekawa Y, Ishii K, Zhang T, Nashed BF, Sakai T, Takashima M, Himeno K (2001) Nippocystatin, a cysteine protease inhibitor from *Nippostrongylus brasiliensis*, inhibits antigen processing and modulates antigen-specific immune response. Infect Immun 69:7380–7386. https://doi.org/10.1128/IAI.69.12.7380-7386.2001

Daniłowicz-Luebert E, Steinfelder S, Kühl AA, Drozdenko G, Lucius R, Worm M, Hamelmann E, Hartmann S (2013) A nematode immunomodulator suppresses grass pollen-specific allergic responses by controlling excessive Th2 inflammation. Int J Parasitol 43:201–210. https://doi.org/10.1016/j.ijpara.2012.10.014

de Araújo CAA, Perini A, Martins MA, Macedo MS, Macedo-Soares MF (2010) PAS-1, an *Ascaris suum* protein, modulates allergic airway inflammation via CD8+γδTCR+ and CD4+CD25+FoxP3+ T cells. Scand J Immunol 72:491–503. https://doi.org/10.1111/j.1365-3083.2010.02465.x

Dittrich AM, Erbacher A, Specht S, Diesner F, Krokowski M, Avagyan A, Stock P, Ahrens B, Hoffmann WH, Hoerauf A, Hamelmann E (2008) Helminth infection with Litomosoides sigmodontis induces regulatory T cells and inhibits allergic sensitization, airway inflammation, and hyperreactivity in a murine asthma model. J Immunol 180(3):1792–1799

Enobe CS, Araújo CA, Perini A, Martins MA, Macedo MS, Macedo-Soares MF (2006) Early stages of *Ascaris suum* induce airway inflammation and hyperreactivity in a mouse model. Parasite Immunol 28:453–461. https://doi.org/10.1111/j.1365-3024.2006.00892.x

Esser-von Bieren J, Mosconi I, Guiet R, Piersgilli A, Volpe B, Chen F, Gause WC, Seitz A, Verbeek JS, Harris NL (2013) Antibodies trap tissue migrating helminth larvae and prevent tissue damage by driving IL-4Rα-independent alternative differentiation of macrophages. PLoS Pathog 9:e1003771. https://doi.org/10.1371/journal.ppat.1003771

Esser-von Bieren J, Volpe B, Sutherland DB, Bürgi J, Verbeek JS, Marsland BJ, Urban JF, Harris NL (2015) Immune antibodies and helminth products drive CXCR2-dependent macrophage-myofibroblast crosstalk to promote intestinal repair. PLoS Pathog 11:e1004778. https://doi.org/10.1371/journal.ppat.1004778

Evans H, Mitre E (2015) Worms as therapeutic agents for allergy and asthma: understanding why benefits in animal studies have not translated into clinical success. J Allergy Clin Immunol 135:343–353. https://doi.org/10.1016/j.jaci.2014.07.007

Everts B, Perona-Wright G, Smits HH, Hokke CH, van der Ham AJ, Fitzsimmons CM, Doenhoff MJ, van der Bosch J, Mohrs K, Haas H, Mohrs M, Yazdanbakhsh M, Schramm G (2009) Omega-1, a glycoprotein secreted by *Schistosoma mansoni* eggs, drives Th2 responses. J Exp Med 206:1673–1680. https://doi.org/10.1084/jem.20082460

Feary JR, Venn AJ, Mortimer K, Brown AP, Hooi D, Falcone FH, Pritchard DI, Britton JR (2010) Experimental hookworm infection: a randomized placebo-controlled trial in asthma. Clin Exp Allergy 40:299–306. https://doi.org/10.1111/j.1365-2222.2009.03433.x

Fleming JO, Weinstock JV (2015) Clinical trials of helminth therapy in autoimmune diseases: rationale and findings. Parasite Immunol 37:277–292. https://doi.org/10.1111/pim.12175

Geering B, Stoeckle C, Conus S, Simon H-U (2013) Living and dying for inflammation: neutrophils, eosinophils, basophils. Trends Immunol 34:398–409. https://doi.org/10.1016/j.it.2013.04.002

Grainger JR, Smith KA, Hewitson JP, McSorley HJ, Harcus Y, Filbey KJ, Finney CAM, Greenwood EJD, Knox DP, Wilson MS, Belkaid Y, Rudensky AY, Maizels RM (2010) Helminth secretions induce de novo T cell Foxp3 expression and regulatory function through the TGF-β pathway. J Exp Med 207:2331–2341. https://doi.org/10.1084/jem.20101074

Hartmann S, Schnoeller C, Dahten A, Avagyan A, Rausch S, Lendner M, Bocian C, Pillai S, Loddenkemper C, Lucius R, Worm M, Hamelmann E (2009) Gastrointestinal nematode infection interferes with experimental allergic airway inflammation but not atopic dermatitis. Clin Exp Allergy 39:1585–1596. https://doi.org/10.1111/j.1365-2222.2009.03290.x

Hayes KS, Bancroft AJ, Goldrick M, Portsmouth C, Roberts IS, Grencis RK (2010) Exploitation of the intestinal microflora by the parasitic nematode *Trichuris muris*. Science 328:1391–1394. https://doi.org/10.1126/science.1187703

Hirsch C, Zouain CS, Alves JB, Goes AM (1997) Induction of protective immunity and modulation of granulomatous hypersensitivity in mice using PIII, an anionic fraction of *Schistosoma mansoni* adult worm. Parasitology 115(Pt 1):21–28

Humphreys NE, Xu D, Hepworth MR, Liew FY, Grencis RK (2008) IL-33, a potent inducer of adaptive immunity to intestinal nematodes. J Immunol Baltim Md 1950(180):2443–2449

Itami DM, Oshiro TM, Araujo CA, Perini A, Martins MA, Macedo MS, Macedo-Soares MF (2005) Modulation of murine experimental asthma by *Ascaris suum* components. Clin Exp Allergy 35:873–879. https://doi.org/10.1111/j.1365-2222.2005.02268.x

Jeong Y-I, Kim SH, Ju JW, Cho SH, Lee WJ, Park JW, Park Y-M, Lee SE (2011) Clonorchis sinensis-derived total protein attenuates airway inflammation in murine asthma model by inducing regulatory T cells and modulating dendritic cell functions. Biochem Biophys Res Commun 407:793–800. https://doi.org/10.1016/j.bbrc.2011.03.102

Ji P, Hu H, Yang X, Wei X, Zhu C, Liu J, Feng Y, Yang F, Okanurak K, Li N, Zeng X, Zheng H, Wu Z, Lv Z (2015) AcCystatin, an immunoregulatory molecule from *Angiostrongylus cantonensis*, ameliorates the asthmatic response in an aluminium hydroxide/ovalbumin-induced rat model of asthma. Parasitol Res 114:613–624. https://doi.org/10.1007/s00436-014-4223-z

Kim HY, Chang Y-J, Subramanian S, Lee H-H, Albacker LA, Matangkasombut P, Savage PB, McKenzie ANJ, Smith DE, Rottman JB, DeKruyff RH, Umetsu DT (2012) Innate lymphoid cells responding to IL-33 mediate airway hyperreactivity independently of adaptive immunity. J Allergy Clin Immunol 129:216–227. https://doi.org/10.1016/j.jaci.2011.10.036

Kitagaki K, Businga TR, Racila D, Elliott DE, Weinstock JV, Kline JN (2006) Intestinal helminths protect in a murine model of asthma. J Immunol Baltim Md 1950(177):1628–1635

Klein Wolterink RGJ, Kleinjan A, van Nimwegen M, Bergen I, de Bruijn M, Levani Y, Hendriks RW (2012) Pulmonary innate lymphoid cells are major producers of IL-5 and IL-13 in murine models of allergic asthma. Eur J Immunol 42:1106–1116. https://doi.org/10.1002/eji. 201142018

Klotz C, Ziegler T, Figueiredo AS, Rausch S, Hepworth MR, Obsivac N, Sers C, Lang R, Hammerstein P, Lucius R, Hartmann S (2011) A helminth immunomodulator exploits host signaling events to regulate cytokine production in macrophages. PLoS Pathog 7:e1001248. https://doi.org/10.1371/journal.ppat.1001248

Kvarnhammar AM, Cardell LO (2012) Pattern-recognition receptors in human eosinophils. Immunology 136:11–20. https://doi.org/10.1111/j.1365-2567.2012.03556.x

Larson D, Hübner MP, Torrero MN, Morris CP, Brankin A, Swierczewski BE, Davies SJ, Vonakis BM, Mitre E (2012) Chronic helminth infection reduces basophil responsiveness in an IL-10-dependent manner. J Immunol Baltim Md 1950. https://doi.org/10.4049/jimmunol. 1101859

Layland LE, Straubinger K, Ritter M, Loffredo-Verde E, Garn H, Sparwasser T, Prazeres da Costa C (2013) *Schistosoma mansoni*-mediated suppression of allergic airway inflammation requires patency and Foxp3+ Treg cells. PLoS Negl Trop Dis 7:e2379. https://doi.org/10.1371/journal. pntd.0002379

Levin M, Muloiwa R, Le Souëf P, Motala C (2012) Ascaris sensitization is associated with aeroallergen sensitization and airway hyperresponsiveness but not allergic disease in urban Africa. J Allergy Clin Immunol 130:265–267. https://doi.org/10.1016/j.jaci.2012.03.033

Liu P, Li J, Yang X, Shen Y, Zhu Y, Wang S, Wu Z, Liu X, An G, Ji W, Gao W, Yang X (2010) Helminth infection inhibits airway allergic reaction and dendritic cells are involved in the modulation process. Parasite Immunol 32:57–66. https://doi.org/10.1111/j.1365-3024.2009. 01161.x

Liu J-Y, Lu P, Hu L-Z, Shen Y-J, Zhu Y-J, Ren J-L, Ji W-H, Zhang X-Z, Wu Z-Q, Yang X-Z, Yang J, Li L-Y, Yang X, Liu P-M (2014) CD8α⁻ DC is the major DC subset which mediates inhibition of allergic responses by Schistosoma infection. Parasite Immunol 36:647–657. https://doi.org/10.1111/pim.12134

Magalhães KG, Almeida PE, Atella GC, Maya-Monteiro CM, Castro-Faria-Neto HC, Pelajo-Machado M, Lenzi HL, Bozza MT, Bozza PT (2010) Schistosomal-derived lysophosphatidylcholine are involved in eosinophil activation and recruitment through Toll-like receptor-2-dependent mechanisms. J Infect Dis 202:1369–1379. https://doi.org/10.1086/ 656477

Maizels RM (2016) Parasitic helminth infections and the control of human allergic and autoimmune disorders. Clin Microbiol Infect 22:481–486. https://doi.org/10.1016/j.cmi.2016.04.024

Mangan NE, van Rooijen N, McKenzie ANJ, Fallon PG (2006) Helminth-modified pulmonary immune response protects mice from allergen-induced airway hyperresponsiveness. J Immunol Baltim Md 1950(176):138–147

Mantovani A, Cassatella MA, Costantini C, Jaillon S (2011) Neutrophils in the activation and regulation of innate and adaptive immunity. Nat Rev Immunol 11:519–531. https://doi.org/10. 1038/nri3024

Massacand JC, Stettler RC, Meier R, Humphreys NE, Grencis RK, Marsland BJ, Harris NL (2009) Helminth products bypass the need for TSLP in Th2 immune responses by directly modulating dendritic cell function. Proc Natl Acad Sci USA 106:13968–13973. https://doi.org/10.1073/ pnas.0906367106

McConchie BW, Norris HH, Bundoc VG, Trivedi S, Boesen A, Urban JF, Keane-Myers AM (2006) *Ascaris suum*-derived products suppress mucosal allergic inflammation in an interleukin-10-independent manner via interference with dendritic cell function. Infect Immun 74:6632–6641. https://doi.org/10.1128/IAI.00720-06

McSorley HJ, O'Gorman MT, Blair N, Sutherland TE, Filbey KJ, Maizels RM (2012) Suppression of type 2 immunity and allergic airway inflammation by secreted products of the helminth Heligmosomoides polygyrus. Eur J Immunol. https://doi.org/10.1002/eji.201142161

McSorley HJ, Blair NF, Smith KA, McKenzie ANJ, Maizels RM (2014) Blockade of IL-33 release and suppression of type 2 innate lymphoid cell responses by helminth secreted products in airway allergy. Mucosal Immunol 7:1068–1078. https://doi.org/10.1038/mi.2013.123

McSorley HJ, Blair NF, Robertson E, Maizels RM (2015) Suppression of OVA-alum induced allergy by heligmosomoides polygyrus products is MyD88-, TRIF-, regulatory t- and b cell-independent, but is associated with reduced innate lymphoid cell activation. Exp Parasitol. https://doi.org/10.1016/j.exppara.2015.02.009

Medeiros M Jr, Figueiredo JP, Almeida MC, Matos MA, Araújo MI, Cruz AA, Atta AM, Rego MAV, de Jesus AR, Taketomi EA, Carvalho EM (2003) Schistosoma mansoni infection is associated with a reduced course of asthma. J Allergy Clin Immunol 111:947–951

Melendez AJ, Harnett MM, Pushparaj PN, Wong WSF, Tay HK, McSharry CP, Harnett W (2007) Inhibition of Fc epsilon RI-mediated mast cell responses by ES-62, a product of parasitic filarial nematodes. Nat Med 13:1375–1381. https://doi.org/10.1038/nm1654

Mitre E, Norwood S, Nutman TB (2005) Saturation of immunoglobulin E (IgE) binding sites by polyclonal IgE does not explain the protective effect of helminth infections against atopy. Infect Immun 73:4106–4111. https://doi.org/10.1128/IAI.73.7.4106-4111.2005

Murray PJ (2016) Macrophage polarization. Annu Rev Physiol. https://doi.org/10.1146/annurev-physiol-022516-034339

Navarro S, Pickering DA, Ferreira IB, Jones L, Ryan S, Troy S, Leech A, Hotez PJ, Zhan B, Laha T, Prentice R, Sparwasser T, Croese J, Engwerda CR, Upham JW, Julia V, Giacomin PR, Loukas A (2016) Hookworm recombinant protein promotes regulatory T cell responses that suppress experimental asthma. Sci Transl Med 8:362ra143. https://doi.org/10.1126/scitranslmed.aaf8807

Neill DR, Wong SH, Bellosi A, Flynn RJ, Daly M, Langford TKA, Bucks C, Kane CM, Fallon PG, Pannell R, Jolin HE, McKenzie ANJ (2010) Nuocytes represent a new innate effector leukocyte that mediates type-2 immunity. Nature 464:1367–1370. https://doi.org/10.1038/nature08900

Nieuwenhuizen NE, Lopata AL (2014) Allergic reactions to Anisakis found in fish. Curr Allergy Asthma Rep 14:455. https://doi.org/10.1007/s11882-014-0455-3

Nogueira DS, Gazzinelli-Guimarães PH, Barbosa FS, Resende NM, Silva CC, de Oliveira LM, Amorim CCO, Oliveira FMS, Mattos MS, Kraemer LR, Caliari MV, Gaze S, Bueno LL, Russo RC, Fujiwara RT (2016) Multiple exposures to Ascaris suum induce tissue injury and mixed Th2/Th17 immune response in mice. PLoS Negl Trop Dis 10:e0004382. https://doi.org/10.1371/journal.pntd.0004382

Nyan OA, Walraven GE, Banya WA, Milligan P, Van Der Sande M, Ceesay SM, Del Prete G, McAdam KP (2001) Atopy, intestinal helminth infection and total serum IgE in rural and urban adult Gambian communities. Clin Exp Allergy 31:1672–1678

Obeng BB, Amoah AS, Larbi IA, de Souza DK, Uh H-W, Fernández-Rivas M, van Ree R, Rodrigues LC, Boakye DA, Yazdanbakhsh M, Hartgers FC (2014) Schistosome infection is negatively associated with mite atopy, but not wheeze and asthma in Ghanaian schoolchildren. Clin Exp Allergy 44:965–975. https://doi.org/10.1111/cea.12307

Obihara CC, Beyers N, Gie RP, Hoekstra MO, Fincham JE, Marais BJ, Lombard CJ, Dini LA, Kimpen JLL (2006) Respiratory atopic disease, Ascaris-immunoglobulin E and tuberculin testing in urban South African children. Clin Exp Allergy 36:640–648. https://doi.org/10.1111/j.1365-2222.2006.02479.x

Park SK, Cho MK, Park H-K, Lee KH, Lee SJ, Choi SH, Ock MS, Jeong HJ, Lee MH, Yu HS (2009) Macrophage migration inhibitory factor homologs of anisakis simplex suppress Th2 response in allergic airway inflammation model via CD4+CD25+Foxp3+ T cell recruitment. J Immunol Baltim Md 1950(182):6907–6914. https://doi.org/10.4049/jimmunol.0803533

Patnode ML, Bando JK, Krummel MF, Locksley RM, Rosen SD (2014) Leukotriene B4 amplifies eosinophil accumulation in response to nematodes. J Exp Med 211:1281–1288. https://doi.org/10.1084/jem.20132336

Perdiguero EG, Geissmann F (2016) The development and maintenance of resident macrophages. Nat Immunol 17:2–8. https://doi.org/10.1038/ni.3341

Pinot de Moira A, Fitzsimmons CM, Jones FM, Wilson S, Cahen P, Tukahebwa E, Mpairwe H, Mwatha JK, Bethony JM, Skov PS, Kabatereine NB, Dunne DW (2014) Suppression of basophil histamine release and other IgE-dependent responses in childhood Schistosoma mansoni/hookworm coinfection. J Infect Dis 210:1198–1206. https://doi.org/10.1093/infdis/jiu234

Salimi M, Barlow JL, Saunders SP, Xue L, Gutowska-Owsiak D, Wang X, Huang L-C, Johnson D, Scanlon ST, McKenzie ANJ, Fallon PG, Ogg GS (2013) A role for IL-25 and IL-33-driven type-2 innate lymphoid cells in atopic dermatitis. J Exp Med 210:2939–2950. https://doi.org/10.1084/jem.20130351

Schierack P, Lucius R, Sonnenburg B, Schilling K, Hartmann S (2003) Parasite-specific immunomodulatory functions of filarial cystatin. Infect Immun 71:2422–2429

Schmiedel Y, Mombo-Ngoma G, Labuda LA, Janse JJ, de Gier B, Adegnika AA, Issifou S, Kremsner PG, Smits HH, Yazdanbakhsh M (2015) CD4+CD25hiFOXP3+ regulatory T cells and cytokine responses in human schistosomiasis before and after treatment with praziquantel. PLoS Negl Trop Dis 9:e0003995. https://doi.org/10.1371/journal.pntd.0003995

Schnoeller C, Rausch S, Pillai S, Avagyan A, Wittig BM, Loddenkemper C, Hamann A, Hamelmann E, Lucius R, Hartmann S (2008) A helminth immunomodulator reduces allergic and inflammatory responses by induction of IL-10-producing macrophages. J Immunol Baltim Md 1950(180):4265–4272

Scrivener S, Yemaneberhan H, Zebenigus M, Tilahun D, Girma S, Ali S, McElroy P, Custovic A, Woodcock A, Pritchard D, Venn A, Britton J (2001) Independent effects of intestinal parasite infection and domestic allergen exposure on risk of wheeze in Ethiopia: a nested case-control study. Lancet 358:1493–1499. https://doi.org/10.1016/S0140-6736(01)06579-5

Smits HH, Hammad H, van Nimwegen M, Soullie T, Willart MA, Lievers E, Kadouch J, Kool M, Kos-van Oosterhoud J, Deelder AM, Lambrecht BN, Yazdanbakhsh M (2007) Protective effect of Schistosoma mansoni infection on allergic airway inflammation depends on the intensity and chronicity of infection. J Allergy Clin Immunol 120:932–940. https://doi.org/10.1016/j.jaci.2007.06.009

Stein M, Greenberg Z, Boaz M, Handzel ZT, Meshesha MK, Bentwich Z (2016) The role of helminth infection and environment in the development of allergy: a prospective study of newly-arrived Ethiopian immigrants in Israel. PLoS Negl Trop Dis 10:e0004208. https://doi.org/10.1371/journal.pntd.0004208

Steinfelder S, Andersen JF, Cannons JL, Feng CG, Joshi M, Dwyer D, Caspar P, Schwartzberg PL, Sher A, Jankovic D (2009) The major component in Schistosome eggs responsible for conditioning dendritic cells for Th2 polarization is a T2 ribonuclease (omega-1). J Exp Med 206:1681–1690. https://doi.org/10.1084/jem.20082462

Straubinger K, Paul S, Prazeres da Costa O, Ritter M, Buch T, Busch DH, Layland LE, Prazeres da Costa CU (2014) Maternal immune response to helminth infection during pregnancy determines offspring susceptibility to allergic airway inflammation. J Allergy Clin Immunol. https://doi.org/10.1016/j.jaci.2014.05.034

Suzuki M, Hara M, Ichikawa S, Kamijo S, Nakazawa T, Hatanaka H, Akiyama K, Ogawa H, Okumura K, Takai T (2016) Presensitization to Ascaris antigens promotes induction of mite-specific IgE upon mite antigen inhalation in mice. Allergol Int 65:44–51. https://doi.org/10.1016/j.alit.2015.07.003

Tamarozzi F, Wright HL, Johnston KL, Edwards SW, Turner JD, Taylor MJ (2014) Human filarial Wolbachia lipopeptide directly activates human neutrophils in vitro. Parasite Immunol 36:494–502. https://doi.org/10.1111/pim.12122

Thomas CJ, Schroder K (2013) Pattern recognition receptor function in neutrophils. Trends Immunol 34:317–328. https://doi.org/10.1016/j.it.2013.02.008

Trujillo-Vargas CM, Werner-Klein M, Wohlleben G, Polte T, Hansen G, Ehlers S, Erb KJ (2007) Helminth-derived products inhibit the development of allergic responses in mice. Am J Respir Crit Care Med 175:336–344. https://doi.org/10.1164/rccm.200601-054OC

van den Biggelaar AH, van Ree R, Rodrigues LC, Lell B, Deelder AM, Kremsner PG, Yazdanbakhsh M (2000) Decreased atopy in children infected with *Schistosoma haematobium*: a role for parasite-induced interleukin-10. Lancet 356:1723–1727. https://doi.org/10.1016/S0140-6736(00)03206-2

van der Kleij D, Latz E, Brouwers JFHM, Kruize YCM, Schmitz M, Kurt-Jones EA, Espevik T, de Jong EC, Kapsenberg ML, Golenbock DT, Tielens AGM, Yazdanbakhsh M (2002) A novel host-parasite lipid cross-talk. Schistosomal lyso-phosphatidylserine activates toll-like receptor 2 and affects immune polarization. J Biol Chem 277:48122–48129. https://doi.org/10.1074/jbc.M206941200

van der Vlugt LEPM, Labuda LA, Ozir-Fazalalikhan A, Lievers E, Gloudemans AK, Liu K-Y, Barr TA, Sparwasser T, Boon L, Ngoa UA, Feugap EN, Adegnika AA, Kremsner PG, Gray D, Yazdanbakhsh M, Smits HH (2012) Schistosomes induce regulatory features in human and mouse CD1d(hi) B cells: inhibition of allergic inflammation by IL-10 and regulatory T cells. PloS One 7:e30883. https://doi.org/10.1371/journal.pone.0030883

van Stijn CMW, Meyer S, van den Broek M, Bruijns SCM, van Kooyk Y, Geyer R, van Die I (2010) *Schistosoma mansoni* worm glycolipids induce an inflammatory phenotype in human dendritic cells by cooperation of TLR4 and DC-SIGN. Mol Immunol 47:1544–1552. https://doi.org/10.1016/j.molimm.2010.01.014

Velupillai P, Harn DA (1994) Oligosaccharide-specific induction of interleukin 10 production by B220+ cells from schistosome-infected mice: a mechanism for regulation of CD4+ T-cell subsets. Proc Natl Acad Sci USA 91:18–22

Wammes LJ, Hamid F, Wiria AE, May L, Kaisar MMM, Prasetyani-Gieseler MA, Djuardi Y, Wibowo H, Kruize YCM, Verweij JJ, de Jong SE, Tsonaka R, Houwing-Duistermaat JJ, Sartono E, Luty AJF, Supali T, Yazdanbakhsh M (2016) Community deworming alleviates geohelminth-induced immune hyporesponsiveness. Proc Natl Acad Sci USA 113:12526–12531. https://doi.org/10.1073/pnas.1604570113

Wang X, Zhou S, Chi Y, Wen X, Hoellwarth J, He L, Liu F, Wu C, Dhesi S, Zhao J, Hu W, Su C (2009) CD4+CD25+ Treg induction by an HSP60-derived peptide SJMHE1 from *Schistosoma japonicum* is TLR2 dependent. Eur J Immunol 39:3052–3065. https://doi.org/10.1002/eji.200939335

Wilson MS, Taylor MD, Balic A, Finney CAM, Lamb JR, Maizels RM (2005) Suppression of allergic airway inflammation by helminth-induced regulatory T cells. J Exp Med 202:1199–1212. https://doi.org/10.1084/jem.20042572

Wilson MS, Taylor MD, O'Gorman MT, Balic A, Barr TA, Filbey K, Anderton SM, Maizels RM (2010) Helminth-induced CD19+CD23hi B cells modulate experimental allergic and autoimmune inflammation. Eur J Immunol 40:1682–1696. https://doi.org/10.1002/eji.200939721

Wohlleben G, Trujillo C, Müller J, Ritze Y, Grunewald S, Tatsch U, Erb KJ (2004) Helminth infection modulates the development of allergen-induced airway inflammation. Int Immunol 16:585–596

Xue J, Schmidt SV, Sander J, Draffehn A, Krebs W, Quester I, De Nardo D, Gohel TD, Emde M, Schmidleithner L, Ganesan H, Nino-Castro A, Mallmann MR, Labzin L, Theis H, Kraut M, Beyer M, Latz E, Freeman TC, Ulas T, Schultze JL (2014) Transcriptome-based network analysis reveals a spectrum model of human macrophage activation. Immunity 40:274–288. https://doi.org/10.1016/j.immuni.2014.01.006

Yang J, Zhao J, Yang Y, Zhang L, Yang X, Zhu X, Ji M, Sun N, Su C (2007) *Schistosoma japonicum* egg antigens stimulate CD4 CD25 T cells and modulate airway inflammation in a murine model of asthma. Immunology 120:8–18. https://doi.org/10.1111/j.1365-2567.2006.02472.x

Yazdanbakhsh M, Kremsner PG, van Ree R (2002) Allergy, parasites, and the hygiene hypothesis. Science 296:490–494. https://doi.org/10.1126/science.296.5567.490

Young MC (2015) Taking the leap earlier: the timing of tolerance. Curr Opin Pediatr 27:736–740. https://doi.org/10.1097/MOP.0000000000000291

Zaiss MM, Rapin A, Lebon L, Dubey LK, Mosconi I, Sarter K, Piersigilli A, Menin L, Walker AW, Rougemont J, Paerewijck O, Geldhof P, McCoy KD, Macpherson AJ, Croese J, Giacomin PR, Loukas A, Junt T, Marsland BJ, Harris NL (2015) The intestinal microbiota contributes to the ability of helminths to modulate allergic inflammation. Immunity 43:998–1010. https://doi.org/10.1016/j.immuni.2015.09.012

Ziegler T, Rausch S, Steinfelder S, Klotz C, Hepworth MR, Kühl AA, Burda P-C, Lucius R, Hartmann S (2015) A novel regulatory macrophage induced by a helminth molecule instructs IL-10 in CD4+ T cells and protects against mucosal inflammation. J Immunol Baltim Md 1950 (194):1555–1564. https://doi.org/10.4049/jimmunol.1401217

Initiation, Persistence and Exacerbation of Food Allergy

Rodrigo Jiménez-Saiz, Derek K. Chu, Susan Waserman, and Manel Jordana

Abstract Th2 humoral immunity (IgE) is protective against venoms and parasites but detrimental when mounted against innocuous proteins such as food allergens. The generation of IgE immunity toward harmless allergens is initiated at the body barriers (i.e. mucosae and skin) where the allergen and the immune system first meet. Epithelial cytokines (such as TSLP, IL-25, and IL-33), damage-associated molecular patterns (DAMPs), alarmins, or barrier disruption at the time of allergen encounter can deviate dendritic cells (DCs) away from the natural tolerogenic response to a food allergen. Then, instructed DCs migrate to draining lymph nodes and facilitate Th2 CD4 T cell polarization by limiting IL-12p40 production and upregulating costimulatory molecules such as OX40L. In this setting, IL-4 production by CD4 Th2 cells is crucial for the emergence of IgE$^+$ B cells and plasma cells. The lifespan of allergen-specific IgE$^+$ plasma cells is short, thereby limiting their ability to sustain IgE titres over time. In contrast, long-lasting immunological memory that includes CD4 T and B cells is imprinted at the time of sensitization. These cells are activated on allergen exposure and replenish the transient IgE$^+$ plasma cell compartment in an IL-4 dependent manner. While immunological memory provides sustainable immunity against pathogens, it underlies persistence and exacerbation of food allergy. Therefore, reaching a better understanding of Th2 immune memory and the cellular and molecular mechanisms driving IgE-generating secondary responses is a major undertaking in the search for novel therapeutic targets in food allergy.

R. Jiménez-Saiz • M. Jordana (✉)
McMaster Immunology Research Centre (MIRC), McMaster University, Hamilton, ON, Canada

Department of Pathology and Molecular Medicine, McMaster University, Hamilton, ON, Canada
e-mail: jordanam@mcmaster.ca

D.K. Chu • S. Waserman
McMaster Immunology Research Centre (MIRC), McMaster University, Hamilton, ON, Canada

Department of Medicine, McMaster University, Hamilton, ON, Canada

© Springer International Publishing AG 2017   121
C.B. Schmidt-Weber (ed.), *Allergy Prevention and Exacerbation*, Birkhäuser
Advances in Infectious Diseases, https://doi.org/10.1007/978-3-319-69968-4_7

1   Introduction

The generation of an adaptive immune response requires an extraordinary amount of energy. Incipient, complex interactions between structural cells and innate cells within the mucosal or skin compartments, migration of instructed dendritic cells (DCs) to secondary lymph nodes, where further interactions take place with B and CD4 T cells, and trafficking of newly generated effector cells to peripheral sites demand an extraordinary metabolic and, hence, energy investment. It would stand to reason that programs are encoded upon a first exposure to antigen such that future encounters with the same antigen require a less energy demanding response. Faced with constant challenges over a lifetime, parsimony is, in the immunological sociolect, an operating principle of the immune system. From this perspective, the generation of memory would be a prime example of this principle because it is an "on demand" response.

Much of the knowledge on the development of memory responses has been generated in the context of infections against viruses and bacteria; these responses generally induce IgG. This knowledge has been transformative, as it is the scaffolding that has led to successful vaccination strategies. Much less is known about the development of humoral memory in Th2 responses. Th2 immunity, IgE in particular, has protective roles against parasites and poisons and, as such, provides a survival advantage to the host. This is not the case for foods. As foodstuff is essential for survival there is no teleological explanation as to why an immune response would be generated directly against food antigens, unless such antigens are encountered at a time when disturbances in the barrier microenvironment subvert homeostasis and innocuous food antigens are "seen" by the immune surveillance system as dangerous. Regardless of the specific entity, whether a poison, a parasite or a food allergen, the generation of memory is an inherent part of the Th2 immune program, just as is for Th1 responses. Arguably, the establishment of IgE memory against food allergens is an undesired effect of the system.

The clinical manifestations of food allergy are largely mediated by IgE. The generation of IgE-expressing cells involves elaborate interactions between allergen, DCs, CD4 T cells and B cells in the draining lymph nodes of barrier sites (i.e. mucosae and skin). Specifically, Th2 cells are a critical requirement for the generation of humoral Th2 immunity. In turn, the differentiation of naïve CD4 T cells into Th2 cells is dependent on the instructions provided by DCs; these instructions are acquired at barrier sites, the initial site of interaction between allergens and the immune system. In this chapter, we will briefly review the nature of the instructional program acquired by DCs in the barrier microenvironment; then, we will discuss the generation of CD4 T and B cell memory with particular reference to the persistence and exacerbation of food allergy.

2   CD4 T Cell Fate Is Determined by DCs Programmed at Barrier Sites: (Do) All Roads Lead to Rome (?)

Dendritic cells (DCs) collect, organize and present information regarding the state of the tissue microenvironment to the adaptive immune system, chiefly T cells. In this context, it is becoming increasingly clear that there is a diversity of modes of DC activation that can launch an immune program leading to allergic sensitization and Th2 memory.

Allergen exposure at barrier sites can activate epithelial cells to provide instructions to DCs that will promote a Th2 response. While it is incompletely understood precisely how epithelial cells become activated after allergen exposure, the production of the cytokines thymic stromal lymphopoietin (TSLP), IL-25, IL-33, and GM-CSF have been shown to be critical initiators of type 2 immunity in various allergic and *helminth* infection systems (see also chapter "Parasite Mediated Protection Against Allergy"). TSLP conditions DCs to promote Th2 polarization by down-regulating IL-12p40 production and up-regulating the co-stimulatory molecule OX40L (Liu et al. 2007). IL-25 and IL-33 can also induce OX40L in DCs (Besnard et al. 2011; Chu et al. 2013), and GM-CSF and IL-33 limit IL-12p40 production (Tada et al. 2000; Mayuzumi et al. 2009). The convergence of epithelial cytokines on DCs provides the immune system a number of avenues to launch a type 2 response. Recent data in a model of *helminth*-induced liver fibrosis reinforce this concept, borne out by the requirement for TSLP, IL-25 and IL-33 to be simultaneously blocked in order to inhibit Th2 immunity (Vannella et al. 2016). In contrast, several models of Th2 immunity to allergens are exquisitely sensitive to individual epithelial cytokines. For example, some models of food allergy require TSLP to induce a Th2 response (Noti et al. 2014; Muto et al. 2014), whereas others require IL-25 (Han et al. 2014; Lee et al. 2016) or IL-33 (Chu et al. 2013; Tordesillas et al. 2014; Noval Rivas et al. 2016). One explanation for the differential requirement of individual epithelial cytokines is that allergens and/or their sensitizing conditions can functionally mimic these cytokines, thereby making them redundant. For example, TSLP can be dispensable to initiate Th2 responses to some *helminths*, house dust mite or peanut because these antigens can supplant TSLP's ability to inhibit IL-12p40 and induce OX40L in DCs (Massacand et al. 2009; Chu et al. 2013). Although we and others have shown that GM-CSF is important for initiating Th2 responses in allergic asthma (Stampfli et al. 1998; Gajewska et al. 2003; Cates et al. 2004; Willart et al. 2012), conditions where GM-CSF or other epithelial-derived signals are critical in driving food allergic sensitization remain to be identified. These data nevertheless show that epithelial cell activation can lead to the production of a plethora of signals that encode nearby DCs with instructions that prime a Th2 response.

In addition to cytokines, damage-associated molecular patterns (DAMPs), also termed alarmins, can be a key requirement in driving DC activation and T cell priming (Heil and Land 2014). Cellular damage and death can lead to the release of intracellular contents and molecules such as uric acid (UA) crystals, IL-1α, ATP, and DNA, which have been shown to be critical for DC activation in models of allergic asthma (Hammad and Lambrecht 2015). IL-1α and GM-CSF have also been shown to feedback onto epithelial cells to stimulate IL-33 release (Willart et al. 2012; Llop-Guevara et al. 2014), thereby linking the pathways of DAMPs and epithelial cytokine secretion during DC activation. In relation to food allergy, we examined the role of UA in allergic sensitization after finding that this molecule was elevated in the blood of peanut allergic patients and mice (Kong et al. 2015). UA crystals activated DCs, including OX40L up-regulation, in a PI3K-dependent manner and facilitated Th2 immunity in peanut allergy independently of TLR2, TLR4, and IL-1 signalling. While the role of many other intracellular DAMPs in the generation of food allergy is not yet fully understood, cellular damage and death clearly can have important implications in determining DC responses.

Alarm signals without cell death can also influence DC behaviour. Oxidative stress/damage generated from epithelial cells, eosinophil granules, or DCs themselves can promote Th2 responses in asthma and food allergy by conditioning DCs to down-regulate the pro-Th1/Th17 molecules IL-12 and CD70, and up-regulate IL-6, CD80, OX40L and CCR7 (Tang et al. 2010; Ckless et al. 2011; Chu et al. 2014). Positive feedback loops might exist between epithelial cell secreted IL-33 and eosinophil degranulation, given the high expression of IL-33 receptor (ST2) on intestinal eosinophils (Cherry et al. 2008; Chu et al. 2014). Together, these data show that DAMPs and alarmins can be one important mechanism in driving DC responses during the incipient events of food allergic sensitization.

A principal function of epithelial barriers is to strictly enforce the compartmentalization of molecules between the host and the external environment. It follows that barrier disruption or dysfunction leads to unregulated micro- and macromolecule movement between sites, and in the context of immunity, aberrant and excessive exposure of DCs to antigen. Reporting on the state of the tissue, DCs promote type 2 immune responses to expel exogenous antigen and return the tissue to homeostasis. That mutations in molecules that make up the skin barrier, such as *filaggrin* and lympho-epithelial Kazal-type-related inhibitor (encoded by *spink5*), are associated with the development of food allergy support this concept (Brown et al. 2011; Thyssen and Maibach 2014; Ashley et al. 2017). Likewise, increased intestinal permeability to ingested allergens has been proposed to facilitate the development of food allergy (Forbes et al. 2008; Perrier and Corthesy 2011). These events may lead to the encounter of DCs with allergens concomitant with endogenous or exogenous adjuvant molecules as well as autoallergens as discussed in chapter "Microbial Triggers in Autoimmunity, Severe Allergy, and Autoallergy", leading to a Th2 biased immune response.

Direct interaction of antigens with DCs can lead to a pro-Th2 phenotype through a number of mechanisms, but they share in common inhibition of DC IL-12 production, and up-regulation of co-stimulatory molecules including OX40L (Pulendran et al. 2010). In specific reference to food allergy, peanut allergens can bind DC-SIGN to cause human DC MHC II, CD83 and CD86 up-regulation in vitro (Shreffler et al. 2006). Similarly, peanut agglutinin induced DC activation (Kamalakannan et al. 2016) and glycosylated ovalbumin (OVA), an allergen from chicken egg white, was superior to OVA alone in inducing DC IL-6 (Hilmenyuk et al. 2010). We have shown that peanut-exposed DCs upregulate OX40L and downregulate IL-12p40 production (Chu et al. 2013). Pollens, which can cause either allergic asthma or food allergy, also limit DC IL-12 production (Traidl-Hoffmann et al. 2005). Though the majority of these data were generated in vitro, and require evidence that they are operative in vivo, they establish plausible mechanisms for DC activation by food allergens.

It is clear that diverse allergen-host interactions are able to launch distinct molecular programs that converge on DCs to induce adaptive Th2 immunity. The emergent paradigm suggests that DCs promote Th2 responses under conditions involving epithelial cytokine secretion, DAMPs/alarmins, or loss of barrier integrity. Whether there are distinct clinicopathologic variants/endotypes in food allergy, driven by different DC programs that elicit subtypes of Th2 cells (e.g., TSLP driven, IL-25 driven, and/or IL-33 driven), each with their own molecular signature and effector armamentarium and, ultimately, distinct therapeutic susceptibilities, requires further investigation. That two different TSLP-dependent models of Th2 immunity induce entirely distinct DC activation patterns (Connor et al. 2017) supports this hypothesis.

Even if "all roads" do not "lead to Rome" (precisely the same type of Th2 response) the next critical step that follows DC activation of T cells is the formation of adaptive immune memory. In the next section, we review mechanisms of adaptive immune memory with a focus on IgE biology and food allergy.

3 Understanding of Memory

The capacity of the immune system to "remember" a pathogen with a high degree of specificity, virtually for a lifetime, has fascinated immunologists for centuries. The Greek historian Thucydides first reported what is currently recognized as immunological memory when describing the plague of Athens in 430 b.c. "the same man was never attacked twice, at least fatally" (Finley 1951). Appreciation for immunological memory was invigorated in the nineteenth century when Dr. Panum reported the results of a natural experiment on the remote Faroe Islands. The investigation began in 1781 during a measles outbreak and lasted the 65 years that the Faroes remained measles-free (1846). He noted that "all the old people who had not gone through with measles in earlier life were attacked when they were exposed to infection" and concluded that "protective immunity could be sustained in the absence of re-exposure to the measles virus" (Panum 1847). This observation intimated the existence of processes that establish persistent long-term memory.

Early studies on the requirements for the generation of long-term memory, particularly those pertaining to humoral immunity, focused on the CD4 T and B cell axis as T cell-independent antibody responses were considered weak (Defrance et al. 2011). The conventional understanding at the time was that plasma cells (PCs) were, in general, short-lived effector cells. Hence, it was proposed that recurrent stimulation of memory B cells was required to drive continuing differentiation to PCs and, ultimately, maintain serum antibody titres. This notion changed with the seminal studies by the Radbruch (Manz et al. 1997) and Ahmed (Slifka et al. 1998) groups in the late 1990s, demonstrating the existence of non-cycling, long-lived PCs in the bone marrow of mice immunized against archetypical T cell-dependent antigens. This population of cells had the ability to maintain neutralizing antibody titres for months, even after depletion of memory B cells. Later, data from a longitudinal analysis of antibody titres and memory B cells specific for common viral antigens (vaccinia, measles, mumps, etc.) suggested the existence of long-lived PCs in humans (Amanna et al. 2007). However, direct proof of their existence in humans remained elusive (Gonzalez-Garcia et al. 2008; Qian et al. 2010; Mei et al. 2015) until recently when viral-induced long-lived PCs in the bone marrow (Halliley et al. 2015), and small intestine (Landsverk et al. 2017) were identified and comprehensively characterized.

The discovery of mitotically-quiescent, long-lived PCs producing thousands of antibodies per second advanced the understanding of the memory compartment, which was previously thought to rely exclusively on memory B and T cells. The rare sub-populations of PCs that were indeed long-lived became recognized not only as effector cells but also as important players of immune memory (i.e. serological memory). Thereafter, the complex machinery of immune memory has been structured into two arms: (i) humoral memory, mediated by antibodies produced by long-lived PCs; and (ii) cellular memory, which involves antigen-experienced B and T cells (Yoshida et al. 2010). The development of these two arms is important for protective immunity or, in the case of autoimmune diseases and allergy, pathology.

3.1 Immune Memory in Food Allergy

Knowledge about immune memory in food allergy, and allergy in general, has been partially extrapolated from the extensive studies in viral infections. Yet, responses against viruses are ontogenically different than those against allergens as they are derived from CD8 and CD4 Th1 cells, and mostly effected by IgG-expressing PCs. Furthermore, memory responses against viral infections must take into account the potential of viruses to mutate; food allergens do not rapidly mutate, although allergen structure and immunogenicity can change significantly due to human processing (Jiménez-Saiz et al. 2015). The role of immune memory in food allergy is intriguing. The heterogeneity of clinical phenotypes insinuates distinct immuno-logical mechanisms at play, such that some patients outgrow their allergy (e.g. typical of cow milk and hen egg allergy) while others remain severely reactive for a lifetime (e.g. shellfish and peanut allergy), with an array of phenotypes in between (Savage et al. 2016). Persistent food allergy has been attributed to the longevity of long-lived IgE$^+$ PCs (Luger et al. 2010; Winter et al. 2012; Moutsoglou and Dreskin 2016). However, the evidence for this is scarce, indirect, and largely limited to experimental models of systemic immunization.

New research on Th2 immunity and IgE biology in particular has shown that IgG- and IgE-expressing cells are fundamentally different. IgE$^+$ B cells have a brief germinal centre (GC) phase (Yang et al. 2012; He et al. 2013) and express low levels of membrane IgE because of suboptimal polyadenylation signals downstream of the exons encoding the cytoplasmic tail of IgE (Karnowski et al. 2006). In addition, IgE$^+$ plasmablasts exhibit inefficient migration towards CXCL12, which affects homing to the bone marrow (Achatz-Straussberger et al. 2008). Furthermore, we recently conducted a long-term study in peanut-allergic mice and determined the half-life of allergen-specific IgE$^+$ PCs to be approximately 60 days, notably shorter than that of IgG1$^+$ PCs of the same specificity (~234 days) (Jiménez-Saiz et al. 2017). The novel findings on IgE$^+$ PC biology highlight the central role of memory B and CD4 T cells in replenishing the effector IgE$^+$ PC compartment and, as such, may transform the current paradigm on the persistence of food allergy.

4 Generation and Nature of Memory CD4 T and B Cells

The generation and nature of memory B and CD4 T cells is regulated at the time of primary immunization by multiple factors including type of pathogen (or adjuvant), route of exposure, antigenic load and persistence. However, immune memory is plastic and its evolution is also influenced by the frequency and context of subsequent antigen exposures (Weisel and Shlomchik 2017).

4.1 Heterogeneity of Memory CD4 T Cells

Whether memory CD4 T cells originate directly from naïve or differentiated CD4 T cells has been the focus of intense research; indeed, a number of models on CD4 T cell differentiation have been proposed (e.g. linear, decreasing potential, divergent differentiation, etc.) (Jaigirdar and MacLeod 2015; Buchholz et al. 2016). Classical understanding (i.e. linear model) states that upon priming, activated CD4 T cells undergo proliferation and become effector cells. After antigen clearance, effector CD4 T cells enter a contraction phase during which the majority of them (>90–95%) undergo apoptosis, while the remaining cells become memory CD4 T cells (Ahmed and Gray 1996; Seder and Ahmed 2003). Yet, none of the proposed models completely conciliate with the heterogeneous findings reported on CD4 T cell memory generation and differentiation (Gasper et al. 2014), thus suggesting the existence of distinct pathways.

The selection of cells that survive the contraction phase, or directly become memory upon activation, is not a stochastic process but, rather, influenced by the priming conditions. The cytokine milieu at the time of priming affects CD4 T cell memory formation. Common γ-chain cytokines promote memory T cell differentiation (Schluns and Lefrancois 2003). For example, IL-2 signalling was required for the formation of lung-resident allergen-specific CD4 Th2 memory cells in an asthma model (Hondowicz et al. 2016). The ability of IL-2 to down-regulate apoptotic pathways and up-regulate IL-7R expression on effector CD4 T cells can promote their survival as memory CD4 T cells (McKinstry et al. 2014). The timing of CD4 T cell priming is also an important consideration. CD4 T cells that are activated toward the end of the immune response are more likely to enter the memory phase; this was attributed to fewer rounds of division of late-arriving CD4 T cells, and to the reduced number of DCs displaying peptide-MHCII complex (Catron et al. 2006). In fact, the nature of the TCR-peptide:MHCII interaction appears to be a key factor in determining CD4 T cell-fate. It has been reported that the CD4 T cells that experience stronger (i.e. affinity and duration) TCR-peptide: MHCII signalling are selected to enter into the memory phase (Fazilleau et al. 2007; Williams et al. 2008; Kim et al. 2013). This mechanism allows for a larger and more diverse number of CD4 T cell clones to participate in the effector response compared to the few clones of high specificity that remain as memory cells. This process, along with the inherently high diversity of TCRs (reported to be $\sim 10^{18}$ vs. $\sim 5 \times 10^{13}$ for BCRs) (Parham 2015) might fulfill the lack of TCR editing to increase affinity, and ensure that highly specific clones are conserved in the memory compartment.

Regardless of their origin, memory CD4 T cells were initially divided into two subsets based on CCR7 and CD62L expression, first described in human peripheral blood mononuclear cells (PBMCs) (Sallusto et al. 1999). Central memory CD4 T cells express high levels of CD62L and CCR7 that allows them to home to lymph nodes. These cells lack immediate effector function but on re-activation differentiate rapidly into effector memory CD4 T cells (CD62L^{low} and CCR7^{low}), which migrate to peripheral tissues and exert effector function (Sallusto and Lanzavecchia 2009). However, since the pioneer work by Lanzavecchia's group in 1999, additional subsets of memory CD4 T cells including follicular, resident and stem cells have been described.

CD4 T follicular helper (Tfh) cells are necessary for the formation and maintenance of GCs where the generation of high affinity and long-lived humoral immunity takes place (Corcoran and Tarlinton 2016). However, there is increasing evidence that Tfh cells may also become memory cells. Markers to identify CD4 Tfh cells are well defined; many of them (e.g. Bcl-6, ICOS, c-MAF, etc.) are downregulated when transitioning into memory cells (Hale and Ahmed 2015). Only CXCR5 expression is retained, which promotes prompt migration to B cell follicles upon reactivation and additional help to the B cell response (MacLeod et al. 2011). The existence of CD4 Tfh memory cells in humans was controversial until the finding by Locci et al. of a population of circulating CXCR5$^+$ CXCR3$^-$ T cells with a resting phenotype that expressed low levels of PD-1 (Locci et al. 2013). This population was maintained in vitro without TCR stimulation and its functionality upon re-activation was similar to that of GC Tfh cells from human tonsils. Moreover, a recent study investigating circulating CD69$^-$ICOS$^-$CXCR5$^+$ CD4 T cells identified a sub-population of Th2-like cells that was increased in patients with juvenile dermatomyositis. These cells lacked expression of CXCR3 and CCR6 and induced high production of IgG and IgE in co-culture assays (Morita et al. 2011). The capacity of CD4 Tfh-2 cells to rapidly aid B cell proliferation and facilitate IgE production requires further research to assess its clinical significance in allergic diseases.

The CD4 T cell effector memory compartment contains a population of non-migratory, tissue resident, memory T cells (Trm) that has attracted considerable interest because of their protective ability against pathogens. Most studies on Trm cells have focused on CD8 T cells, but CD4 Trm cells have now been reported in both mice and humans. Interestingly, as compared to their CD8 counterparts, the TCR repertoires of CD4 Trm cells seem more compartmentalized and/or less cross-reactive, which may indicate greater regulation. The mechanisms underlying retention of CD4 Trm cells within tissues vary depending on the tissue but CD69, which antagonizes the tissue egression receptor S1PR1, and CD103, which mediates interaction of CD4 T cells with epithelial cells, have been proposed to play a major role (Schenkel and Masopust 2014). CD4 T cells preferentially migrate back to the original site of priming where they can differentiate into Trm cells (Gebhardt et al. 2011). Ugur et al. tracked a persistent population of $CD62L^{low}$ CD4 T cells in Peyer's patches and mesenteric lymph nodes in a model of intestinal sensitization to OVA and a classical Th2 adjuvant, cholera toxin. Upon oral antigen exposure this population increased in the Peyer's patches, but remained stable in the mesenteric lymph nodes (Ugur et al. 2014). However, the generation of Trm CD4 T cells does not necessarily require antigen exposure and inflammation in situ. In this regard, Von Adrian's group showed development of CD4 Trm cells in the uterus after mucosal (nasal or vaginal), but not subcutaneous, immunization against *Chlamydia trachomatis* (Stary et al. 2015). Altogether, these data suggest the existence of a migration program of CD4 Trm cells that is specific to the site of sensitization.

Stem cell T memory cells have been studied for the most part in human PMBCs. They are defined by the expression of CD45RA, CCR7 and CD95. A novel transcriptional analysis of human stem cell memory cells ($CD45RO^+CCR7^+CD95^+$) placed these cells with a profile somewhat in-between naïve and central memory CD4 T cells (Takeshita et al. 2015). The same study showed that these cells gradually acquire diverse polarities, including a Th2 phenotype when cultured in the presence of IL-4 and anti-INF-γ. There may be a potential pathogenic role for these cells in allergy that remains to be explored.

In reference to Th2 immunity, memory Th2 CD4 T cells are considered a population with a central role in the pathogenesis of allergic diseases. These cells have been mostly studied in respiratory allergic diseases, particularly in asthma, where their ability to produce IL-5 and associated eosinophil-driven pathology is a key feature (Nakayama et al. 2016). However, the clinical manifestations of other allergic diseases such as food allergy are largely mediated by IgE. Therefore, identifying what type of CD4 Th2 memory cells produce IL-4, aid B cell class-switching to IgE and mediate plasmablast proliferation has considerable implications. The recent discovery of new subsets of memory CD4 T (Th2) cells with differential functionalities, survival requirements, migration and homing patterns, etc. brings to light a previously unappreciated heterogeneity and complex regulation of the CD4 T cell memory compartment. While all CD4 T cells are "born equal", they carry multiple potentials. The activation of a given potential likely by environmental cues confers to the cell a degree of specialization that ultimately expresses its identity. The challenge is to understand the biological and clinical relevance of these different identities in allergic diseases.

4.2 The Elusive IgE⁺ Memory B Cell

In general, the generation of memory B cells is better understood than that of memory CD4 T cells because the identification of antigen-specific B cells is less technically demanding and, thus, has allowed for comprehensive and in depth analyses of specific populations of memory B cells. Also, the evolution of the B cell lineage is linear and thought to be irreversible; albeit, there is active debate on the capacity of terminally-differentiated PCs to re-enter the cell cycle (Tooze 2013). Naïve B cells egress the bone marrow expressing IgM and IgD BCRs and patrol lymphoid tissues in the search for their cognate antigen. Upon its encounter, entry into an extrafollicular or GC pathway is the first major fate-decision step during a CD4 T cell-dependent B cell response. Initially, this decision was thought to be aleatory (Dal Porto et al. 1998; Hasbold et al. 2004) but this notion has changed as a result of the discovery of factors that regulate this process.

The strength of the initial interaction between BCR and antigen is a primary determinant of B cell fate. Paus et al. immunized mice expressing BCRs specific for lysozyme, an allergen from hen's egg white, against sheep red blood cells coupled to a series of engineered lysozymes that bound this BCR over a 10,000-fold affinity range (Paus et al. 2006). They found that B cells reactive against either high affinity or abundant epitopes were preferentially selected toward the extrafollicular pathway. Conversely, responding clones with weaker antigen reactivity were primarily directed to GCs where they would undergo affinity maturation. Another important determinant of B cell fate is the nature and duration of B-Tfh cell interactions. Germain's group (Qi et al. 2008) used two-photon intravital imaging and found that signalling lymphocyte activation molecule (SLAM) associated protein (SAP) deficiency impaired the ability of CD4 T cells to stably interact with cognate B cells, which resulted in defective GC formation. Among the different signals provided by CD4 T cells, CD40 signalling alone is sufficient to induce activated B cells to differentiate extrafollicularly into memory B cells (Taylor et al. 2012). On the other hand, IL-21 signalling and associated up-regulation of the transcription factor Bcl-6 in B cells seems to be crucial for entry into the GC (Fukuda et al. 1997; Linterman et al. 2010). Altogether, these data suggest that activated B cells that are able to interact in a stable and durable manner with Tfh cells and receive adequate T cell help are licensed to enter the GC. However, if the duration of this interaction is rather short, B cells are more likely to contribute to the GC-independent memory B cell pool (Kurosaki et al. 2015). While somatic hypermutation and affinity maturation are thought to occur exclusively in GCs, class-switch recombination and memory B cell differentiation can take place in a follicular or extrafollicular manner (Kurosaki et al. 2015; Weisel and Shlomchik 2017). Therefore, both pathways (i.e. GC-dependent or independent) may contribute to the production of IgE^+ B cells (Gould and Ramadani 2015).

Historically, the study of IgE^+ B cells, and in particular of IgE^+ memory B cells, has been challenging (Gould and Ramadani 2015). These cells are extremely rare and heavily outnumbered by B cells expressing different isotypes. Also, many cells, including B cells, can bind IgE via FcεRII (CD23), thus hindering the detection of bona fide IgE^+ B cells. In order to study IgE-expressing cells, three laboratories independently generated IgE-reporter mice; these mouse strains encode fluorescent proteins for membrane IgE by targeting the endogenous IgE locus (Talay et al. 2012; Yang et al. 2012; He et al. 2013); these approaches have been recently reviewed (Yang et al. 2014; Gould and Ramadani 2015). An alternative to the use of transgenic mice to reliably detect these cells is the use of sophisticated flow cytometry staining methods. One approach is to acid-wash the cells before extracellular staining to reduce cell-bound IgE and facilitate the staining of membrane IgE (Erazo et al. 2007); although this method seems to lack of sufficient sensitivity and may affect other surface molecules (Yang et al. 2014). Recently, Christopher C. D. Allen's group published a detection method that blocks surface IgE and, thus, exclusively detects intracellular IgE. It was validated in Verigem (Venus reporter for membrane IgE) mice (Yang et al. 2012) and we recently adapted it to detect allergen-specific IgE^+ B cells and PCs in a preclinical model of food allergy (Jiménez-Saiz et al. 2017). These innovative developments to detect IgE^+ B cells in mouse models have brought considerable insight into the biology of these cells.

Recent studies in models of Th2 immunity against helminths (*N. brasiliensis*) or haptenated-proteins (NP-KLH, NP-CGG) have shown that IgE^+ B cells are predisposed to differentiate into PCs. The emergence of these IgE-secreting cells peaked early in the immune response without signs of somatic hypermutation, which is indicative of an extrafollicular response (Talay et al. 2012; Yang et al. 2012; He et al. 2013). In addition, there is evidence that mice with a deficiency in Bcl-6 restricted to the B-cell lineage (Mb1-cre × $Bcl^{fl/fl}$), which lack GCs, produced similar serum levels of peanut-specific IgE early in the immune response compared with wild type mice on mucosal sensitization (Jiménez-Saiz et al. 2017). Similarly, IgE^+ GC B cells are present only briefly in the GC as they have a unique propensity to upregulate Blimp-1, a key transcription factor in PC differentiation (Yang et al. 2012). In addition, IgE^+ GC B cells show higher rates of apoptosis, compared to their IgG counterparts, which was attributed to impaired BCR functioning (He et al. 2013). New studies have provided more insights into the changes in B cell fate that result from membrane IgE-expression per se (Laffleur et al. 2015; Haniuda et al. 2016; Yang et al. 2016). While the data generated in these studies support that autonomous IgE-BCR signaling promotes PC differentiation via CD19-PI3K-Akt-IRF4 signalling (Haniuda et al. 2016; Yang et al. 2016), the findings on autonomous IgE-BCR signalling and activation of apoptotic pathways via BLNK-Jnk/p38 signalling remain controversial (Laffleur et al. 2015; Haniuda et al. 2016; Yang et al. 2016). Taken together, these findings suggest that regardless of their origin (follicular vs. extrafollicular), the scarce population of IgE$^+$ B cells appears to be intrinsically predisposed to differentiate into short-lived IgE-secreting cells. IgE-BCR signalling may also promote apoptotic pathways that regulate the IgE response. This might explain the virtual absence of IgE$^+$ memory B cells in mice. Indeed, these cells have not been readily detected in vivo under physiological conditions with one exception (Talay et al. 2012). This group used IgE reporter mice (M1$'$) that express the M1$'$ segment of human IgE (Talay et al. 2012). However, there is controversy regarding M1$'$ mice as they contain additional sequences including an exogenous polyadenylation signal sequence that might alter normal IgE responses (Brightbill et al. 2010; Lafaille et al. 2012).

In humans, the existence of IgE$^+$ memory B cells has also been a matter of debate (Davies et al. 2013) until the recent work by Van Zelm's group (Berkowska et al. 2014). They developed an 8-color flow cytometry based strategy to reliably detect IgE$^+$ memory B cells after stepwise exclusion of IgM$^+$, IgD$^+$, IgG$^+$ and IgA$^+$ B cells. Using this approach, they detected bona fide IgE$^+$ memory B cells in the blood of healthy donors. The phenotype of these cells was characterized by the higher expression of CD80 and CD86, the TNF superfamily member transmembrane activator and CAML interactor (TACI) compared with naïve B cells. In contrast to other memory B cells, IgE$^+$ memory B cells did not show upregulation of toll-like receptor-related CD180 and showed the lowest expression levels of the B-cell antigen receptor complex member CD79b. The authors found that the differential expression of CD27 was associated with a GC-dependent (CD27$^+$) or -independent (CD27$^-$) origin based on their replication history, somatic hypermutation level and frequencies in patients deficient in CD40L. Interestingly, GC-dependent IgE$^+$ memory B cells were scarce in children <6 years old, with a median frequency of 0.03% of total B cells; this frequency increased with age and reached 0.1% of total B cells in adults aged 31–40 years. On the other hand, GC independent IgE$^+$ memory B cells were more abundant in children than adults and the highest frequency of 0.12% of total B cells was reported in children from 6 to 15 years of age. These results may insinuate the existence of unexplored biological roles for IgE$^+$ memory B cells, which exist in healthy individuals and increase affinity over time. They also found that the frequency of GC-independent IgE$^+$ memory B cells were increased in patients with atopic dermatitis. This population exhibited increased somatic hypermutation, hypothesized to reflect either the mainly local manifestation of this allergic disease or an abnormal phenotype (Berkowska et al. 2014). In this context, a recent study by the Shlomchik's group in mice infected with *Salmonella typhimurium* proposed the existence of a somatic hypermutation program outside the GC (Di Niro et al. 2015) that may be at play in patients with atopic dermatitis or other allergic diseases.

The ability to reliably identify and study bona fide IgE$^+$ memory B cells by flow cytometry has advanced the understanding of human IgE biology in general. Yet, the challenge remains to identify, within this very rare population, those cells that are allergen-specific. The phenotypic, molecular, and functional analysis of allergen-specific IgE$^+$ memory B cells would provide a deeper and unpreceded understanding on the immunobiology underlying these pathogenic cells.

5 Activation of Memory Responses in Food Allergy

The dynamics that control the fate of memory B cells on re-call responses remain largely unexplored (McHeyzer-Williams et al. 2015). Recent research has focused on IgE-generating memory responses, for the most part in the context of immunity against helminths or systemic Th2 immunization to haptenated proteins (Erazo et al. 2007; Yang et al. 2012; Xiong et al. 2012; Turqueti-Neves et al. 2015). The body of data generated from these studies has advanced the concept that secondary responses resulting in high-affinity IgE$^+$ PC generation originate from GC-derived IgG1$^+$ memory B cells. This concept is supported by several lines of evidence: (i) IgE$^+$ memory B cells have not been readily detected in vivo (reviewed in the previous section); (ii) IgE$^+$ B cells do not thrive in GCs (where somatic hypermutation and affinity maturation takes place) (Erazo et al. 2007; Yang et al. 2012); yet, the generation of high-affinity IgE has been reported along with a number of IgE mutations in the CDR3 region comparable with IgG1 (Erazo et al. 2007); (iii) a large proportion of IgE-expressing B cells emerging during a Th2 immune response contained IgG1 switch-region (Sγ1) remnants, and the presence of Sγ1 remnants (sequential class-switching) was associated with increased affinity (Xiong et al. 2012); and (iv) IgE production on Th2 re-call responses was lost in mice in which the extracellular part of IgG1 had been replaced with IgE sequences (Turqueti-Neves et al. 2015).

The notion that the reservoir of IgE$^+$ PCs resides in the IgG1$^+$ memory B cell compartment has also been proposed in a murine model of mucosal sensitization to peanut (Jiménez-Saiz et al. 2017). We reported, using flow cytometry, a population of allergen-specific IgG1$^+$, but not IgE$^+$, memory B cells, present both at the mucosal and the systemic compartments 9 months post allergic sensitization. On oral allergen exposure, a population of allergen-specific IgE-expressing B cells emerged along with the appearance of allergen-specific IgE$^+$ PCs in the bone marrow and high levels of allergen-specific IgE in circulation. While allergen-specific GC activity was induced during sensitization, it was absent during recall responses (Jiménez-Saiz et al. 2017). These findings suggest that the strength of BCR-allergen binding is a major decision checkpoint, not only for naïve B cells but also memory B cells (Paus et al. 2006). Furthermore, culture of splenocytes from mice 15 months post-sensitization to peanut resulted in memory B cell and plasmablast proliferation, suggesting that these memory cells last for virtually a lifetime. In sum, there is compelling evidence that long-lived IgG1$^+$ memory B cells might serve as "a", or even "the only", reservoir of IgE$^+$ plasma cells on murine secondary responses in Th2 immunity in general, and food allergy in particular, but this requires further investigation.

In humans, evidence for sequential class-switching from an intermediate IgG-expressing B cell to IgE has been reported in patients with atopic dermatitis (Berkowska et al. 2014) and tonsil B-cell cultures (Ramadani et al. 2017). However, the contribution of the allergen-specific IgE$^+$ versus IgG1$^+$ memory B cell pool to the IgE$^+$ PC compartment during secondary responses to food allergens remains to be fully elucidated (Davies et al. 2013; Otte et al. 2016). From a therapeutic perspective, defining the contribution of each type of memory (IgE vs. IgG) has substantial implications. Strategies to deplete IgG-expressing cells for the treatment of allergic diseases would be complicated by the need to retain protective immunity against pathogens. As an alternative, depletion of IgE-expressing cells with antibodies specific for human membrane IgE has been investigated (Chowdhury et al. 2012; Chu et al. 2012; Gauvreau et al. 2014; Harris et al. 2016). However, the replenishment of IgE$^+$ PCs from IgG$^+$ memory B cells would limit the potential benefits of this approach.

There are other populations of allergen-specific memory B cells beyond those expressing IgE and IgG. Zuccarino-Catania et al. identified in mice immunized to haptenated proteins, various populations of IgM$^+$ memory B cells whose functionality on re-call was defined by the differential expression of CD80 and PD-L2 (Zuccarino-Catania et al. 2014). Double negative (CD80$^-$PD-L2$^-$) IgM$^+$ memory B cells showed a low number of mutations and were able to generate GCs and class-switched PCs on recall. The existence of these non-switched (IgM$^+$) memory B cells exhibiting no increased affinity (compared with naïve B cells) challenges the classical definition of a memory B cell, which has been recently re-defined as a B cell that has responded to an antigen and becomes quiescent (Weisel and Shlomchik 2017). On the other hand, double positive (CD80$^+$PD-L2$^+$) IgM$^+$ memory B cells were mutated and did not produce GCs but quickly generated a large isotype-switched secondary PC response. Interestingly, diverse functionalities were also observed in IgG1$^+$ memory B cells according to the expression of CD80 and PD-L2. It is then possible that low-affinity memory B cells participate in secondary responses to allergens which structure has changed from that at the time of sensitization; for example, food processing is known to alter allergen's structure and binding capacity to IgE (Jiménez-Saiz et al. 2015). Also, these cells could serve as a last reservoir of memory in the event of high-affinity memory B cell exhaustion (Weisel and Shlomchik 2017). The IgE-generating capacity of these subpopulations on re-call in allergic disease, particularly food allergy, remains unknown.

5.1 Therapeutic Implications/Prospects

The idea that the IgE$^+$ PC compartment can be replenished from memory B cells expressing different isotypes, whether it is through sequential or direct class-switching, strengthens the search for depletion strategies based on allergen specificity rather than BCR isotype. In this regard, a liposome-based preparation containing Ara h 2 (a major peanut allergen) and CD22L, to prevent B cell activation and induce apoptosis in peanut-specific cells, has been recently reported (Orgel et al. 2017). While this preparation was successfully tested in a prophylactic model, it may face challenges on a therapeutic setting because of safety (allergic reactions on systemic administration) and/or delivery (stability of the liposomes on oral administration) concerns. Recent findings on the cellular and molecular requirements for successful secondary IgE generation have shed light on new therapeutic targets with disease transforming potential in food allergy. Motsoglou and Dreskin showed that in a system involving transfer of splenic B cells from systemically sensitized mice along with CD4 T cells, followed by two systemic challenges, food allergy was elicited in naïve recipients in a CD4 T cell-dependent manner (Moutsoglou and Dreskin 2016). More recently, we have found memory CD4 T cell proliferation and associated Th2 cytokine production in supernatants of spleen cells from peanut allergic mice 15 months post-sensitization when cultured with peanut (Jiménez-Saiz et al. 2017). Importantly, the depletion of CD4 T cells (prior to co-culture) significantly reduced both memory B cell and plasmablast proliferation. In addition, IL-4 was central for plasmablast proliferation. Of note, the reduction of memory cells and/or plasmablast proliferation was significant but not complete suggesting several, non-mutually exclusive interpretations: (i) the blocking and depleting capacity of anti-IL-4 and anti-CD4 respectively may have been incomplete; (ii) allergen-specific memory B cells (generated in a CD4 T cell-dependent response) could have a limited ability to drive secondary responses in the absence of CD4 T cell help (Zuccarino-Catania et al. 2014); (iii) molecules other than IL-4, such as IL-13, may aid memory B cells in IgE class-switching and plasmablast expansion on recall (Turqueti-Neves et al. 2015).

The understanding of the cellular and molecular machinery that takes place in secondary responses to food allergens is still in its infancy. However, the idea of interfering with IL-4, or both IL-4 and IL-13, signaling via blocking its common receptor (IL-4Rα) has merit. First, blocking IL-4/IL-13 signaling would prevent IgE-regeneration from any memory B cell reservoir that requires IgE class-switching. Secondly, the commitment of pathogenic memory B cells to an IgE$^+$ PC lineage, and hence IgE production, would be impaired. In addition, de novo Th2 polarization would be prevented. In this regard, a recent study in peanut-allergic patients undergoing oral immunotherapy reported the presence of a population of peanut-reactive Th2 cells that persisted despite evidence of clinical desensitization (12–24 months of oral immunotherapy). The authors hypothesized that this residual population of IL-4$^+$ CD4 T cells, while in low numbers, could serve to undermine the durability of non-responsiveness to peanut (Wisniewski et al. 2015). Hence, concomitant interference with IL-4 and IL-13, derived from CD4 T cells, in IgE-mediated allergic conditions such as food allergy, may not only effectively impair the machinery re-generating IgE but also potentiate regulatory pathways leading to oral tolerance.

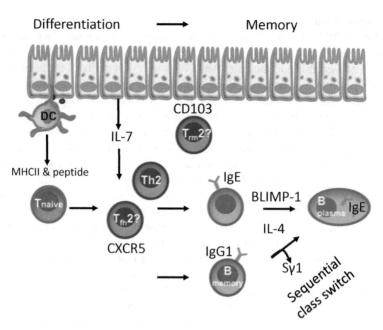

Fig. 1 Heterogeneous Th2 memory formation and perpetuation of humoral IgE immunity

6   Conclusion

The discovery of allergic memory responses as a pivotal mechanism in the perpetuation and exacerbation of allergic diseases identifies allergen-specific memory B and CD4 T cells as key therapeutic targets (Fig. 1). Therefore, efforts to disable the machinery that maintains and activates memory responses may lead to the development of therapies for IgE-mediated food allergies beyond the symptomatic drugs currently available. However, further research is needed to fully understand the cellular and molecular events mediating secondary responses resulting in IgE production.

Acknowledgements Research by the Jordana-Waserman lab cited in this work has been supported by the Canadian Institutes of Health Research (CIHR), MedImmune LLC (USA), the National Institutes of Health (NIH, USA), AllerGen NCE, Food Allergy Canada, the Delaney family and the Walter and Maria Schroeder Foundation. MJ is a senior Canada Research Chair in Immunobiology of Respiratory Diseases and Allergy. RJ holds a MITACS Postdoctoral Fellowship. DKC is a Vanier Scholar. We thank Joshua Koenig for critical review of this chapter.

References

Achatz-Straussberger G, Zaborsky N, Konigsberger S, Luger EO, Lamers M, Crameri R, Achatz G (2008) Migration of antibody secreting cells towards CXCL12 depends on the isotype that forms the BCR. Eur J Immunol 38(11):3167–3177. https://doi.org/10.1002/eji.200838456

Ahmed R, Gray D (1996) Immunological memory and protective immunity: understanding their relation. Science 272(5258):54–60

Amanna IJ, Carlson NE, Slifka MK (2007) Duration of humoral immunity to common viral and vaccine antigens. N Engl J Med 357(19):1903–1915. https://doi.org/10.1056/NEJMoa066092

Ashley SE, Tan HT, Vuillermin P, Dharmage SC, Tang ML, Koplin J, Gurrin LC, Lowe A, Lodge C, Ponsonby AL, Molloy J, Martin P, Matheson MC, Saffery R, Allen KJ, Ellis JA, Martino D, HealthNuts team, Barwon Infant Study tMACstPATS, the Peanut Oral Immuno-Therapy Study (2017) The skin barrier function gene SPINK5 is associated with challenge-proven IgE-mediated food allergy in infants. Allergy. https://doi.org/10.1111/all.13143

Berkowska MA, Heeringa JJ, Hajdarbegovic E, van der Burg M, Thio HB, van Hagen PM, Boon L, Orfao A, van Dongen JJ, van Zelm MC (2014) Human IgE(+) B cells are derived from T cell-dependent and T cell-independent pathways. J Allergy Clin Immunol 134 (3):688–697.e686. https://doi.org/10.1016/j.jaci.2014.03.036

Besnard AG, Togbe D, Guillou N, Erard F, Quesniaux V, Ryffel B (2011) IL-33-activated dendritic cells are critical for allergic airway inflammation. Eur J Immunol 41 (6):1675–1686. https://doi.org/10.1002/eji.201041033

Brightbill HD, Jeet S, Lin Z, Yan D, Zhou M, Tan M, Nguyen A, Yeh S, Delarosa D, Leong SR, Wong T, Chen Y, Ultsch M, Luis E, Ramani SR, Jackman J, Gonzalez L, Dennis MS, Chuntharapai A, DeForge L, Meng YG, Xu M, Eigenbrot C, Lee WP, Refino CJ, Balazs M, Wu LC (2010) Antibodies specific for a segment of human membrane IgE deplete IgE-producing B cells in humanized mice. J Clin Invest 120(6):2218–2229. https://doi.org/10.1172/JCI40141

Brown SJ, Asai Y, Cordell HJ, Campbell LE, Zhao Y, Liao H, Northstone K, Henderson J, Alizadehfar R, Ben-Shoshan M, Morgan K, Roberts G, Masthoff LJ, Pasmans SG, van den Akker PC, Wijmenga C, Hourihane JO, Palmer CN, Lack G, Clarke A, Hull PR, Irvine AD, McLean WH (2011) Loss-of-function variants in the filaggrin gene are a significant risk factor for peanut allergy. J Allergy Clin Immunol 127(3):661–667. https://doi.org/10.1016/j.jaci.2011.01.031

Buchholz VR, Schumacher TN, Busch DH (2016) T cell fate at the single-cell level. Annu Rev Immunol 34:65–92. https://doi.org/10.1146/annurev-immunol-032414-112014

Cates EC, Fattouh R, Wattie J, Inman MD, Goncharova S, Coyle AJ, Gutierrez-Ramos JC, Jordana M (2004) Intranasal exposure of mice to house dust mite elicits allergic airway inflammation via a GM-CSF-mediated mechanism. J Immunol 173(10):6384–6392

Catron DM, Rusch LK, Hataye J, Itano AA, Jenkins MK (2006) CD4+ T cells that enter the draining lymph nodes after antigen injection participate in the primary response and become central-memory cells. J Exp Med 203(4):1045–1054. https://doi.org/10.1084/jem.20051954

Cherry WB, Yoon J, Bartemes KR, Iijima K, Kita H (2008) A novel IL-1 family cytokine, IL-33, potently activates human eosinophils. J Allergy Clin Immunol 121(6):1484–1490. https://doi.org/10.1016/j.jaci.2008.04.005

Chowdhury PS, Chen Y, Yang C, Cook KE, Nyborg AC, Ettinger R, Herbst R, Kiener PA, Wu H (2012) Targeting the junction of CvarepsilonmX and varepsilon-migis for the specific deple-tion of mIgE-expressing B cells. Mol Immunol 52(3–4):279–288. https://doi.org/10.1016/j.molimm.2012.06.004

Chu DK, Llop-Guevara A, Walker TD, Flader K, Goncharova S, Boudreau JE, Moore CL, Seunghyun In T, Waserman S, Coyle AJ, Kolbeck R, Humbles AA, Jordana M (2013) IL-33, but not thymic stromal lymphopoietin or IL-25, is central to mite and peanut allergic sensitization. J Allergy Clin Immunol 131(1):187–200.e181–188. https://doi.org/10.1016/j.jaci.2012.08.002

Chu DK, Jimenez-Saiz R, Verschoor CP, Walker TD, Goncharova S, Llop-Guevara A, Shen P, Gordon ME, Barra NG, Bassett JD, Kong J, Fattouh R, McCoy KD, Bowdish DM, Erjefalt JS, Pabst O, Humbles AA, Kolbeck R, Waserman S, Jordana M (2014) Indigenous enteric eosinophils control DCs to initiate a primary Th2 immune response in vivo. J Exp Med 211 (8):1657–1672. https://doi.org/10.1084/jem.20131800

Chu SY, Horton HM, Pong E, Leung IW, Chen H, Nguyen DH, Bautista C, Muchhal US, Bernett MJ, Moore GL, Szymkowski DE, Desjarlais JR (2012) Reduction of total IgE by targeted coengagement of IgE B-cell receptor and FcgammaRIIb with Fc-engineered antibody. J Allergy Clin Immunol 129(4):1102–1115. https://doi.org/10.1016/j.jaci.2011.11.029

Ckless K, Hodgkins SR, Ather JL, Martin R, Poynter ME (2011) Epithelial, dendritic, and CD4(+) T cell regulation of and by reactive oxygen and nitrogen species in allergic sensitization. Biochim Biophys Acta 1810(11):1025–1034. https://doi.org/10.1016/j.bbagen.2011.03.005

Connor LM, Tang SC, Cognard E, Ochiai S, Hilligan KL, Old SI, Pellefigues C, White RF, Patel D, Smith AA, Eccles DA, Lamiable O, McConnell MJ, Ronchese F (2017) Th2 responses are primed by skin dendritic cells with distinct transcriptional profiles. J Exp Med 214 (1):125–142. https://doi.org/10.1084/jem.20160470

Corcoran LM, Tarlinton DM (2016) Regulation of germinal center responses, memory B cells and plasma cell formation-an update. Curr Opin Immunol 39:59–67. https://doi.org/10.1016/j.coi.2015.12.008

Dal Porto JM, Haberman AM, Shlomchik MJ, Kelsoe G (1998) Antigen drives very low affinity B cells to become plasmacytes and enter germinal centers. J Immunol 161(10):5373–5381

Davies JM, Platts-Mills TA, Aalberse RC (2013) The enigma of IgE+ B-cell memory in human subjects. J Allergy Clin Immunol 131(4):972–976. https://doi.org/10.1016/j.jaci.2012.12.1569

Defrance T, Taillardet M, Genestier L (2011) T cell-independent B cell memory. Curr Opin Immunol 23(3):330–336. https://doi.org/10.1016/j.coi.2011.03.004

Di Niro R, Lee SJ, Vander Heiden JA, Elsner RA, Trivedi N, Bannock JM, Gupta NT, Kleinstein SH, Vigneault F, Gilbert TJ, Meffre E, McSorley SJ, Shlomchik MJ (2015) Salmonella infection drives promiscuous B cell activation followed by extrafollicular affinity maturation. Immunity 43(1):120–131. https://doi.org/10.1016/j.immuni.2015.06.013

Erazo A, Kutchukhidze N, Leung M, Christ AP, Urban JF Jr, Curotto de Lafaille MA, Lafaille JJ (2007) Unique maturation program of the IgE response in vivo. Immunity 26(2):191–203. https://doi.org/10.1016/j.immuni.2006.12.006

Fazilleau N, Eisenbraun MD, Malherbe L, Ebright JN, Pogue-Caley RR, McHeyzer-Williams LJ, McHeyzer-Williams MG (2007) Lymphoid reservoirs of antigen-specific memory T helper cells. Nat Immunol 8(7):753–761. https://doi.org/10.1038/ni1472

Finley JJ (1951) The complete writings of thucydides: the Peloponesian War. Modern Library, New York

Forbes EE, Groschwitz K, Abonia JP, Brandt EB, Cohen E, Blanchard C, Ahrens R, Seidu L, McKenzie A, Strait R, Finkelman FD, Foster PS, Matthaei KI, Rothenberg ME, Hogan SP (2008) IL-9- and mast cell-mediated intestinal permeability predisposes to oral antigen hypersensitivity. J Exp Med 205(4):897–913. https://doi.org/10.1084/jem.20071046

Fukuda T, Yoshida T, Okada S, Hatano M, Miki T, Ishibashi K, Okabe S, Koseki H, Hirosawa S, Taniguchi M, Miyasaka N, Tokuhisa T (1997) Disruption of the Bcl6 gene results in an impaired germinal center formation. J Exp Med 186(3):439–448

Gajewska BU, Wiley RE, Jordana M (2003) GM-CSF and dendritic cells in allergic airway inflammation: basic mechanisms and prospects for therapeutic intervention. Curr Drug Targets Inflamm Allergy 2(4):279–292

Gasper DJ, Tejera MM, Suresh M (2014) CD4 T-cell memory generation and maintenance. Crit Rev Immunol 34(2):121–146

Gauvreau GM, Harris JM, Boulet LP, Scheerens H, Fitzgerald JM, Putnam WS, Cockcroft DW, Davis BE, Leigh R, Zheng Y, Dahlen B, Wang Y, Maciuca R, Mayers I, Liao XC, Wu LC, Matthews JG, O'Byrne PM (2014) Targeting membrane-expressed IgE B cell receptor with an antibody to the M1 prime epitope reduces IgE production. Sci Transl Med 6(243):243ra285. https://doi.org/10.1126/scitranslmed.3008961

Gebhardt T, Whitney PG, Zaid A, Mackay LK, Brooks AG, Heath WR, Carbone FR, Mueller SN (2011) Different patterns of peripheral migration by memory CD4+ and CD8+ T cells. Nature 477(7363):216–219. https://doi.org/10.1038/nature10339

Gonzalez-Garcia I, Rodriguez-Bayona B, Mora-Lopez F, Campos-Caro A, Brieva JA (2008) Increased survival is a selective feature of human circulating antigen-induced plasma cells synthesizing high-affinity antibodies. Blood 111(2):741–749. https://doi.org/10.1182/blood-2007-08-108118

Gould HJ, Ramadani F (2015) IgE responses in mouse and man and the persistence of IgE memory. Trends Immunol 36(1):40–48. https://doi.org/10.1016/j.it.2014.11.002

Hale JS, Ahmed R (2015) Memory T follicular helper CD4 T cells. Front Immunol 6:16. https://doi.org/10.3389/fimmu.2015.00016

Halliley JL, Tipton CM, Liesveld J, Rosenberg AF, Darce J, Gregoretti IV, Popova L, Kaminiski D, Fucile CF, Albizua I, Kyu S, Chiang KY, Bradley KT, Burack R, Slifka M, Hammarlund E, Wu H, Zhao L, Walsh EE, Falsey AR, Randall TD, Cheung WC, Sanz I, Lee FE (2015) Long-lived plasma cells are contained within the CD19(-)CD38(hi)CD138(+) subset in human bone marrow. Immunity 43(1):132–145. https://doi.org/10.1016/j.immuni.2015.06.016

Hammad H, Lambrecht BN (2015) Barrier epithelial cells and the control of type 2 immunity. Immunity 43(1):29–40. https://doi.org/10.1016/j.immuni.2015.07.007

Han H, Thelen TD, Comeau MR, Ziegler SF (2014) Thymic stromal lymphopoietin-mediated epicutaneous inflammation promotes acute diarrhea and anaphylaxis. J Clin Invest 124 (12):5442–5452. https://doi.org/10.1172/JCI77798

Haniuda K, Fukao S, Kodama T, Hasegawa H, Kitamura D (2016) Autonomous membrane IgE signaling prevents IgE-memory formation. Nat Immunol 17(9):1109–1117. https://doi.org/10.1038/ni.3508

Harris JM, Maciuca R, Bradley MS, Cabanski CR, Scheerens H, Lim J, Cai F, Kishnani M, Liao XC, Samineni D, Zhu R, Cochran C, Soong W, Diaz JD, Perin P, Tsukayama M, Dimov D, Agache I, Kelsen SG (2016) A randomized trial of the efficacy and safety of quilizumab in adults with inadequately controlled allergic asthma. Respir Res 17:29. https://doi.org/10.1186/s12931-016-0347-2

Hasbold J, Corcoran LM, Tarlinton DM, Tangye SG, Hodgkin PD (2004) Evidence from the generation of immunoglobulin G-secreting cells that stochastic mechanisms regulate lymphocyte differentiation. Nat Immunol 5(1):55–63. https://doi.org/10.1038/ni1016

He JS, Meyer-Hermann M, Xiangying D, Zuan LY, Jones LA, Ramakrishna L, de Vries VC, Dolpady J, Aina H, Joseph S, Narayanan S, Subramaniam S, Puthia M, Wong G, Xiong H, Poidinger M, Urban JF, Lafaille JJ, Curotto de Lafaille MA (2013) The distinctive germinal center phase of IgE+ B lymphocytes limits their contribution to the classical memory response. J Exp Med 210(12):2755–2771. https://doi.org/10.1084/jem.20131539

Heil M, Land WG (2014) Danger signals – damaged-self recognition across the tree of life. Front Plant Sci 5:578. https://doi.org/10.3389/fpls.2014.00578

Hilmenyuk T, Bellinghausen I, Heydenreich B, Ilchmann A, Toda M, Grabbe S, Saloga J (2010) Effects of glycation of the model food allergen ovalbumin on antigen uptake and presentation by human dendritic cells. Immunology 129(3):437–445. https://doi.org/10.1111/j.1365-2567.2009.03199.x

Hondowicz BD, An D, Schenkel JM, Kim KS, Steach HR, Krishnamurty AT, Keitany GJ, Garza EN, Fraser KA, Moon JJ, Altemeier WA, Masopust D, Pepper M (2016) Interleukin-2-dependent allergen-specific tissue-resident memory cells drive asthma. Immunity 44 (1):155–166. https://doi.org/10.1016/j.immuni.2015.11.004

Jaigirdar SA, MacLeod MK (2015) Development and function of protective and pathologic memory CD4 T cells. Front Immunol 6:456. https://doi.org/10.3389/fimmu.2015.00456

Jiménez-Saiz R, Benede S, Molina E, Lopez-Exposito I (2015) Effect of processing technologies on the allergenicity of food products. Crit Rev Food Sci Nutr 55(13):1902–1917. https://doi.org/10.1080/10408398.2012.736435

Jiménez-Saiz R, Chu DK, Mandur TS, Walker TD, Gordon ME, Chaudhary R, Koenig J, Saliba S, Galipeau HJ, Utley A, King IL, Lee K, Ettinger R, Waserman S, Kolbeck R, Jordana M (2017) Lifelong memory responses perpetuate humoral TH2 immunity and anaphylaxis in food allergy. J Allergy Clin Immunol. https://doi.org/10.1016/j.jaci.2017.01.018

Kamalakannan M, Chang LM, Grishina G, Sampson HA, Masilamani M (2016) Identification and characterization of DC-SIGN-binding glycoproteins in allergenic foods. Allergy 71 (8):1145–1155. https://doi.org/10.1111/all.12873

Karnowski A, Achatz-Straussberger G, Klockenbusch C, Achatz G, Lamers MC (2006) Inefficient processing of mRNA for the membrane form of IgE is a genetic mechanism to limit recruitment of IgE-secreting cells. Eur J Immunol 36(7):1917–1925. https://doi.org/10.1002/eji.200535495

Kim C, Wilson T, Fischer KF, Williams MA (2013) Sustained interactions between T cell receptors and antigens promote the differentiation of CD4(+) memory T cells. Immunity 39 (3):508–520. https://doi.org/10.1016/j.immuni.2013.08.033

Kong J, Chalcraft K, Mandur TS, Jimenez-Saiz R, Walker TD, Goncharova S, Gordon ME, Naji L, Flader K, Larche M, Chu DK, Waserman S, McCarry B, Jordana M (2015) Comprehensive metabolomics identifies the alarmin uric acid as a critical signal for the induction of peanut allergy. Allergy 70(5):495–505. https://doi.org/10.1111/all.12579

Kurosaki T, Kometani K, Ise W (2015) Memory B cells. Nat Rev Immunol 15(3):149–159. https://doi.org/10.1038/nri3802

Lafaille JJ, Xiong H, Curotto de Lafaille MA (2012) On the differentiation of mouse IgE(+) cells. Nat Immunol 13 (7):623; author reply 623–624. https://doi.org/10.1038/ni.2313

Laffleur B, Duchez S, Tarte K, Denis-Lagache N, Peron S, Carrion C, Denizot Y, Cogne M (2015) Self-restrained B cells arise following membrane IgE expression. Cell Rep. https://doi.org/10.1016/j.celrep.2015.01.023

Landsverk OJ, Snir O, Casado RB, Richter L, Mold JE, Reu P, Horneland R, Paulsen V, Yaqub S, Aandahl EM, Oyen OM, Thorarensen HS, Salehpour M, Possnert G, Frisen J, Sollid LM, Baekkevold ES, Jahnsen FL (2017) Antibody-secreting plasma cells persist for decades in human intestine. J Exp Med 214(2):309–317. https://doi.org/10.1084/jem.20161590

Lee JB, Chen CY, Liu B, Mugge L, Angkasekwinai P, Facchinetti V, Dong C, Liu YJ, Rothenberg ME, Hogan SP, Finkelman FD, Wang YH (2016) IL-25 and CD4(+) TH2 cells enhance type 2 innate lymphoid cell-derived IL-13 production, which promotes IgE-mediated experimental food allergy. J Allergy Clin Immunol 137(4):1216–1225.e1211–1215. https://doi.org/10.1016/j.jaci.2015.09.019

Linterman MA, Beaton L, Yu D, Ramiscal RR, Srivastava M, Hogan JJ, Verma NK, Smyth MJ, Rigby RJ, Vinuesa CG (2010) IL-21 acts directly on B cells to regulate Bcl-6 expression and germinal center responses. J Exp Med 207(2):353–363. https://doi.org/10.1084/jem.20091738

Liu YJ, Soumelis V, Watanabe N, Ito T, Wang YH, Malefyt Rde W, Omori M, Zhou B, Ziegler SF (2007) TSLP: an epithelial cell cytokine that regulates T cell differentiation by conditioning dendritic cell maturation. Annu Rev Immunol 25:193–219. https://doi.org/10.1146/annurev.immunol.25.022106.141718

Llop-Guevara A, Chu DK, Walker TD, Goncharova S, Fattouh R, Silver JS, Moore CL, Xie JL, O'Byrne PM, Coyle AJ, Kolbeck R, Humbles AA, Stampfli MR, Jordana M (2014) A GM-CSF/IL-33 pathway facilitates allergic airway responses to sub-threshold house dust mite exposure. PLoS One 9(2):e88714. https://doi.org/10.1371/journal.pone.0088714

Locci M, Havenar-Daughton C, Landais E, Wu J, Kroenke MA, Arlehamn CL, Su LF, Cubas R, Davis MM, Sette A, Haddad EK, International AVIPCPI, Poignard P, Crotty S (2013) Human circulating PD-1+CXCR3-CXCR5+ memory Tfh cells are highly functional and correlate with broadly neutralizing HIV antibody responses. Immunity 39(4):758–769. https://doi.org/10.1016/j.immuni.2013.08.031

Luger EO, Wegmann M, Achatz G, Worm M, Renz H, Radbruch A (2010) Allergy for a lifetime? Allergol Int 59(1):1–8. https://doi.org/10.2332/allergolint.10-RAI-0175

MacLeod MK, David A, McKee AS, Crawford F, Kappler JW, Marrack P (2011) Memory CD4 T cells that express CXCR5 provide accelerated help to B cells. J Immunol 186(5):2889–2896. https://doi.org/10.4049/jimmunol.1002955

Manz RA, Thiel A, Radbruch A (1997) Lifetime of plasma cells in the bone marrow. Nature 388 (6638):133–134. https://doi.org/10.1038/40540

Massacand JC, Stettler RC, Meier R, Humphreys NE, Grencis RK, Marsland BJ, Harris NL (2009) Helminth products bypass the need for TSLP in Th2 immune responses by directly modulating dendritic cell function. Proc Natl Acad Sci USA 106(33):13968–13973. https://doi.org/10.1073/pnas.0906367106

Mayuzumi N, Matsushima H, Takashima A (2009) IL-33 promotes DC development in BM culture by triggering GM-CSF production. Eur J Immunol 39(12):3331–3342. https://doi.org/10.1002/eji.200939472

McHeyzer-Williams LJ, Milpied PJ, Okitsu SL, McHeyzer-Williams MG (2015) Class-switched memory B cells remodel BCRs within secondary germinal centers. Nat Immunol 16 (3):296–305. https://doi.org/10.1038/ni.3095

McKinstry KK, Strutt TM, Bautista B, Zhang W, Kuang Y, Cooper AM, Swain SL (2014) Effector CD4 T-cell transition to memory requires late cognate interactions that induce autocrine IL-2. Nat Commun 5:5377. https://doi.org/10.1038/ncomms6377

Mei HE, Wirries I, Frolich D, Brisslert M, Giesecke C, Grun JR, Alexander T, Schmidt S, Luda K, Kuhl AA, Engelmann R, Durr M, Scheel T, Bokarewa M, Perka C, Radbruch A, Dorner T (2015) A unique population of IgG-expressing plasma cells lacking CD19 is enriched in human bone marrow. Blood 125(11):1739–1748. https://doi.org/10.1182/blood-2014-02-555169

Morita R, Schmitt N, Bentebibel SE, Ranganathan R, Bourdery L, Zurawski G, Foucat E, Dullaers M, Oh S, Sabzghabaei N, Lavecchio EM, Punaro M, Pascual V, Banchereau J, Ueno H (2011) Human blood CXCR5(+)CD4(+) T cells are counterparts of T follicular cells and contain specific subsets that differentially support antibody secretion. Immunity 34 (1):108–121. https://doi.org/10.1016/j.immuni.2010.12.012

Moutsoglou DM, Dreskin SC (2016) B cells establish, but do not maintain, long-lived murine anti-peanut IgE(a). Clin Exp Allergy 46(4):640–653. https://doi.org/10.1111/cea.12715

Muto T, Fukuoka A, Kabashima K, Ziegler SF, Nakanishi K, Matsushita K, Yoshimoto T (2014) The role of basophils and proallergic cytokines, TSLP and IL-33, in cutaneously sensitized food allergy. Int Immunol 26(10):539–549. https://doi.org/10.1093/intimm/dxu058

Nakayama T, Hirahara K, Onodera A, Endo Y, Hosokawa H, Shinoda K, Tumes DJ, Okamoto Y (2016) Th2 cells in health and disease. Annu Rev Immunol. https://doi.org/10.1146/annurev-immunol-051116-052350

Noti M, Kim BS, Siracusa MC, Rak GD, Kubo M, Moghaddam AE, Sattentau QA, Comeau MR, Spergel JM, Artis D (2014) Exposure to food allergens through inflamed skin promotes intestinal food allergy through the thymic stromal lymphopoietin-basophil axis. J Allergy Clin Immunol 133(5):1390–1399.e1391–1396. https://doi.org/10.1016/j.jaci.2014.01.021

Noval Rivas M, Burton OT, Oettgen HC, Chatila T (2016) IL-4 production by group 2 innate lymphoid cells promotes food allergy by blocking regulatory T-cell function. J Allergy Clin Immunol 138(3):801–811.e809. https://doi.org/10.1016/j.jaci.2016.02.030

Orgel KA, Duan S, Wright BL, Maleki SJ, Wolf JC, Vickery BP, Burks AW, Paulson JC, Kulis MD, Macauley MS (2017) Exploiting CD22 on antigen-specific B cells to prevent allergy to the major peanut allergen Ara h 2. J Allergy Clin Immunol 139(1):366–369.e362. https://doi.org/10.1016/j.jaci.2016.06.053

Otte M, Mahler V, Kerpes A, Pabst O, Voehringer D (2016) Persistence of the IgE repertoire in birch pollen allergy. J Allergy Clin Immunol 137(6):1884–1887.e1888. https://doi.org/10.1016/j.jaci.2015.12.1333

Panum P (1847) Beobachtungen über das Maserncontagium. Virchows Arch 1(3):492–512. https://doi.org/10.1007/BF02114472

Parham P (2015) The immune system. Garland Science, New York

Paus D, Phan TG, Chan TD, Gardam S, Basten A, Brink R (2006) Antigen recognition strength regulates the choice between extrafollicular plasma cell and germinal center B cell differentiation. J Exp Med 203(4):1081–1091. https://doi.org/10.1084/jem.20060087

Perrier C, Corthesy B (2011) Gut permeability and food allergies. Clin Exp Allergy 41(1):20–28. https://doi.org/10.1111/j.1365-2222.2010.03639.x

Pulendran B, Tang H, Manicassamy S (2010) Programming dendritic cells to induce T(H)2 and tolerogenic responses. Nat Immunol 11(8):647–655. https://doi.org/10.1038/ni.1894

Qi H, Cannons JL, Klauschen F, Schwartzberg PL, Germain RN (2008) SAP-controlled T-B cell interactions underlie germinal centre formation. Nature 455(7214):764–769. https://doi.org/10.1038/nature07345

Qian Y, Wei C, Eun-Hyung Lee F, Campbell J, Halliley J, Lee JA, Cai J, Kong YM, Sadat E, Thomson E, Dunn P, Seegmiller AC, Karandikar NJ, Tipton CM, Mosmann T, Sanz I, Scheuermann RH (2010) Elucidation of seventeen human peripheral blood B-cell subsets and quantification of the tetanus response using a density-based method for the automated identification of cell populations in multidimensional flow cytometry data. Cytometry B Clin Cytom 78(Suppl 1):S69–S82. https://doi.org/10.1002/cyto.b.20554

Ramadani F, Bowen H, Upton N, Hobson PS, Chan YC, Chen JB, Chang TW, McDonnell JM, Sutton BJ, Fear DJ, Gould HJ (2017) Ontogeny of human IgE-expressing B cells and plasma cells. Allergy 72(1):66–76. https://doi.org/10.1111/all.12911

Sallusto F, Lanzavecchia A (2009) Heterogeneity of CD4+ memory T cells: functional modules for tailored immunity. Eur J Immunol 39(8):2076–2082. https://doi.org/10.1002/eji.200939722

Sallusto F, Lenig D, Forster R, Lipp M, Lanzavecchia A (1999) Two subsets of memory T lymphocytes with distinct homing potentials and effector functions. Nature 401(6754):708–712. https://doi.org/10.1038/44385

Savage J, Sicherer S, Wood R (2016) The natural history of food allergy. J Allergy Clin Immunol Pract 4 (2):196–203; quiz 204. https://doi.org/10.1016/j.jaip.2015.11.024

Schenkel JM, Masopust D (2014) Tissue-resident memory T cells. Immunity 41(6):886–897. https://doi.org/10.1016/j.immuni.2014.12.007

Schluns KS, Lefrancois L (2003) Cytokine control of memory T-cell development and survival. Nat Rev Immunol 3(4):269–279. https://doi.org/10.1038/nri1052

Seder RA, Ahmed R (2003) Similarities and differences in CD4+ and CD8+ effector and memory T cell generation. Nat Immunol 4(9):835–842. https://doi.org/10.1038/ni969

Shreffler WG, Castro RR, Kucuk ZY, Charlop-Powers Z, Grishina G, Yoo S, Burks AW, Sampson HA (2006) The major glycoprotein allergen from Arachis hypogaea, Ara h 1, is a ligand of dendritic cell-specific ICAM-grabbing nonintegrin and acts as a Th2 adjuvant in vitro. J Immunol 177(6):3677–3685

Slifka MK, Antia R, Whitmire JK, Ahmed R (1998) Humoral immunity due to long-lived plasma cells. Immunity 8(3):363–372

Stampfli MR, Wiley RE, Neigh GS, Gajewska BU, Lei XF, Snider DP, Xing Z, Jordana M (1998) GM-CSF transgene expression in the airway allows aerosolized ovalbumin to induce allergic sensitization in mice. J Clin Invest 102(9):1704–1714. https://doi.org/10.1172/JCI4160

Stary G, Olive A, Radovic-Moreno AF, Gondek D, Alvarez D, Basto PA, Perro M, Vrbanac VD, Tager AM, Shi J, Yethon JA, Farokhzad OC, Langer R, Starnbach MN, von Andrian UH (2015) VACCINES. A mucosal vaccine against Chlamydia trachomatis generates two waves of protective memory T cells. Science 348(6241):aaa8205. https://doi.org/10.1126/science.aaa8205

Tada Y, Asahina A, Nakamura K, Tomura M, Fujiwara H, Tamaki K (2000) Granulocyte/macrophage colony-stimulating factor inhibits IL-12 production of mouse Langerhans cells. J Immunol 164(10):5113–5119

Takeshita M, Suzuki K, Kassai Y, Takiguchi M, Nakayama Y, Otomo Y, Morita R, Miyazaki T, Yoshimura A, Takeuchi T (2015) Polarization diversity of human CD4+ stem cell memory T cells. Clin Immunol 159(1):107–117. https://doi.org/10.1016/j.clim.2015.04.010

Talay O, Yan D, Brightbill HD, Straney EE, Zhou M, Ladi E, Lee WP, Egen JG, Austin CD, Xu M, Wu LC (2012) IgE(+) memory B cells and plasma cells generated through a germinal-center pathway. Nat Immunol 13(4):396–404. https://doi.org/10.1038/ni.2256

Tang H, Cao W, Kasturi SP, Ravindran R, Nakaya HI, Kundu K, Murthy N, Kepler TB, Malissen B, Pulendran B (2010) The T helper type 2 response to cysteine proteases requires dendritic cell-basophil cooperation via ROS-mediated signaling. Nat Immunol 11(7):608–617. https://doi.org/10.1038/ni.1883

Taylor JJ, Pape KA, Jenkins MK (2012) A germinal center-independent pathway generates unswitched memory B cells early in the primary response. J Exp Med 209(3):597–606. https://doi.org/10.1084/jem.20111696

Thyssen JP, Maibach HI (2014) Filaggrin: basic science, epidemiology, clinical aspects and management. Springer, Berlin

Tooze RM (2013) A replicative self-renewal model for long-lived plasma cells: questioning irreversible cell cycle exit. Front Immunol 4:460. https://doi.org/10.3389/fimmu.2013.00460

Tordesillas L, Goswami R, Benede S, Grishina G, Dunkin D, Jarvinen KM, Maleki SJ, Sampson HA, Berin MC (2014) Skin exposure promotes a Th2-dependent sensitization to peanut allergens. J Clin Invest 124(11):4965–4975. https://doi.org/10.1172/JCI75660

Traidl-Hoffmann C, Mariani V, Hochrein H, Karg K, Wagner H, Ring J, Mueller MJ, Jakob T, Behrendt H (2005) Pollen-associated phytoprostanes inhibit dendritic cell interleukin-12 production and augment T helper type 2 cell polarization. J Exp Med 201(4):627–636. https://doi.org/10.1084/jem.20041065

Turqueti-Neves A, Otte M, Schwartz C, Schmitt ME, Lindner C, Pabst O, Yu P, Voehringer D (2015) The extracellular domains of IgG1 and T cell-derived IL-4/IL-13 are critical for the polyclonal memory IgE response in vivo. PLoS Biol 13(11):e1002290. https://doi.org/10.1371/journal.pbio.1002290

Ugur M, Schulz O, Menon MB, Krueger A, Pabst O (2014) Resident CD4+ T cells accumulate in lymphoid organs after prolonged antigen exposure. Nat Commun 5:4821. https://doi.org/10.1038/ncomms5821

Vannella KM, Ramalingam TR, Borthwick LA, Barron L, Hart KM, Thompson RW, Kindrachuk KN, Cheever AW, White S, Budelsky AL, Comeau MR, Smith DE, Wynn TA (2016) Combinatorial targeting of TSLP, IL-25, and IL-33 in type 2 cytokine-driven inflammation and fibrosis. Sci Transl Med 8(337):337ra365. https://doi.org/10.1126/scitranslmed.aaf1938

Weisel F, Shlomchik M (2017) Memory B cells of mice and humans. Annu Rev Immunol. https://doi.org/10.1146/annurev-immunol-041015-055531

Willart MA, Deswarte K, Pouliot P, Braun H, Beyaert R, Lambrecht BN, Hammad H (2012) Interleukin-1alpha controls allergic sensitization to inhaled house dust mite via the epithelial release of GM-CSF and IL-33. J Exp Med 209(8):1505–1517. https://doi.org/10.1084/jem.20112691

Williams MA, Ravkov EV, Bevan MJ (2008) Rapid culling of the CD4+ T cell repertoire in the transition from effector to memory. Immunity 28(4):533–545. https://doi.org/10.1016/j.immuni.2008.02.014

Winter O, Dame C, Jundt F, Hiepe F (2012) Pathogenic long-lived plasma cells and their survival niches in autoimmunity, malignancy, and allergy. J Immunol 189(11):5105–5111. https://doi.org/10.4049/jimmunol.1202317

Wisniewski JA, Commins SP, Agrawal R, Hulse KE, Yu MD, Cronin J, Heymann PW, Pomes A, Platts-Mills TA, Workman L, Woodfolk JA (2015) Analysis of cytokine production by peanut-reactive T cells identifies residual Th2 effectors in highly allergic children who received peanut oral immunotherapy. Clin Exp Allergy 45(7):1201–1213. https://doi.org/10.1111/cea.12537

Xiong H, Dolpady J, Wabl M, Curotto de Lafaille MA, Lafaille JJ (2012) Sequential class switching is required for the generation of high affinity IgE antibodies. J Exp Med 209(2):353–364. https://doi.org/10.1084/jem.20111941

Yang Z, Sullivan BM, Allen CD (2012) Fluorescent in vivo detection reveals that IgE(+) B cells are restrained by an intrinsic cell fate predisposition. Immunity 36(5):857–872. https://doi.org/10.1016/j.immuni.2012.02.009

Yang Z, Robinson MJ, Allen CD (2014) Regulatory constraints in the generation and differentiation of IgE-expressing B cells. Curr Opin Immunol 28:64–70. https://doi.org/10.1016/j.coi.2014.02.001

Yang Z, Robinson MJ, Chen X, Smith GA, Taunton J, Liu W, Allen CD (2016) Regulation of B cell fate by chronic activity of the IgE B cell receptor. Elife 5. https://doi.org/10.7554/eLife.21238

Yoshida T, Mei H, Dorner T, Hiepe F, Radbruch A, Fillatreau S, Hoyer BF (2010) Memory B and memory plasma cells. Immunol Rev 237(1):117–139. https://doi.org/10.1111/j.1600-065X.2010.00938.x

Zuccarino-Catania GV, Sadanand S, Weisel FJ, Tomayko MM, Meng H, Kleinstein SH, Good-Jacobson KL, Shlomchik MJ (2014) CD80 and PD-L2 define functionally distinct memory B cell subsets that are independent of antibody isotype. Nat Immunol 15(7):631–637. https://doi.org/10.1038/ni.2914

The Role of the Gut in Type 2 Immunity

Caspar Ohnmacht

Abstract Allergic and autoimmune disorders have been on the raise in the last decades in westernized countries while infectious diseases could be dramatically reduced due to efficient vaccination campaigns, improvement of personal hygiene and use of medication. Alteration of immunological tolerance may be implicated in both autoimmune and allergic inflammation yet underlying mechanisms remain poorly understood. Microbes colonizing all barrier sites are now known to have strong impact on our general immune status. The intestinal tract harbors the densest community of microbes and in combination with a huge surface area, the micro-environment in the gut implies strong tolerogenic properties. While a healthy symbiosis between intestinal microbes and the immune system is beneficial for the host, dysbiosis of microbial communities can heavily impact on immune responses and immunological tolerance. A systemic effect on the general susceptibility to allergic and autoimmune disorders is difficult to study and remains controversial. As an alternative explanation, allergic sensibilisation via the skin or lung is thought to be an important pathway but lacks the capability to explain the increase of allergic disorders on an epidemiological level. In this chapter, I will highlight a possible impact of a beneficial host-microbiota relationship in both the intestinal tract and other mucosal surfaces.

1 Introduction

The gut harbors the largest contact area to our immediate environment (up to 200 qm^2) and is the organ with the highest colonization density by commensal microbes. These microbes including bacteria, fungi and viruses are known to have a fundamental impact on the host physiology and the immune system. This so-called

C. Ohnmacht
Center of Allergy and Environment (ZAUM), Technical University Munich and Helmholtz
Zentrum Munich, Munich, Germany
e-mail: caspar.ohnmacht@helmholtz-muenchen.de

© Springer International Publishing AG 2017   145
C.B. Schmidt-Weber (ed.), *Allergy Prevention and Exacerbation*, Birkhäuser
Advances in Infectious Diseases, https://doi.org/10.1007/978-3-319-69968-4_8

microbiota co-evolved with its host in all vertebrates and non-vertebrates for a very long time and has established a mutual beneficial symbiosis. The concomitant emergence of genome-encoded and somatically recombined receptors of the innate and adaptive immune system enabled probably an even more fine-tuned symbiosis between the host and its inner self. Noteworthy, the immune system is constantly influenced by the composition of this microbiota and at the same time also impacts on this composition mainly through the secretion of anti-microbial peptides, defensines and immunoglobulin A. Given the high density of microbes and foreign antigens (both food- and microbe-derived) as well as the presence of potential danger ligands, e.g. TLR ligands, the symbiosis of host and microbiota is based on both passive and active tolerance mechanisms. In recent years, it became clear that a loss of tolerance to symbiotic microbes can have fatal consequences and result in chronic inflammatory disorders in the intestinal tract (e.g. Colitis Ulcerosa). Furthermore, a number of recent epidemiological studies indicate a strong association between the risk of allergic disorders and different lifestyle factors. Most of these studies found an association between the presence of different microbial patterns in the environment or directly at different body sites early in life. This finding implicates that the immune system is particularly vulnerable to environmental influences in a critical time window when the immune system is built up, immune memory is formed and a stable host-microbiota is established. Studies have shown that at younger age the intestinal microbiota shows a high degree of variability until a stable community is formed. Once this community is established it can only be disturbed upon massive antibiotic treatment (Lozupone et al. 2012; Ubeda and Pamer 2012).

The establishment of the microbiota also includes the induction of antigen-specific regulatory T cells (Tregs) and accordingly, microbial colonization induces a steep increase in the frequency of Tregs (Atarashi et al. 2011). It is generally believed that allergens in non-susceptible people are either ignored by the adaptive immune system, induce a state of non-responsiveness by inducing antibody isotypes that provide negative signals upon FcR binding or directly induce Tregs that in turn negatively regulate the accumulation of allergen-specific Th2 cells. Given the highly tolerogenic environment in the gut one possible explanation for enhanced susceptibility to allergic disorders may be the loss of a beneficial host-microbiota relationship in the gut and a subsequent general state of lower immune tolerance even at distant barrier sites. This may occur due to factors associated to a 'westernized' life style (e.g. chemicals, diet or altered oral exposure to microbes in childhood) leading to a non-healthy microbiota and predispose individuals for allergic disorders that may manifest later in life at different body sites. As an alternative explanation, other barrier sites such as the lung or the skin may be at the origin of high susceptibility to allergic disorders. In favor of this theory various cytokines that have been linked to the beginning of the allergenic cascade (e.g. TSLP, IL-33 or IL-25) are expressed at high levels at all barrier sites either at steady state or upon injury. A lot of current research efforts have been focused on the role of Tregs and I will illustrate how microbes and Tregs are thought to affect each other and how this may impact type 2 immunity.

2 The Gut: Site of Tolerance or Induction of Allergic Disorders?

As mentioned above the gut is the organ with the highest load of bacterial coloni-zation and at the same time harbors the highest frequencies of Forkhead box P3-expressing Tregs (Foxp3$^+$ Tregs). In general, Tregs can either differentiate in the thymus (tTregs) or be induced in the periphery (iTregs) and the latter are thought to play a key role for the tolerance of colonizing microbes but possibly also regulate distinct inflammatory responses (Ohnmacht 2016; Tanoue et al. 2016). In light of the frequent migration of cells to and from the gut to distal organs (Morton et al. 2014) it seems very likely that T cells integrating signals from the intestinal microenvironment may have a strong impact at other sites.

2.1 The Fundamental Role of Microbes

The most striking results describing a role of the microbiota for protection from type 2 immune disorders are derived from the analysis of germfree or antibiotic-treated mice. Germfree mice do have an intrinsic bias to mount stronger type 2 immune responses in many different disease models. The absence of microbes has a tremendous impact on immune maturation that may advice caution in the interpretation of such studies. During the last decade it became obvious that microbial colonization is tightly associated with a sharp increase of Tregs homing to the intestinal lamina propria in both the small and the large bowel. Tregs are thought to directly counteract Th2 cells and it is believed that the ratio between Treg and Th2 cells decides whether an allergy will develop or not. Experimental models have revealed now some factors on both the microbial and the host side that play an important role in the tight balance between Th2 and Tregs and therefore with a possible risk of allergic predisposition.

Germfree or antibiotic treated mice show increased sensitization to food aller-gens and this can be partially attributed to a reduced barrier function of the intestinal epithelium (Stefka et al. 2014). Similarly, allergic airway responses in such mice show a more severe phenotype (Herbst et al. 2011; Hill et al. 2012; Russell et al. 2012). Different antibiotic treatments have a selective effect on the induction of allergic airway inflammation and hypersensitivity pneumonitis arguing for microbial-specific effects in prevention of allergic airway inflammation (Russell et al. 2012, 2015). The protective effect of microbes may be at least partly attributed to production of short chain fatty acids (SCFA) from fermentation of dietary fibers (Trompette et al. 2014). SCFA bind to G-protein coupled receptors and exert multiple functions on the immune system. Most importantly, SCFA have been shown to directly promote the development of intestinal regulatory T cells through inhibition of histone deacetylases and consequently enhanced acetylation of the *foxp3* locus. At the same time SCFA can limit the maturation and activation of dendritic cells that results also in increased differentiation and accumulation of intestinal Tregs (Furusawa et al. 2013; Arpaia et al. 2013; Singh et al. 2010). The SCFA propionate was also shown to impact hematopoiesis of dendritic cell precursors and increased generation of phagocytic dendritic cells in the lung expressing reduced CD40 and MHC-II molecules but more FcεRI (Trompette et al. 2014). Human FcεRI$^+$ DCs in humanized mouse models have been associated with tolerogenic function in allergic disorders (Platzer et al. 2015) while murine FcεRI$^+$ DCs are rather associated with a pro-inflammatory role for the initiation and maintenance of allergic inflammation (Plantinga et al. 2013; Hammad et al. 2010).

Besides microbes, also parasites in the gut can substantially impact on the host's immune system. One beneficial immunomodulatory role of intestinal parasite infections on allergic inflammation relies on the manipulation of the intestinal microbiota leading to higher SCFA levels and subsequent secretion of the immunomodulatory cytokine IL-10 in the lungs of infected mice (Zaiss et al. 2015). Furthermore, the intestinal parasite *Heligmosomosoides polygyrus* is able to directly manipulate immunological tolerance through secretion of a TGF-β-like molecule resulting directly in the induction of Foxp3$^+$ Tregs (Wilson et al. 2005). These parasite-induced Tregs have been shown to protect from allergic airway inflammation even though the question of how antigen-specificity is integrated in such a general mechanism remains to be discovered.

In addition to effects on allergic inflammation, germfree mice have also been shown to react more drastically in a murine model of ulcerative colitis (Olszak et al. 2012) which is also dependent on typical Th2-related cytokines (Boirivant et al. 1998). Perhaps most strikingly, germfree mice show increased hematopoiesis of basophil progenitors and develop spontaneously increased levels of serum IgE possibly due to IgE class switch recombination preferentially in gut-associated lymphoid tissues (Hill et al. 2012; Cahenzli et al. 2013). Interestingly, this phenomenon is reversible until a certain age by microbial colonization illustrating a key role of microbes for the prevention of high IgE titers in the gut at younger age (Cahenzli et al. 2013). Even a transient alteration of microbial colonization and associated metabolic alterations at younger age may result in an increased risk allergic asthma since reduced microbes from stool samples of babies which will develop in the future signs of allergic asthma are able to ameliorate allergic airway inflammation in a mouse model (Arrieta et al. 2015). Altogether, a general absence of microbial colonization clearly shifts the immune system to a bias in which type 2 immunity is dominating and may therefore enhance different parameters of allergy.

As mentioned before microbial colonization has a massive impact on the intestinal immune system and induces a steep increase in the frequency of Tregs (Atarashi et al. 2011; Geuking et al. 2011). Similarly, oral exposure to allergens leads to the *de novo* induction of Tregs (iTregs) that are able to regulate allergic airway inflammation in an antigen-specific manner even in the absence of naturally occurring Tregs (Curotto de Lafaille et al. 2008; Mucida et al. 2005). tTregs differentiate in the thymus and are probably selected on the basis of self antigen recognition. The different origin of Tregs may be useful to understand some properties of tTregs and iTregs—the latter probably playing a major role for the regulation of allergic disorders—but one may keep in mind that both subsets act together and that Treg origin does not necessarily reflect origin of antigen (Legoux et al. 2015). Some aspects of the allergic cascade e.g. the accumulation of eosinophils can also be counteracted by the induction of IFN-γ after contact to the allergen (Curotto de Lafaille et al. 2008). This mechanism is also exploited by the subcutaneous allergen immune therapy (Secrist et al. 1993; Varney et al. 1993; Durham et al. 1996) but it remains to be discovered whether a similar mechanism constantly occurs in the gut under physiological conditions or whether oral exposure uniquely induces professional regulatory cells such as iTregs.

2.2 Regulatory T Cells: Not Only Foxp3

Certain bacterial types are particularly effective to prevent sensitization to food allergens (Stefka et al. 2014) and in the induction of Foxp3$^+$ Tregs including *Clostridia* Cluster IV, XIVa and XVIII and (Atarashi et al. 2011, 2013) and *Bacteroides fragilis* (Round and Mazmanian 2010). Importantly, colonization with *Clostridium* was shown to limit the secretion of IL-4 and induction of IgE while increasing the generation of IL-10 from restimulated splenocytes after induction of a systemic sensitization with the adjuvant Alum (Atarashi et al. 2011). Therefore, a systemic influence of microbial-induced Tregs may also play a role in the regulation of type 2 immune responses. Given that foreign antigens are presumably absent during T cell selection in the thymus [even though exceptions from this rule may be possible (Hadeiba et al. 2012)], it is not surprising that iTregs are an essential component to establish a beneficial microbiota-host relationship in the gut (Geuking et al. 2011). Interestingly, Tregs have been shown to contribute to microbial diversification in the intestinal tract via regulation inflammatory responses and IgA production resulting in intestinal homeostasis (Kawamoto et al. 2014). However, when iTreg differentiation is prevented through the knockout of the CNS1 region next to the *Foxp3* promoter, iTreg-deficient mice develop a spontaneous type 2 inflammation at mucosal barrier sites including high IgE levels and sensitization to known food antigens (Josefowicz et al. 2012b). A fairly high percentage of intestinal Tregs has a TCR repertoire that recognizes microbial-derived antigens (Lathrop et al. 2011) and constant signaling via the TCR has been shown to be essential for maintenance of the Treg pool and their function (Levine et al. 2014, 2017; Vahl et al. 2014). Interestingly, iTregs in the gut are able to transdifferentiate into CD4$^+$ intraepithelial cells in a microbiota-dependent manner where they exert complementary roles with iTregs to mediate intestinal tolerance (Sujino et al. 2016). Thus, iTregs may constantly receive signals from the intestinal microbiota—a process that seems to be necessary for their functional capacity to adequately react to changes in microbial compositions.

Although Tregs in general have been known to prevent excessive inflammation of type 2 immune responses (Lin et al. 2005), distinct Treg subpopulations and their associated molecular mechanisms draw the attention only recently. For instance, the specific knockout of IRF4 in Foxp3$^+$ Tregs was shown to result in a systemic autoinflammation dominated by increased accumulation of Th2 cells (Zheng et al. 2009). Since then, additional transcription factors have been identified that control the fate decision of naïve T cells between Treg- and Th2 differentiation programs: The inflammasome component Nlrp3 but not other inflammasome members binds together with IRF4 to the *Il4* promoter to transactivate this locus and may thus serve as a positive feedback loop during Th2 differentiation (Bruchard et al. 2015). In contrast, Musculin suppresses the Th2 transcriptional program by preventing the binding of Gata3 to Th2-cell-related cytokines and this process is essential for the proper differentiation of iTregs (Wu et al. 2017). In line with CNS1-deficient mice [(Josefowicz et al. 2012a), see below], Musculin-deficient mice develop with a certain age a spontaneous Th2-dominated inflammation in the gut and the lung (Wu et al. 2017). A similar role was recently described for the autophagy gene *Atg16l1*: In the absence of this gene, iTregs do have a selective disadvantage in survival and mice with a loss of Atg16l1 in T cells or Tregs develop a spontaneous Th2-driven intestinal inflammation (Kabat et al. 2016).

IRF4 is also among the transcription factors that have been shown to play a pivotal role in DCs capable to induce Th2-dominated immune responses (Williams et al. 2013; Gao et al. 2013; Kumamoto et al. 2013). Yet IRF4 is not unique in its ability to confer Th2-induction potential to DC subsets but also regulates the induction of Th17 immune responses. The DC potential is additionally governed by factors such as KLF4 that enables IRF4-expressing DCs to mount full-blown Th2 immune responses (Tussiwand et al. 2015). DCs and Tregs have been shown to strongly interdepend and regulate the function of each other (Darrasse-Jeze et al. 2009). One example relevant for the regulation of type 2 immunity is the knockout of the kinase CK2 in Tregs. Such mice develop a spontaneous inflammation of the lung that is dominated by the accumulation of Th2 cells (Ulges et al. 2015). Tregs deficient for CK2 were shown to be unable to regulate IRF4$^+$ DCs and may thus cause DC phenotypes associated with type 2 inflammations. A second example for the interaction of Tregs and DCs in the context of Th2-driven inflammation is the specific deletion of CTLA-4 in Tregs: Such mice develop a spontaneous autoinflammation characterized by accumulation of Th2 cells and high serum IgE levels (Wing et al. 2008). CTLA-4 expression may be one of the mechanisms by which Tregs exert their regulatory function on T effector cells but at the same time also limit the expression of costimulatory receptors on DCs (Wing et al. 2008).

Another striking example for the essential role of DCs in the Th2/Treg balance is the conditional knockout of TRAF6 in DCs: Such mice develop a spontaneous Th2-domianted inflammation in the gut and show reduced tolerance induction to model antigens as evidence by reduced *de novo* Treg induction (Han et al. 2013). One possible explanation highlighted by the authors was the reduced Il-2 secretion by TRAF6-deficient DCs. IL-2 is known to have an important role on Treg homeostasis in general (Chinen et al. 2016; Fontenot et al. 2005). In line with this concept, classical DCs were shown to be necessary and sufficient to induce oral tolerance to model allergens (Esterhazy et al. 2016).

Originally, the expression of master transcription factors was associated with distinct T helper lineages. More recently, the expression of those transcription factors has been shown to regulate the function of Tregs particularly in the periphery. For instance, normal microbial colonization induces the appearance of a subset of Foxp3$^+$ Tregs expressing the transcription factor ROR(γt) in the lamina propria (Sefik et al. 2015; Ohnmacht et al. 2015; Lochner et al. 2011). Expression of ROR(γt) in peripheral T helper cells has been previously associated rather with Th17 cells that are induced particularly by commensals that adhere to the intestinal epithelium (Ivanov et al. 2006, 2009; Atarashi et al. 2015). The specific knockout of ROR(γt) in Tregs regulates the degree of type 2 immune responses after infection with *H. polygyrus* and in a model of Colitis Ulcerosa (Ohnmacht et al. 2015). This surprising counterregulation of type 2 inflammation by 'type 3' Tregs was recently confirmed both at steady state and in a model of systemic lupus erythematosus (Kluger et al. 2017). Interestingly, different and non-related bacterial species are able to induce the accumulation of ROR(γt)$^+$ Tregs which may underline the importance and notably the evolutionary stability of this immune-microbe crosstalk (Geva-Zatorsky et al. 2017; Sefik et al. 2015). Thus, ROR(γt)$^+$ cells and particularly ROR(γt)$^+$ Tregs may have a key role in establishing effective tolerance towards physiological colonization of the intestinal tract (Ohnmacht 2016).

In man, the Hyper-IgE syndrome may be linked to a failure to generate these 'type 3' Tregs: Different mutations in man affect the STAT3 pathway and some of these Hyper-IgE patients specifically lack the CCR6$^+$ subsets of Tregs (Kluger et al. 2014). In the murine system, STAT3-deficiency reduces the frequency of ROR(γt)$^+$ Tregs and ROR(γt)$^+$ Tregs show high expression of CCR6 (Kluger et al. 2014; Ohnmacht et al. 2015; Sefik et al. 2015). These ROR(γt)$^+$ Tregs develop as a consequence of a contact to harmless microbes and in their absence immune responses to microbes can not be adequately regulated and become exaggerated (Sefik et al. 2015; Barthels et al. 2017; Yang et al. 2016). Thus, a failure of iTreg generation or maintenance does not always result *per se* in a type 2 immune bias but only in combination with a certain microbial composition.

The prototypic transcription factor driving Th2 differentiation is Gata3 and its initial IL-4-induced expression during T cell priming was demonstrated to limit the differentiation of induced Tregs (Mantel et al. 2007; Wu et al. 2017). Note-worthy, a subset of Tregs co-expressed the transcription factors Foxp3 and Gata3 in the skin and the intestinal tract (Wohlfert et al. 2011). Gata3 expression in Tregs was shown to depend on the ST2 receptor (the receptor for IL-33) signaling and was suggested to enable the maintenance of Tregs under inflammatory conditions (Wohlfert et al. 2011; Schiering et al. 2014). Interestingly, IL-33 is also thought to be at the origin of the allergic cascade by activating and inducing the proliferation of innate lymphoid cells type 2 (ILC2s) (de Kleer et al. 2016; Barlow et al. 2013; Bartemes et al. 2012). Nevertheless, IL-33 has also a considerable impact on Tregs and injection of recombinant IL-33 leads to the accumulation of ST2/Gata3$^+$ Tregs (Chen et al. 2017). Such Tregs have been shown to express prototypic type 2 cytokines (Il-4, IL-5 and IL-13) yet ST2-expressing Tregs are fully capable of suppressing proliferation of naïve T helper cells (Chen et al. 2017; Siede et al. 2016). ST2/Gata3-expressing cells are particularly abundant in a murine model of a human mutation in the IL-4R leading to enhanced signaling in such mice (IL-4Ra^{F709} mice) and have therefore been termed 'Th2-like Tregs' (Noval Rivas et al. 2015). In line with their endogenous bias towards Th2 cells (Siede et al. 2016) Tregs isolated from IL-4Ra^{F709} mice with food allergy may even be able to transdifferentiate towards Th2 cells. Additionally, the conditional knockout of Th2 pathways in Tregs protects such mice from excessive food allergy demonstrating the pathological role of these Tregs in this model (Noval Rivas et al. 2015). Similarly, allergen tolerance in the lungs can be broken upon dysregulated IL-33 exposure in combination with contact to the respective allergen at least partially due to secretion of typical Th2 cytokines by Tregs (Chen et al. 2017). When IL-4Ra^{F709} mice are challenged with a model of allergic airway inflammation, ablation of *Rorc* [the gene encoding for ROR(γt)] specifically in Tregs, the exaggerated allergic inflammation including high IL-17 levels observed in the lungs of IL-4Ra^{F709} mice is attenuated (Massoud et al. 2016). Thus, 'type 3 Tregs'—even though in general confined to the intestinal tract (Ohnmacht et al. 2015; Sefik et al. 2015; Yang et al. 2016)—may also contribute to inflammation under some circumstances particularly in cases of mixed Th2/Th17-driven allergic immune responses.

Gata3$^+$ and ROR(γt)$^+$ Tregs are mutually exclusive (Wohlfert et al. 2011; Ohnmacht et al. 2015) and it is interesting to note that IL-23 can reduce the impact of IL-33 on Tregs in a cell intrinsic manner (Schiering et al. 2014). In addition, Gata3$^+$ Tregs developing under physiological conditions are not dependent on the microbiota and may rather originate from the thymus (Ohnmacht et al. 2015). This implies that they have been initially selected on the basis of self-antigen recognition. In line with this concept, ablation of Gata3 alone or in combination with T-bet in Tregs has been associated with enhanced Th17-dominated inflammation and autoimmune features (Wang et al. 2011; Yu et al. 2014).

In analogy to the link of some Hyper-IgE patients with a lack of CCR6$^+$ Tregs (Kluger et al. 2014), Gata3$^+$ Tregs have been associated to failed oral tolerance. Patients suffering from mutation in WASP (Wiskott-Aldrich Syndrome Protein) show high titers of IgE that are able to recognize different food antigens (Ozcan et al. 2008). In a murine model, the conditional ablation of WASP in Tregs was able to phenocopy this intestinal Th2 bias and this was again accompanied by an increase of Gata3$^+$ Tregs in gut-draining secondary lymphoid organs (Lexmond et al. 2016). Whether other Treg subsets are also affected in WASP-deficient mice has not yet been addressed but the development of food antigen specific IgE titers was shown to be independent of the microbiota.

Microbial antigens are able to trigger a subset of iTregs responsible to limit a spontaneous intestinal type 2 immune bias observed in germfree mice (Josefowicz et al. 2012b; Ohnmacht et al. 2015; Cahenzli et al. 2013). This spontaneous type 2 inflammation that develop without any 'artificial' trigger may thus be the most extreme but nevertheless may illustrate how a delicate microbial-host relationship sets the rheostat to prevent a systemic 'pro-allergic' immune bias. Interestingly, when both microbial and food antigens are absent, the intestinal immune system shows a bias towards Th1-driven inflammation including the accumulation of high amounts of T-bet expressing Tregs (Kim et al. 2016). In summary, intestinal microbes are particularly effective in regulating Th2-driven inflammation while food antigen-induced Tregs—even though mostly the cause for allergic responses in the gut—may rather control Th1-driven immune responses.

3 Microbial Impact on Allergy on an Epidemiological Perspective

A variety of epidemiologic studies have addressed the question how microbes impact the risk of different types of allergic disorders. One of the pioneer works by von Mutius described that children growing up on a farm are largely protected from the later development of childhood asthma (Ege et al. 2011). This protection correlated with the diversity of bacteria and fungi that could be sampled in the staples of this protective farm environment. Later on, this was confirmed in studies with Amish and Hutterite farming populations that still have a very traditional lifestyle (Stein et al. 2016). Again, the traditional lifestyle was protective for allergic disorders and correlated with increased bacterial diversity found in the dust of the Amish homes. Importantly, LPS as one of the main TLR ligands impacting predominantly innate immune cells was present at much higher levels in the homes of the Amish people and protected excessive inflammation in a murine model of allergic asthma (Stein et al. 2016). The importance of microbial components is further illustrated by the observation that defined exposure of purified LPS via the airways protects from allergic airway disease (Schuijs et al. 2015). This protection is at least partially mediated by the gene *Tnfaip3* (encoding for the NFκB limiting ubiquitinase A20) in airway epithelial cells (Schuijs et al. 2015). Importantly, SNPs of *Tnfaip3* have been linked with an enhanced risk of allergic airway disease in humans. Thus, limiting exposure to endotoxin and other bacterial ligands that were initially thought to have a unique role for the induction of immune responses may be a risk factor for the later development of atopic diseases. Given the high bacterial load in the gut one may speculate that exposure to such bacterial ligands may also contribute to the education of the immune system. Indeed, a cross-country study in Finland, Estonia and Russia revealed marked differences in microbial compositions in the feces of young children that was associated with a high risk of autoimmune disorders and allergies (Vatanen et al. 2016). Interestingly, genes encoding for Lipid A and LPS synthesis were among the most differently expressed genes in stool samples of children that were protected from the later development of allergies or type 1 diabetes (Vatanen et al. 2016). Different structures of LPS encoded by different bacterial strains correlated with immunogenicity or tolerogenic properties in both human and mice (Vatanen et al. 2016).

Other studies assessed retrospectively whether bacterial communities in the feces of pre-atopic and healthy children differed in their composition. Indeed, Finlay and colleagues found differences in the microbial composition and metabolic profile of such children and correlated these differences with the later development of wheeze and atopy (Arrieta et al. 2015). In a separate study, neonatal microbial composition could be categorized into different risk categories for the later development of multisensitized atopy (Fujimura et al. 2016). Dysbiotic microbial communities associated with a high-risk for atopy produced metabolites that were able to induce higher frequencies of IL-4[+] T helper cells while limiting the induction of Tregs (Fujimura et al. 2016).

Further evidence for an important role of the gut is based on the idea that oral exposure to allergens induces a state of tolerance (also known as oral tolerance) (Julia et al. 2015). This oral tolerance can be established in experimental models by allergens present in breast milk (Verhasselt et al. 2008) and children that have been breastfed are thought to have an overall lower risk for asthma and certain allergies (Lodge et al. 2015). While most of the mentioned studies focused on an impact of the intestinal tract on allergic airway diseases or the skin there is now direct evidence that early exposure to dietary allergens can prevent the development of food allergies: Feeding defined amounts of peanut extracts to young children at high risk of peanut allergy prevented the later development of allergic reactions to peanuts while prevention of contact to peanuts resulted in high sensitization rates (Du Toit et al. 2015). This protection via the oral route is clearly not possible for all food allergens but it illustrates a critical time window to establish oral tolerance towards common allergens.

4 Non-intestinal Barrier Sites at the Origin of Allergic Disorders

In principal, all anatomical sites ever exposed to the diverse array of known and unknown allergens may be able to contribute to allergic sensitization. One currently discussed question is which anatomical site—and therefore which environmental factor with a dominant impact on the microenvironment at this site—could be responsible for a systemic influence of the immune system and thus contribute to allergic predisposition of a subject. The answer of this question may hold the key to better understand protecting factors within our closest environment. Besides the gut, obviously other barrier sites with a huge surface area in contact to the environment such as the lung or the skin may be important. One of the best-studied organs is the lung due to the high numbers of patients suffering from allergic asthma. For a very long time it was thought that the lower airways are sterile and not influenced by microbes. However, Marsland and colleagues showed in murine models that very young animals develop a more drastic allergic immune response in the lung compared to adult mice and this was associated to tolerogenic DC function and subsequent induction of iTregs in the lungs (Gollwitzer et al. 2014).

In humans, harmless aeroallergens (e.g. house dust mite, fungal spores or pollen) have been recently shown to elicit the expected ratios of tolerogenic Tregs that are fully capable of immune suppression (Bacher et al. 2016). However, allergenic Th2 responses are still induced probably due to divergent antigen specificities as a result of allergen dissociation from a particulate aeroallergen (Bacher et al. 2016). This study therefore highlights the important role of antigen specificity for Treg-mediated suppression of allergic responses: Testing Treg specificity on crude allergen extracts containing several potential allergens may not be sufficient to prove intact Treg function. This concept is in line with data from murine studies highlighting a critical role of the TCR and therefore also specificity for full Treg function (Vahl et al. 2014; Levine et al. 2014). A microbial impact on general allergic predisposition is therefore most likely not antigen specific but acts via metabolic compounds or small molecular compounds indirectly in favor of a general easier antigen specific induction of Tregs as a secondary consequence e.g. of innate immune cells.

The human skin harbors substantial frequencies of Tregs and these Tregs have an activated/memory phenotype (Sanchez Rodriguez et al. 2014). Hair follicles are known to harbor a big fraction of the cutaneous microbiome (Lange-Asschenfeldt et al. 2011) and one may speculate that a considerable fraction of these activated/ memory Tregs promote tolerance against these harmless microbes. Indeed, a wave of skin-homing Tregs following neonatal colonization of the skin is critical to induce tolerance against commensal microbes (Scharschmidt et al. 2015). The migration and accumulation of Tregs to hair follicles is driven by the local expression of CCL20 that act as a chemokine for the recruitment of $CCR6^+$ Tregs (Scharschmidt et al. 2017). Noteworthy, activated keratinocytes seem to regulate Tregs via release of TSLP and the conditional knockout of the TSLP receptor in Tregs results in the progression from a pro-inflammatory condition in the skin to a fatal systemic inflammation (Kashiwagi et al. 2017). In humans, a very prominent example for break of tolerance was recently described in the alpha-gal syndrome: This syndrome is characterized by induction of IgE antibodies against the oligosaccharide galactose-α-1,3-galactose (alpha-gal) which is present in red meat (Commins et al. 2009). Humans have lost their ability to produce this oligosaccharide and usually develop a strong immunolgical tolerance against the alpha-gal epitope including large quantities of IgG and IgM antibodes specific for alpha-gal. However, tick bites are able to break this tolerance, induce anti-alpha-gal IgE antibodies and increase the risk of anaphylaxis after consumption of red meat or systemic infusion of therapeutic antibodies carrying the alpha-gal epitope (Commins et al. 2011; Chung et al. 2008). Thus, break of tolerance via the skin is possible but it remains to be established whether this is an exception or rather the standard in the etiology of allergic patients.

In mouse models, sustained exposure of an allergen to the (shaved) skin is able to induce allergic sensitization and allergic symptoms in the lung upon a single exposure to the allergen (Spergel et al. 1998). Similarly, exposure of food allergens via the skin results in a Th2-dependent sensitization via activation of the innate immune system and notably secretion of IL-6 and IL-33 (Tordesillas et al. 2014). The inflamed skin potentiates allergic sensibilization and is able to trigger intestinal food allergy by triggering secretion of IL-33 and TSLP and activation of basophils (Galand et al. 2016; Tordesillas et al. 2014; Noti et al. 2014). Depending on the site of allergen challenge the cutaneous exposure of allergens on allergic dermatitis-like skin lesions can also lead to the development of TSLP-dependent eosinophilic esophagitis (Noti et al. 2013). Interestingly, IL-33 is associated with Treg function (Schiering et al. 2014) even though excessive signaling may contribute to Treg instability and acquisition of a Th2 phenotype in the lung during allergen-driven allergic airway inflammation (Chen et al. 2017). Thus, cytokines associated with barrier sites, e.g. IL-33 and TSLP, are able to induce typical allergic reactions but at the same time interact also with immunological tolerance notably by affecting Treg biology.

The question what distinguishes healthy and allergy-prone subjects in terms of reactivity to foreign harmless antigens remains open. In addition, allergen-specific immunotherapy relies on injections of high allergen doses into the skin with the aim to induce non-IgE antibody subtypes and allergen-specific Tregs to induce tolerance to future allergen exposure. It seems very unlikely that subjects suffering from allergic reactions simply had enhanced contact to allergens via the skin in the past or an 'unlucky combination' of skin injury and allergen exposure at the same time. Such a scenario cannot explain the recent raise of allergic disorders and it is even likely that skin injuries and associated exposure to harmless antigens has been more frequent in children grown up before the allergic raise was observed. Therefore, environmental factors must be involved to explain this change on an epidemiological level.

5 Conclusion

Altogether the current literature suggests an important role of the gut for the regulation of immune responses yet the underlying mechanisms are only beginning to be elucidated. It is clear that induction of Tregs is tightly associated with microbial colonization and iTregs clearly are important players in regulating allergic inflammation. However, much less is known about what drives intestinal Th2 differentiation apart from parasite infections. A pro-allergic microbiome (Fig. 1) could be hypothesized that lacks anti-inflammatory triggers supporting a physiologic steady-state condition. The delicate balance between iTregs and Th2 cells in all barrier sites may be essential to prevent a general 'type 2' immune bias of the immune system and limit thereby the risk of allergic sensibilisation at different body sites. When ignoring infections by huge parasites one may also ask the question why Th2-dominated immune responses develop in the gut at all. Two possible explanations come to mind: First, type 2 immunity may play a crucial role for tissue repair and healing which may be necessary in the gut due to constant renewal of the intestinal epithelium and potential bacterial or chemical damage even in the absence of parasite-induced tissue damage (Gause et al. 2013). Second, the 'toxin theory' by Galli proposes toxin neutralization after snake or insect stings as the physiological role of IgE and associated mast cell activation at the forefront of type 2 immunity (Metz et al. 2006). This concept has so far been applied to toxins derived from arthropods or reptiles (Tsai et al. 2015) but it would be interesting to apply a similar concept to bacterial toxins that are frequently present in the

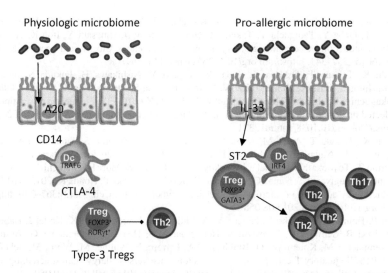

Fig. 1 A physiologic microbiome induces ROR(γt)⁺ Tregs via classical antigen-presenting cells, e.g. dendritic cells (Dc). These 'Type 3 Tregs' are able to counteract Th2 cells and prevent allergic predisposition (*left*). Upon exposure to a pro-allergic microbiome (*right*), Dc are enabled to induce the differentiation of Th2 cells. Gata3-expressing Tregs may accumulate e.g. through the effect of epithelial-derived IL-33. These Tregs secrete type 2 cytokines themselves and possibly even transdifferentiate into Th2 cells thereby predisposing individuals to allergic disorders

intestinal tract during severe gastrointestinal infections. More generally, co-evolution the intestinal microbiota and the associated risk of bacterial toxin production at this site may necessitate the capability to generate Th2-driven immune responses even in the absence of helminth infections or tissue injury. We are now beginning to understand how microbes modulate the intestinal immune system and this knowledge may hopefully also be used in the future to prevent allergic disorders. In the mean time, understanding the physiological role of type 2 immunity may help us to gain mechanistic insight of allergic disorders.

References

Arpaia N, Campbell C, Fan X, Dikiy S, van der Veeken J, deRoos P, Liu H, Cross JR, Pfeffer K, Coffer PJ, Rudensky AY (2013) Metabolites produced by commensal bacteria promote peripheral regulatory T-cell generation. Nature 504(7480):451–455. https://doi.org/10.1038/nature12726

Arrieta MC, Stiemsma LT, Dimitriu PA, Thorson L, Russell S, Yurist-Doutsch S, Kuzeljevic B, Gold MJ, Britton HM, Lefebvre DL, Subbarao P, Mandhane P, Becker A, McNagny KM, Sears MR, Kollmann T, Investigators CS, Mohn WW, Turvey SE, Brett Finlay B (2015) Early infancy microbial and metabolic alterations affect risk of childhood asthma. Sci Transl Med 7(307):307ra152. https://doi.org/10.1126/scitranslmed.aab2271

Atarashi K, Tanoue T, Shima T, Imaoka A, Kuwahara T, Momose Y, Cheng G, Yamasaki S, Saito T, Ohba Y, Taniguchi T, Takeda K, Hori S, Ivanov II, Umesaki Y, Itoh K, Honda K (2011) Induction of colonic regulatory T cells by indigenous Clostridium species. Science 331(6015):337–341. https://doi.org/10.1126/science.1198469

Atarashi K, Tanoue T, Oshima K, Suda W, Nagano Y, Nishikawa H, Fukuda S, Saito T, Narushima S, Hase K, Kim S, Fritz JV, Wilmes P, Ueha S, Matsushima K, Ohno H, Olle B, Sakaguchi S, Taniguchi T, Morita H, Hattori M, Honda K (2013) Treg induction by a rationally selected mixture of Clostridia strains from the human microbiota. Nature 500(7461):232–236. https://doi.org/10.1038/nature12331

Atarashi K, Tanoue T, Ando M, Kamada N, Nagano Y, Narushima S, Suda W, Imaoka A, Setoyama H, Nagamori T, Ishikawa E, Shima T, Hara T, Kado S, Jinnohara T, Ohno H, Kondo T, Toyooka K, Watanabe E, Yokoyama S, Tokoro S, Mori H, Noguchi Y, Morita H, Ivanov II, Sugiyama T, Nunez G, Camp JG, Hattori M, Umesaki Y, Honda K (2015) Th17 cell induction by adhesion of microbes to intestinal epithelial cells. Cell 163(2):367–380. https://doi.org/10.1016/j.cell.2015.08.058

Bacher P, Heinrich F, Stervbo U, Nienen M, Vahldieck M, Iwert C, Vogt K, Kollet J, Babel N, Sawitzki B, Schwarz C, Bereswill S, Heimesaat MM, Heine G, Gadermaier G, Asam C, Assenmacher M, Kniemeyer O, Brakhage AA, Ferreira F, Wallner M, Worm M, Scheffold A (2016) Regulatory T cell specificity directs tolerance versus allergy against aeroantigens in humans. Cell 167(4):1067–1078.e1016. https://doi.org/10.1016/j.cell.2016.09.050

Barlow JL, Peel S, Fox J, Panova V, Hardman CS, Camelo A, Bucks C, Wu X, Kane CM, Neill DR, Flynn RJ, Sayers I, Hall IP, McKenzie AN (2013) IL-33 is more potent than IL-25 in provoking IL-13-producing nuocytes (type 2 innate lymphoid cells) and airway contraction. J Allergy Clin Immunol 132(4):933–941. https://doi.org/10.1016/j.jaci.2013.05.012

Bartemes KR, Iijima K, Kobayashi T, Kephart GM, McKenzie AN, Kita H (2012) IL-33-responsive lineage- CD25+ CD44(hi) lymphoid cells mediate innate type 2 immunity and allergic inflammation in the lungs. J Immunol 188(3):1503–1513. https://doi.org/10.4049/jimmunol.1102832

Barthels C, Ogrinc A, Steyer V, Meier S, Simon F, Wimmer M, Blutke A, Straub T, Zimber-Strobl U, Lutgens E, Marconi P, Ohnmacht C, Garzetti D, Stecher B, Brocker T (2017) CD40-signalling abrogates induction of RORgammat+ Treg cells by intestinal CD103+ DCs and causes fatal colitis. Nat Commun 8:14715. https://doi.org/10.1038/ncomms14715

Boirivant M, Fuss IJ, Chu A, Strober W (1998) Oxazolone colitis: a murine model of T helper cell type 2 colitis treatable with antibodies to interleukin 4. J Exp Med 188(10):1929–1939

Bruchard M, Rebe C, Derangere V, Togbe D, Ryffel B, Boidot R, Humblin E, Hamman A, Chalmin F, Berger H, Chevriaux A, Limagne E, Apetoh L, Vegran F, Ghiringhelli F (2015) The receptor NLRP3 is a transcriptional regulator of TH2 differentiation. Nat Immunol 16(8):859–870. https://doi.org/10.1038/ni.3202

Cahenzli J, Koller Y, Wyss M, Geuking MB, McCoy KD (2013) Intestinal microbial diversity during early-life colonization shapes long-term IgE levels. Cell Host Microbe 14(5):559–570. https://doi.org/10.1016/j.chom.2013.10.004

Chen CC, Kobayashi T, Iijima K, Hsu FC, Kita H (2017) IL-33 dysregulates regulatory T cells and impairs established immunologic tolerance in the lungs. J Allergy Clin Immunol. https://doi.org/10.1016/j.jaci.2017.01.015

Chinen T, Kannan AK, Levine AG, Fan X, Klein U, Zheng Y, Gasteiger G, Feng Y, Fontenot JD, Rudensky AY (2016) An essential role for the IL-2 receptor in Treg cell function. Nat Immunol 17(11):1322–1333. https://doi.org/10.1038/ni.3540

Chung CH, Mirakhur B, Chan E, Le QT, Berlin J, Morse M, Murphy BA, Satinover SM, Hosen J, Mauro D, Slebos RJ, Zhou Q, Gold D, Hatley T, Hicklin DJ, Platts-Mills TA (2008) Cetuximab-induced anaphylaxis and IgE specific for galactose-alpha-1,3-galactose. N Engl J Med 358(11):1109–1117. https://doi.org/10.1056/NEJMoa074943

Commins SP, Satinover SM, Hosen J, Mozena J, Borish L, Lewis BD, Woodfolk JA, Platts-Mills TA (2009) Delayed anaphylaxis, angioedema, or urticaria after consumption of red meat in patients with IgE antibodies specific for galactose-alpha-1,3-galactose. J Allergy Clin Immunol 123(2):426–433. https://doi.org/10.1016/j.jaci.2008.10.052

Commins SP, James HR, Kelly LA, Pochan SL, Workman LJ, Perzanowski MS, Kocan KM, Fahy JV, Nganga LW, Ronmark E, Cooper PJ, Platts-Mills TA (2011) The relevance of tick bites to the production of IgE antibodies to the mammalian oligosaccharide galactose-alpha-1,3-galactose. J Allergy Clin Immunol 127(5):1286–1293.e1286. https://doi.org/10.1016/j.jaci.2011.02.019

Curotto de Lafaille MA, Kutchukhidze N, Shen S, Ding Y, Yee H, Lafaille JJ (2008) Adaptive Foxp3+ regulatory T cell-dependent and -independent control of allergic inflammation. Immunity 29(1):114–126. https://doi.org/10.1016/j.immuni.2008.05.010

Darrasse-Jeze G, Deroubaix S, Mouquet H, Victora GD, Eisenreich T, Yao KH, Masilamani RF, Dustin ML, Rudensky A, Liu K, Nussenzweig MC (2009) Feedback control of regulatory T cell homeostasis by dendritic cells in vivo. J Exp Med 206(9):1853–1862. https://doi.org/10.1084/jem.20090746

de Kleer IM, Kool M, de Bruijn MJ, Willart M, van Moorleghem J, Schuijs MJ, Plantinga M, Beyaert R, Hams E, Fallon PG, Hammad H, Hendriks RW, Lambrecht BN (2016) Perinatal activation of the interleukin-33 pathway promotes type 2 immunity in the developing lung. Immunity 45(6):1285–1298. https://doi.org/10.1016/j.immuni.2016.10.031

Du Toit G, Roberts G, Sayre PH, Bahnson HT, Radulovic S, Santos AF, Brough HA, Phippard D, Basting M, Feeney M, Turcanu V, Sever ML, Gomez Lorenzo M, Plaut M, Lack G, LEAP Study Team (2015) Randomized trial of peanut consumption in infants at risk for peanut allergy. N Engl J Med 372(9):803–813. https://doi.org/10.1056/NEJMoa1414850

Durham SR, Ying S, Varney VA, Jacobson MR, Sudderick RM, Mackay IS, Kay AB, Hamid QA (1996) Grass pollen immunotherapy inhibits allergen-induced infiltration of CD4+ T lymphocytes and eosinophils in the nasal mucosa and increases the number of cells expressing messenger RNA for interferon-gamma. J Allergy Clin Immunol 97(6):1356–1365

Ege MJ, Mayer M, Normand AC, Genuneit J, Cookson WO, Braun-Fahrlander C, Heederik D, Piarroux R, von Mutius E, GABRIELA Transregio 22 Study Group (2011) Exposure to environmental microorganisms and childhood asthma. N Engl J Med 364(8):701–709. https://doi.org/10.1056/NEJMoa1007302

Esterhazy D, Loschko J, London M, Jove V, Oliveira TY, Mucida D (2016) Classical dendritic cells are required for dietary antigen-mediated induction of peripheral T(reg) cells and tolerance. Nat Immunol 17(5):545–555. https://doi.org/10.1038/ni.3408

Fontenot JD, Rasmussen JP, Gavin MA, Rudensky AY (2005) A function for interleukin 2 in Foxp3-expressing regulatory T cells. Nat Immunol 6(11):1142–1151. https://doi.org/10.1038/ni1263

Fujimura KE, Sitarik AR, Havstad S, Lin DL, Levan S, Fadrosh D, Panzer AR, LaMere B, Rackaityte E, Lukacs NW, Wegienka G, Boushey HA, Ownby DR, Zoratti EM, Levin AM, Johnson CC, Lynch SV (2016) Neonatal gut microbiota associates with childhood multisensitized atopy and T cell differentiation. Nat Med 22(10):1187–1191. https://doi.org/10.1038/nm.4176

Furusawa Y, Obata Y, Fukuda S, Endo TA, Nakato G, Takahashi D, Nakanishi Y, Uetake C, Kato K, Kato T, Takahashi M, Fukuda NN, Murakami S, Miyauchi E, Hino S, Atarashi K, Onawa S, Fujimura Y, Lockett T, Clarke JM, Topping DL, Tomita M, Hori S, Ohara O, Morita T, Koseki H, Kikuchi J, Honda K, Hase K, Ohno H (2013) Commensal microbe-derived butyrate induces the differentiation of colonic regulatory T cells. Nature 504(7480):446–450. https://doi.org/10.1038/nature12721

Galand C, Leyva-Castillo JM, Yoon J, Han A, Lee MS, McKenzie AN, Stassen M, Oyoshi MK, Finkelman FD, Geha RS (2016) IL-33 promotes food anaphylaxis in epicutaneously sensitized mice by targeting mast cells. J Allergy Clin Immunol 138(5):1356–1366. https://doi.org/10.1016/j.jaci.2016.03.056

Gao Y, Nish SA, Jiang R, Hou L, Licona-Limon P, Weinstein JS, Zhao H, Medzhitov R (2013) Control of T helper 2 responses by transcription factor IRF4-dependent dendritic cells. Immunity 39(4):722–732. https://doi.org/10.1016/j.immuni.2013.08.028

Gause WC, Wynn TA, Allen JE (2013) Type 2 immunity and wound healing: evolutionary refinement of adaptive immunity by helminths. Nat Rev Immunol 13(8):607–614. https://doi.org/10.1038/nri3476

Geuking MB, Cahenzli J, Lawson MA, Ng DC, Slack E, Hapfelmeier S, McCoy KD, Macpherson AJ (2011) Intestinal bacterial colonization induces mutualistic regulatory T cell responses. Immunity 34(5):794–806. https://doi.org/10.1016/j.immuni.2011.03.021

Geva-Zatorsky N, Sefik E, Kua L, Pasman L, Tan TG, Ortiz-Lopez A, Yanortsang TB, Yang L, Jupp R, Mathis D, Benoist C, Kasper DL (2017) Mining the human gut microbiota for immunomodulatory organisms. Cell 168(5):928–943.e911. https://doi.org/10.1016/j.cell.2017.01.022

Gollwitzer ES, Saglani S, Trompette A, Yadava K, Sherburn R, McCoy KD, Nicod LP, Lloyd CM, Marsland BJ (2014) Lung microbiota promotes tolerance to allergens in neonates via PD-L1. Nat Med. https://doi.org/10.1038/nm.3568

Hadeiba H, Lahl K, Edalati A, Oderup C, Habtezion A, Pachynski R, Nguyen L, Ghodsi A, Adler S, Butcher EC (2012) Plasmacytoid dendritic cells transport peripheral antigens to the thymus to promote central tolerance. Immunity 36(3):438–450. https://doi.org/10.1016/j.immuni.2012.01.017

Hammad H, Plantinga M, Deswarte K, Pouliot P, Willart MA, Kool M, Muskens F, Lambrecht BN (2010) Inflammatory dendritic cells—not basophils—are necessary and sufficient for induction of Th2 immunity to inhaled house dust mite allergen. J Exp Med 207(10):2097–2111. https://doi.org/10.1084/jem.20101563

Han D, Walsh MC, Cejas PJ, Dang NN, Kim YF, Kim J, Charrier-Hisamuddin L, Chau L, Zhang Q, Bittinger K, Bushman FD, Turka LA, Shen H, Reizis B, Defranco AL, Wu GD, Choi Y (2013) Dendritic cell expression of the signaling molecule TRAF6 is critical for gut microbiota-dependent immune tolerance. Immunity 38(6):1211–1222. https://doi.org/10.1016/j.immuni.2013.05.012

Herbst T, Sichelstiel A, Schar C, Yadava K, Burki K, Cahenzli J, McCoy K, Marsland BJ, Harris NL (2011) Dysregulation of allergic airway inflammation in the absence of microbial colonization. Am J Respir Crit Care Med 184(2):198–205. https://doi.org/10.1164/rccm.201010-1574OC

Hill DA, Siracusa MC, Abt MC, Kim BS, Kobuley D, Kubo M, Kambayashi T, Larosa DF, Renner ED, Orange JS, Bushman FD, Artis D (2012) Commensal bacteria-derived signals regulate basophil hematopoiesis and allergic inflammation. Nat Med 18(4):538–546. https://doi.org/10.1038/nm.2657

Ivanov II, McKenzie BS, Zhou L, Tadokoro CE, Lepelley A, Lafaille JJ, Cua DJ, Littman DR (2006) The orphan nuclear receptor RORgammat directs the differentiation program of proinflammatory IL-17+ T helper cells. Cell 126(6):1121–1133. https://doi.org/10.1016/j.cell.2006.07.035

Ivanov II, Atarashi K, Manel N, Brodie EL, Shima T, Karaoz U, Wei D, Goldfarb KC, Santee CA, Lynch SV, Tanoue T, Imaoka A, Itoh K, Takeda K, Umesaki Y, Honda K, Littman DR (2009) Induction of intestinal Th17 cells by segmented filamentous bacteria. Cell 139(3):485–498. https://doi.org/10.1016/j.cell.2009.09.033

Josefowicz SZ, Lu LF, Rudensky AY (2012a) Regulatory T cells: mechanisms of differentiation and function. Annu Rev Immunol 30:531–564. https://doi.org/10.1146/annurev.immunol.25.022106.141623

Josefowicz SZ, Niec RE, Kim HY, Treuting P, Chinen T, Zheng Y, Umetsu DT, Rudensky AY (2012b) Extrathymically generated regulatory T cells control mucosal TH2 inflammation. Nature 482(7385):395–399. https://doi.org/10.1038/nature10772

Julia V, Macia L, Dombrowicz D (2015) The impact of diet on asthma and allergic diseases. Nat Rev Immunol 15(5):308–322. https://doi.org/10.1038/nri3830

Kabat AM, Harrison OJ, Riffelmacher T, Moghaddam AE, Pearson CF, Laing A, Abeler-Dorner L, Forman SP, Grencis RK, Sattentau Q, Simon AK, Pott J, Maloy KJ (2016) The autophagy gene Atg16l1 differentially regulates Treg and TH2 cells to control intestinal inflammation. elife 5: e12444. https://doi.org/10.7554/eLife.12444

Kashiwagi M, Hosoi J, Lai JF, Brissette J, Ziegler SF, Morgan BA, Georgopoulos K (2017) Direct control of regulatory T cells by keratinocytes. Nat Immunol 18(3):334–343. https://doi.org/10.1038/ni.3661

Kawamoto S, Maruya M, Kato LM, Suda W, Atarashi K, Doi Y, Tsutsui Y, Qin H, Honda K, Okada T, Hattori M, Fagarasan S (2014) Foxp3(+) T cells regulate immunoglobulin a selection and facilitate diversification of bacterial species responsible for immune homeostasis. Immunity 41(1):152–165. https://doi.org/10.1016/j.immuni.2014.05.016

Kim KS, Hong SW, Han D, Yi J, Jung J, Yang BG, Lee JY, Lee M, Surh CD (2016) Dietary antigens limit mucosal immunity by inducing regulatory T cells in the small intestine. Science 351(6275):858–863. https://doi.org/10.1126/science.aac5560

Kluger MA, Luig M, Wegscheid C, Goerke B, Paust HJ, Brix SR, Yan I, Mittrucker HW, Hagl B, Renner ED, Tiegs G, Wiech T, Stahl RA, Panzer U, Steinmetz OM (2014) Stat3 programs Th17-specific regulatory T cells to control GN. J Am Soc Nephrol 25(6):1291–1302. https://doi.org/10.1681/ASN.2013080904

Kluger MA, Nosko A, Ramcke T, Goerke B, Meyer MC, Wegscheid C, Luig M, Tiegs G, Stahl RA, Steinmetz OM (2017) RORgammat expression in Tregs promotes systemic lupus erythematosus via IL-17 secretion, alteration of Treg phenotype and suppression of Th2 responses. Clin Exp Immunol 188(1):63–78. https://doi.org/10.1111/cei.12905

Kumamoto Y, Linehan M, Weinstein JS, Laidlaw BJ, Craft JE, Iwasaki A (2013) CD301b(+) dermal dendritic cells drive T helper 2 cell-mediated immunity. Immunity 39(4):733–743. https://doi.org/10.1016/j.immuni.2013.08.029

Lange-Asschenfeldt B, Marenbach D, Lang C, Patzelt A, Ulrich M, Maltusch A, Terhorst D, Stockfleth E, Sterry W, Lademann J (2011) Distribution of bacteria in the epidermal layers and hair follicles of the human skin. Skin Pharmacol Physiol 24(6):305–311. https://doi.org/10.1159/000328728

Lathrop SK, Bloom SM, Rao SM, Nutsch K, Lio CW, Santacruz N, Peterson DA, Stappenbeck TS, Hsieh CS (2011) Peripheral education of the immune system by colonic commensal microbiota. Nature 478(7368):250–254. https://doi.org/10.1038/nature10434

Legoux FP, Lim JB, Cauley AW, Dikiy S, Ertelt J, Mariani TJ, Sparwasser T, Way SS, Moon JJ (2015) CD4(+) T cell tolerance to tissue-restricted self antigens is mediated by antigen-specific regulatory T cells rather than deletion. Immunity 43(5):896–908. https://doi.org/10.1016/j.immuni.2015.10.011

Levine AG, Arvey A, Jin W, Rudensky AY (2014) Continuous requirement for the TCR in regulatory T cell function. Nat Immunol 15(11):1070–1078. https://doi.org/10.1038/ni.3004

Levine AG, Hemmers S, Baptista AP, Schizas M, Faire MB, Moltedo B, Konopacki C, Schmidt-Supprian M, Germain RN, Treuting PM, Rudensky AY (2017) Suppression of lethal autoimmunity by regulatory T cells with a single TCR specificity. J Exp Med 214(3):609–622. https://doi.org/10.1084/jem.20161318

Lexmond WS, Goettel JA, Lyons JJ, Jacobse J, Deken MM, Lawrence MG, DiMaggio TH, Kotlarz D, Garabedian E, Sackstein P, Nelson CC, Jones N, Stone KD, Candotti F, Rings EH, Thrasher AJ, Milner JD, Snapper SB, Fiebiger E (2016) FOXP3+ Tregs require WASP to restrain Th2-mediated food allergy. J Clin Invest 126(10):4030–4044. https://doi.org/10.1172/JCI85129

Lin W, Truong N, Grossman WJ, Haribhai D, Williams CB, Wang J, Martin MG, Chatila TA (2005) Allergic dysregulation and hyperimmunoglobulinemia E in Foxp3 mutant mice. J Allergy Clin Immunol 116(5):1106–1115. https://doi.org/10.1016/j.jaci.2005.08.046

Lochner M, Berard M, Sawa S, Hauer S, Gaboriau-Routhiau V, Fernandez TD, Snel J, Bousso P, Cerf-Bensussan N, Eberl G (2011) Restricted microbiota and absence of cognate TCR antigen leads to an unbalanced generation of Th17 cells. J Immunol 186(3):1531–1537. https://doi.org/10.4049/jimmunol.1001723

Lodge CJ, Tan DJ, Lau MX, Dai X, Tham R, Lowe AJ, Bowatte G, Allen KJ, Dharmage SC (2015) Breastfeeding and asthma and allergies: a systematic review and meta-analysis. Acta Paediatr 104(467):38–53. https://doi.org/10.1111/apa.13132

Lozupone CA, Stombaugh JI, Gordon JI, Jansson JK, Knight R (2012) Diversity, stability and resilience of the human gut microbiota. Nature 489(7415):220–230. https://doi.org/10.1038/nature11550

Mantel PY, Kuipers H, Boyman O, Rhyner C, Ouaked N, Ruckert B, Karagiannidis C, Lambrecht BN, Hendriks RW, Crameri R, Akdis CA, Blaser K, Schmidt-Weber CB (2007) GATA3-driven Th2 responses inhibit TGF-beta1-induced FOXP3 expression and the formation of regulatory T cells. PLoS Biol 5(12):e329. https://doi.org/10.1371/journal.pbio.0050329

Massoud AH, Charbonnier LM, Lopez D, Pellegrini M, Phipatanakul W, Chatila TA (2016) An asthma-associated IL4R variant exacerbates airway inflammation by promoting conversion of regulatory T cells to TH17-like cells. Nat Med 22(9):1013–1022. https://doi.org/10.1038/nm.4147

Metz M, Piliponsky AM, Chen CC, Lammel V, Abrink M, Pejler G, Tsai M, Galli SJ (2006) Mast cells can enhance resistance to snake and honeybee venoms. Science 313(5786):526–530. https://doi.org/10.1126/science.1128877

Morton AM, Sefik E, Upadhyay R, Weissleder R, Benoist C, Mathis D (2014) Endoscopic photo-conversion reveals unexpectedly broad leukocyte trafficking to and from the gut. Proc Natl Acad Sci USA 111(18):6696–6701. https://doi.org/10.1073/pnas.1405634111

Mucida D, Kutchukhidze N, Erazo A, Russo M, Lafaille JJ, Curotto de Lafaille MA (2005) Oral tolerance in the absence of naturally occurring Tregs. J Clin Invest 115(7):1923–1933. https://doi.org/10.1172/JCI24487

Noti M, Wojno ED, Kim BS, Siracusa MC, Giacomin PR, Nair MG, Benitez AJ, Ruymann KR, Muir AB, Hill DA, Chikwava KR, Moghaddam AE, Sattentau QJ, Alex A, Zhou C, Yearley JH, Menard-Katcher P, Kubo M, Obata-Ninomiya K, Karasuyama H, Comeau MR, Brown-Whitehorn T, de Waal Malefyt R, Sleiman PM, Hakonarson H, Cianferoni A, Falk GW, Wang ML, Spergel JM, Artis D (2013) Thymic stromal lymphopoietin-elicited basophil responses promote eosinophilic esophagitis. Nat Med 19(8):1005–1013. https://doi.org/10.1038/nm.3281

Noti M, Kim BS, Siracusa MC, Rak GD, Kubo M, Moghaddam AE, Sattentau QA, Comeau MR, Spergel JM, Artis D (2014) Exposure to food allergens through inflamed skin promotes intestinal food allergy through the thymic stromal lymphopoietin-basophil axis. J Allergy Clin Immunol 133(5):1390–1399, 1399.e1391–1396. https://doi.org/10.1016/j.jaci.2014.01.021

Noval Rivas M, Burton OT, Wise P, Charbonnier LM, Georgiev P, Oettgen HC, Rachid R, Chatila TA (2015) Regulatory T cell reprogramming toward a Th2-Cell-like lineage impairs oral tolerance and promotes food allergy. Immunity 42(3):512–523. https://doi.org/10.1016/j.immuni.2015.02.004

Ohnmacht C (2016) Tolerance to the Intestinal Microbiota Mediated by ROR(gammat)(+) Cells. Trends Immunol 37(7):477–486. https://doi.org/10.1016/j.it.2016.05.002

Ohnmacht C, Park JH, Cording S, Wing JB, Atarashi K, Obata Y, Gaboriau-Routhiau V, Marques R, Dulauroy S, Fedoseeva M, Busslinger M, Cerf-Bensussan N, Boneca IG, Voehringer D, Hase K, Honda K, Sakaguchi S, Eberl G (2015) MUCOSAL IMMUNOLOGY. The microbiota regulates type 2 immunity through RORgammat(+) T cells. Science 349 (6251):989–993. https://doi.org/10.1126/science.aac4263

Olszak T, An D, Zeissig S, Vera MP, Richter J, Franke A, Glickman JN, Siebert R, Baron RM, Kasper DL, Blumberg RS (2012) Microbial exposure during early life has persistent effects on natural killer T cell function. Science 336(6080):489–493. https://doi.org/10.1126/science.1219328

Ozcan E, Notarangelo LD, Geha RS (2008) Primary immune deficiencies with aberrant IgE production. J Allergy Clin Immunol 122(6):1054–1062. https://doi.org/10.1016/j.jaci.2008.10.023

Plantinga M, Guilliams M, Vanheerswynghels M, Deswarte K, Branco-Madeira F, Toussaint W, Vanhoutte L, Neyt K, Killeen N, Malissen B, Hammad H, Lambrecht BN (2013) Conventional and monocyte-derived CD11b(+) dendritic cells initiate and maintain T helper 2 cell-mediated immunity to house dust mite allergen. Immunity 38(2):322–335. https://doi.org/10.1016/j.immuni.2012.10.016

Platzer B, Baker K, Vera MP, Singer K, Panduro M, Lexmond WS, Turner D, Vargas SO, Kinet JP, Maurer D, Baron RM, Blumberg RS, Fiebiger E (2015) Dendritic cell-bound IgE functions to restrain allergic inflammation at mucosal sites. Mucosal Immunol 8(3):516–532. https://doi.org/10.1038/mi.2014.85

Round JL, Mazmanian SK (2010) Inducible Foxp3+ regulatory T-cell development by a commensal bacterium of the intestinal microbiota. Proc Natl Acad Sci USA 107(27):12204–12209. https://doi.org/10.1073/pnas.0909122107

Russell SL, Gold MJ, Hartmann M, Willing BP, Thorson L, Wlodarska M, Gill N, Blanchet MR, Mohn WW, McNagny KM, Finlay BB (2012) Early life antibiotic-driven changes in microbiota enhance susceptibility to allergic asthma. EMBO Rep 13(5):440–447. https://doi.org/10.1038/embor.2012.32

Russell SL, Gold MJ, Reynolds LA, Willing BP, Dimitriu P, Thorson L, Redpath SA, Perona-Wright G, Blanchet MR, Mohn WW, Finlay BB, McNagny KM (2015) Perinatal antibiotic-induced shifts in gut microbiota have differential effects on inflammatory lung diseases. J Allergy Clin Immunol 135(1):100–109. https://doi.org/10.1016/j.jaci.2014.06.027

Sanchez Rodriguez R, Pauli ML, Neuhaus IM, Yu SS, Arron ST, Harris HW, Yang SH, Anthony BA, Sverdrup FM, Krow-Lucal E, Mackenzie TC, Johnson DS, Meyer EH, Lohr A, Hsu A, Koo J, Liao W, Gupta R, Debbaneh MG, Butler D, Huynh M, Levin EC, Leon A, Hoffman WY, McGrath MH, Alvarado MD, Ludwig CH, Truong HA, Maurano MM, Gratz IK, Abbas AK, Rosenblum MD (2014) Memory regulatory T cells reside in human skin. J Clin Invest 124 (3):1027–1036. https://doi.org/10.1172/jci72932

Scharschmidt TC, Vasquez KS, Truong HA, Gearty SV, Pauli ML, Nosbaum A, Gratz IK, Otto M, Moon JJ, Liese J, Abbas AK, Fischbach MA, Rosenblum MD (2015) A wave of regulatory T cells into neonatal skin mediates tolerance to commensal microbes. Immunity 43(5):1011–1021. https://doi.org/10.1016/j.immuni.2015.10.016

Scharschmidt TC, Vasquez KS, Pauli ML, Leitner EG, Chu K, Truong HA, Lowe MM, Sanchez Rodriguez R, Ali N, Laszik ZG, Sonnenburg JL, Millar SE, Rosenblum MD (2017) Commensal microbes and hair follicles morphogenesis coordinately drive treg migration into neonatal skin. Cell Host Microbe 21(4):467–477.e5. https://doi.org/10.1016/j.chom.2017.03.001

Schiering C, Krausgruber T, Chomka A, Frohlich A, Adelmann K, Wohlfert EA, Pott J, Griseri T, Bollrath J, Hegazy AN, Harrison OJ, Owens BM, Lohning M, Belkaid Y, Fallon PG, Powrie F (2014) The alarmin IL-33 promotes regulatory T-cell function in the intestine. Nature. https://doi.org/10.1038/nature13577

Schuijs MJ, Willart MA, Vergote K, Gras D, Deswarte K, Ege MJ, Madeira FB, Beyaert R, van Loo G, Bracher F, von Mutius E, Chanez P, Lambrecht BN, Hammad H (2015) Farm dust and endotoxin protect against allergy through A20 induction in lung epithelial cells. Science 349(6252):1106–1110. https://doi.org/10.1126/science.aac6623

Secrist H, Chelen CJ, Wen Y, Marshall JD, Umetsu DT (1993) Allergen immunotherapy decreases interleukin 4 production in CD4+ T cells from allergic individuals. J Exp Med 178(6): 2123–2130

Sefik E, Geva-Zatorsky N, Oh S, Konnikova L, Zemmour D, McGuire AM, Burzyn D, Ortiz-Lopez A, Lobera M, Yang J, Ghosh S, Earl A, Snapper SB, Jupp R, Kasper D, Mathis D, Benoist C (2015) MUCOSAL IMMUNOLOGY. Individual intestinal symbionts induce a distinct population of RORgamma(+) regulatory T cells. Science 349(6251):993–997. https://doi.org/10.1126/science.aaa9420

Siede J, Frohlich A, Datsi A, Hegazy AN, Varga DV, Holecska V, Saito H, Nakae S, Lohning M (2016) IL-33 receptor-expressing regulatory T cells are highly activated, Th2 biased and suppress CD4 T cell proliferation through IL-10 and TGFbeta release. PLoS One 11(8): e0161507. https://doi.org/10.1371/journal.pone.0161507

Singh N, Thangaraju M, Prasad PD, Martin PM, Lambert NA, Boettger T, Offermanns S, Ganapathy V (2010) Blockade of dendritic cell development by bacterial fermentation products butyrate and propionate through a transporter (Slc5a8)-dependent inhibition of histone deacetylases. J Biol Chem 285(36):27601–27608. https://doi.org/10.1074/jbc.M110.102947

Spergel JM, Mizoguchi E, Brewer JP, Martin TR, Bhan AK, Geha RS (1998) Epicutaneous sensitization with protein antigen induces localized allergic dermatitis and hyperresponsiveness to methacholine after single exposure to aerosolized antigen in mice. J Clin Invest 101(8):1614–1622. https://doi.org/10.1172/JCI1647

Stefka AT, Feehley T, Tripathi P, Qiu J, McCoy K, Mazmanian SK, Tjota MY, Seo GY, Cao S, Theriault BR, Antonopoulos DA, Zhou L, Chang EB, Fu YX, Nagler CR (2014) Commensal bacteria protect against food allergen sensitization. Proc Natl Acad Sci USA 111(36): 13145–13150. https://doi.org/10.1073/pnas.1412008111

Stein MM, Hrusch CL, Gozdz J, Igartua C, Pivniouk V, Murray SE, Ledford JG, Marques dos Santos M, Anderson RL, Metwali N, Neilson JW, Maier RM, Gilbert JA, Holbreich M, Thorne PS, Martinez FD, von Mutius E, Vercelli D, Ober C, Sperling AI (2016) Innate immunity and asthma risk in amish and hutterite farm children. N Engl J Med 375(5):411–421. https://doi.org/10.1056/NEJMoa1508749

Sujino T, London M, Hoytema van Konijnenburg DP, Rendon T, Buch T, Silva HM, Lafaille JJ, Reis BS, Mucida D (2016) Tissue adaptation of regulatory and intraepithelial CD4(+) T cells controls gut inflammation. Science 352(6293):1581–1586. https://doi.org/10.1126/science.aaf3892

Tanoue T, Atarashi K, Honda K (2016) Development and maintenance of intestinal regulatory T cells. Nat Rev Immunol 16(5):295–309. https://doi.org/10.1038/nri.2016.36

Tordesillas L, Goswami R, Benede S, Grishina G, Dunkin D, Jarvinen KM, Maleki SJ, Sampson HA, Berin MC (2014) Skin exposure promotes a Th2-dependent sensitization to peanut allergens. J Clin Invest 124(11):4965–4975. https://doi.org/10.1172/JCI75660

Trompette A, Gollwitzer ES, Yadava K, Sichelstiel AK, Sprenger N, Ngom-Bru C, Blanchard C, Junt T, Nicod LP, Harris NL, Marsland BJ (2014) Gut microbiota metabolism of dietary fiber influences allergic airway disease and hematopoiesis. Nat Med 20(2):159–166. https://doi.org/10.1038/nm.3444

Tsai M, Starkl P, Marichal T, Galli SJ (2015) Testing the 'toxin hypothesis of allergy': mast cells, IgE, and innate and acquired immune responses to venoms. Curr Opin Immunol 36:80–87. https://doi.org/10.1016/j.coi.2015.07.001

Tussiwand R, Everts B, Grajales-Reyes GE, Kretzer NM, Iwata A, Bagaitkar J, Wu X, Wong R, Anderson DA, Murphy TL, Pearce EJ, Murphy KM (2015) Klf4 expression in conventional dendritic cells is required for T helper 2 cell responses. Immunity 42(5):916–928. https://doi.org/10.1016/j.immuni.2015.04.017

Ubeda C, Pamer EG (2012) Antibiotics, microbiota, and immune defense. Trends Immunol 33 (9):459–466. https://doi.org/10.1016/j.it.2012.05.003

Ulges A, Klein M, Reuter S, Gerlitzki B, Hoffmann M, Grebe N, Staudt V, Stergiou N, Bohn T, Bruhl TJ, Muth S, Yurugi H, Rajalingam K, Bellinghausen I, Tuettenberg A, Hahn S, Reissig S, Haben I, Zipp F, Waisman A, Probst HC, Beilhack A, Buchou T, Filhol-Cochet O, Boldyreff B, Breloer M, Jonuleit H, Schild H, Schmitt E, Bopp T (2015) Protein kinase CK2 enables regulatory T cells to suppress excessive T2 responses in vivo. Nat Immunol. https://doi.org/10.1038/ni.3083

Vahl JC, Drees C, Heger K, Heink S, Fischer JC, Nedjic J, Ohkura N, Morikawa H, Poeck H, Schallenberg S, Riess D, Hein MY, Buch T, Polic B, Schonle A, Zeiser R, Schmitt-Graff A, Kretschmer K, Klein L, Korn T, Sakaguchi S, Schmidt-Supprian M (2014) Continuous T cell receptor signals maintain a functional regulatory T cell pool. Immunity 41(5):722–736. https://doi.org/10.1016/j.immuni.2014.10.012

Varney VA, Hamid QA, Gaga M, Ying S, Jacobson M, Frew AJ, Kay AB, Durham SR (1993) Influence of grass pollen immunotherapy on cellular infiltration and cytokine mRNA expression during allergen-induced late-phase cutaneous responses. J Clin Invest 92(2):644–651. https://doi.org/10.1172/JCI116633

Vatanen T, Kostic AD, d'Hennezel E, Siljander H, Franzosa EA, Yassour M, Kolde R, Vlamakis H, Arthur TD, Hamalainen AM, Peet A, Tillmann V, Uibo R, Mokurov S, Dorshakova N, Ilonen J, Virtanen SM, Szabo SJ, Porter JA, Lahdesmaki H, Huttenhower C, Gevers D, Cullen TW, Knip M, DIABIMMUNE Study Group, Xavier RJ (2016) Variation in microbiome LPS immunogenicity contributes to autoimmunity in humans. Cell 165(4):842–853. https://doi.org/10.1016/j.cell.2016.04.007

Verhasselt V, Milcent V, Cazareth J, Kanda A, Fleury S, Dombrowicz D, Glaichenhaus N, Julia V (2008) Breast milk-mediated transfer of an antigen induces tolerance and protection from allergic asthma. Nat Med 14(2):170–175. https://doi.org/10.1038/nm1718

Wang Y, Su MA, Wan YY (2011) An essential role of the transcription factor GATA-3 for the function of regulatory T cells. Immunity 35(3):337–348. https://doi.org/10.1016/j.immuni.2011.08.012

Williams JW, Tjota MY, Clay BS, Vander Lugt B, Bandukwala HS, Hrusch CL, Decker DC, Blaine KM, Fixsen BR, Singh H, Sciammas R, Sperling AI (2013) Transcription factor IRF4 drives dendritic cells to promote Th2 differentiation. Nat Commun 4(2990). https://doi.org/10.1038/ncomms3990

Wilson MS, Taylor MD, Balic A, Finney CA, Lamb JR, Maizels RM (2005) Suppression of allergic airway inflammation by helminth-induced regulatory T cells. J Exp Med 202(9):1199–1212. https://doi.org/10.1084/jem.20042572

Wing K, Onishi Y, Prieto-Martin P, Yamaguchi T, Miyara M, Fehervari Z, Nomura T, Sakaguchi S (2008) CTLA-4 control over Foxp3+ regulatory T cell function. Science 322(5899):271–275. https://doi.org/10.1126/science.1160062

Wohlfert EA, Grainger JR, Bouladoux N, Konkel JE, Oldenhove G, Ribeiro CH, Hall JA, Yagi R, Naik S, Bhairavabhotla R, Paul WE, Bosselut R, Wei G, Zhao K, Oukka M, Zhu J, Belkaid Y (2011) GATA3 controls Foxp3(+) regulatory T cell fate during inflammation in mice. J Clin Invest 121(11):4503–4515. https://doi.org/10.1172/JCI57456

Wu C, Chen Z, Dardalhon V, Xiao S, Thalhamer T, Liao M, Madi A, Franca RF, Han T, Oukka M, Kuchroo V (2017) The transcription factor musculin promotes the unidirectional development of peripheral Treg cells by suppressing the TH2 transcriptional program. Nat Immunol 18(3):344–353. https://doi.org/10.1038/ni.3667

Yang BH, Hagemann S, Mamareli P, Lauer U, Hoffmann U, Beckstette M, Fohse L, Prinz I, Pezoldt J, Suerbaum S, Sparwasser T, Hamann A, Floess S, Huehn J, Lochner M (2016) Foxp3(+) T cells expressing RORgammat represent a stable regulatory T-cell effector lineage with enhanced suppressive capacity during intestinal inflammation. Mucosal Immunol 9(2):444–457. https://doi.org/10.1038/mi.2015.74

Yu F, Sharma S, Edwards J, Feigenbaum L, Zhu J (2014) Dynamic expression of transcription factors T-bet and GATA-3 by regulatory T cells maintains immunotolerance. Nat Immunol. https://doi.org/10.1038/ni.3053

Zaiss MM, Rapin A, Lebon L, Dubey LK, Mosconi I, Sarter K, Piersigilli A, Menin L, Walker AW, Rougemont J, Paerewijck O, Geldhof P, McCoy KD, Macpherson AJ, Croese J, Giacomin PR, Loukas A, Junt T, Marsland BJ, Harris NL (2015) The Intestinal Microbiota Contributes to the Ability of Helminths to Modulate Allergic Inflammation. Immunity 43(5):998–1010. https://doi.org/10.1016/j.immuni.2015.09.012

Zheng Y, Chaudhry A, Kas A, deRoos P, Kim JM, Chu TT, Corcoran L, Treuting P, Klein U, Rudensky AY (2009) Regulatory T-cell suppressor program co-opts transcription factor IRF4 to control T(H)2 responses. Nature 458(7236):351–356. https://doi.org/10.1038/nature07674

Aryl Hydrocarbon Receptor: An Environmental Sensor in Control of Allergy Outcomes

Marco Gargaro, Matteo Pirro, Giorgia Manni, Antonella De Luca, Teresa Zelante, and Francesca Fallarino

Abstract The mechanisms how environmental compounds influence the human immune system are unknown. The environmentally sensitive transcription factor aryl hydrocarbon receptor (AhR) has immune-modulating functions and responds to a wide variety of small molecules. Since AhR is highly expressed in cells at body surfaces, such as skin, gut mucosa and particularly in mucosal-associated lymphocytes, this molecule is perfectly positioned to be a sensor of external environmental signals. The role of AhR in the balance of immunity and tolerance and in the control of local homeostasis has been clearly demonstrated in recent years [Kiss et al., Science (New York, NY) 334(6062):1561–1565, 2011; Li et al., Cell 147 (3):629–640, 2011]. Deletion of AhR in mice resulted in altered composition of gut microbiota, impaired function and inflammatory immune activation of gut epithelium. In addition to xenobiotics, AhR ligands now include endogenous metabolites, dietary derivatives and bacterial metabolites (Denison and Nagy. Annu Rev Pharmacol Toxicol 43:309–334, 2003). Xenobiotics such as dietary components, products of microbiota, and ubiquitous environmental pollutants may have shaped the AhR system in intestinal epithelia or other body surfaces during millions of years of evolution. Thus, the crosstalk among the dietary components/xenobiotics, AhR and the gut microbiome appears to be an important factor in the maintenance of the mucosal immunity and immune homeostasis. These exciting discoveries provide a novel perspective for the biological role of AhR, which has been originally studied only as a sensor of toxicants, but which has been now implicated in a wide range of human conditions, including autoimmune and allergic disorders.

M. Gargaro • G. Manni • A. De Luca • T. Zelante • F. Fallarino (✉)
Department of Experimental Medicine, University of Perugia, Perugia, Italy
e-mail: francesca.fallarino@unipg.it

M. Pirro
Department of Medicine, University of Perugia, Perugia, Italy

© Springer International Publishing AG 2017   167
C.B. Schmidt-Weber (ed.), *Allergy Prevention and Exacerbation*, Birkhäuser
Advances in Infectious Diseases, https://doi.org/10.1007/978-3-319-69968-4_9

1 AhR: A Pleiotropic Immune Regulator

The aryl hydrocarbon receptor (AhR) is a ligand-activated transcription factor that mediates numerous cellular responses, playing critical roles in several cells involved in innate and adaptive immune processes. Although AhR activity was originally investigated in toxicology, because of its ability to bind to environmental contaminants, such as 2,3,7,8-tetrachlorodibenzo-p-dioxin (TCDD), recently it has attracted enormous attention also in immunology. Studies in AhR-null mice indicate that AhR deficiency impaired several physiological progresses, such as the development of specific cell types of the immune system and protection against bacterial infections (Fernandez-Salguero et al. 1997; Kiss et al. 2011) or cohabition, as indicated in the previous chapters. Moreover, the recent discovery of endogenous and plant-derived ligands, point to a crucial role of AhR in normal cell physiology, in addition to its roles in sensing some environmental chemicals. Thus, AhR is emerging as a key protein that mediates both toxicological and physiological effects mostly upon sensing exogenous and endogenous molecules.

Several studies have examined AhR effects in immune cells including dendritic cells (DCs), T cells and macrophages (Di Meglio et al. 2014). AhR is expressed in almost all mouse tissues, and in human highest expression is observed in lung, thymus, kidney and liver (Puga et al. 2009). The selective forces that led to the high degree of conservation of the AhR amino acid sequence are unknown and its physiological functions, are still being elucidated. At present, it is clear that AhR is both an activator of metabolism of small molecules and a key player in many cell functions, including several of the immune system. In this regard, AhR might link adaptive immune responses to environmental factors playing an important role in the balance of immune responses. In these regard, recent studies have unraveled unsuspected physiological roles and novel alternative ligand-specific pathways for this receptor. Thus, the immunomodulatory role of AhR has become a major area of investigation. The implications of this environmentally triggered feedback pathways are of great importance, as they may contribute to new options in immune modulation or in tolerance-promoting treatment strategies.

1.1 AhR Structure and Conventional AhR Signaling Pathway

The transcription factor AhR is a cytosolic sensor of small synthetic xenobiotic molecules and natural compounds that acting as AhR ligands are also called as AhR agonists. AhR is a member of the basic-loop-helix transcription factors (bHLH) that contain a basic Helix-Loop-Helix motif in the NH2-terminus and a PER/ARNT/SIM-homology domain in the COOH-adjacent region. Structurally, the N-terminal half of the mammalian AhR is well-conserved over 600 million years, suggesting critical importance in specific physiological activities (Hahn et al. 2017). The bHLH motif is shared with other transcription factors such as Myc and MyoD and is responsible for DNA-binding and dimerization. It includes a nuclear localisation signal (NLS) and a nuclear export signal (NES) (Hahn et al. 2017). In contrast, the receptor's C-terminal region exhibits species differences reflected by polymorphisms and variations in protein length, which largely accounts for the size differences observed between species (Hahn 2002). AhR polymorphism among species, especially differences in the key amino acids in the ligand-binding pocket, explain differential sensitivity of AhRs response to dioxins (Karchner et al. 2006). The PAS domain contains two imperfect repeats of 50 amino acids, PAS A and PAS B, and is thought as whole to serve as an interactive surface for dimer formation, while PAS B partially corresponds to the ligand-binding domain. bHLH-PAS proteins are biological sensors for a variety of stimuli, controlling neurogenesis, vascularization, circadian rhythms, metabolism and stress responses to hypoxia, among others (Kewley et al. 2004). The COOH-terminal half of the amino acidic sequence mediates transactivation activity through several specific sites, for instance a glutamine-rich box.

The inactive, cytosolic AhR is part of a protein complex that includes the 90-kDa heat shock protein (HSP90), the c-SRC protein kinase, and the AhR-interacting protein Ara9, which keeps AhR structurally competent for binding to ligands and prevents untimely translocation to the nucleus (Mimura and Fujii-Kuriyama 2003). Ligand binding induces a conformational change in AhR, thereby exposing a nuclear translocation site, initiating the so called 'canonical AhR pathway'. AhR translocates to the nucleus and binds to its dimerization partner aryl receptor nuclear translocator ARNT (i.e. another bHLH-PAS protein), then the AhR-ARNT complex initiates transcription of genes containing in their promoters dioxin or xenobiotic responsive elements, known as DRE and XRE respectively. These genes include those encoding members of the cytochrome P450 family 1 (CYP1) such as CYP1A1, CYP1A2 and CYP1B1 (Sogawa and Fujii-Kuriyama 1997). Ligand-activated AhR also induces other enzymes involved in phase I and II xenobiotic metabolism, including UDP-glucuronosyl transferase 1A6, NAD(P)H-dependent quinone oxidoreductase-1, aldehyde dehydrogenase 3A1 and glutathione transferases (Hankinson 1995). Nevertheless, the list of AhR-regulated genes has grown considerably over the recent years and it includes also a variety of genes unrelated to xenobiotic metabolism such as genes encoding proteins involved in proliferation/apoptosis control (TGF-α, TGF-β2, Bax, p27[kip1]) (Yin et al. 1994), cytokines (IL-1β, IL-2) (Hoffer et al. 1996) or proteins forming transcription factor complexes (c-fos, Jun-B, c-Jun, JunD) (Tian et al. 2003). Notably, ligands only need to meet minimal requirements for size and planar shape to fit into the AhR binding pocket. Consequently, a broad range of low- molecular-weight chemicals can activate AhR, albeit at different affinities ranging between 10^{-12} and 10^{-3} M. Among known AhR ligands several are characterized by two carbon ring systems, such as tryptophan derivatives, flavonoids and biphenyls (Denison and Nagy 2003). The AhR system is genetically polymorphic and different alleles influence responsiveness to AhR ligands (Hahn 2002). AhR directly controls transcriptional activity of its targets by interacting with specific subunits of the positive transcription elongation factor (P-TEFb complex) to regulate transcriptional elongation, as well as with subunits of the transcriptional coactivator mediator complex (Kagey et al. 2010; Wang et al. 2004). This interaction could be of particular importance for communication of AhR-regulated transcriptional enhancers with their respective target promoters, as mediator associates with cohesin complexes to regulate chromatin looping and cell type-specific gene expression patterns (Okino and Whitlock 1995). Apart from direct interactions with the general transcription machinery, it has been shown that AhR can affect local chromatin architecture (Wang and Hankinson 2002) by directly interacting with the Brahma/SWI2-related gene 1 (Brg1) subunit of the SWI/SNF chromatin-remodeling complex (Fujii-Kuriyama and Mimura 2005).

The regulation of AhR function includes various feedback mechanisms, which may attenuate AhR response and these signals implies AhR degradation (Ma and Baldwin 2000). Notably, this mechanism has been studied for AhR-mediated TCDD activation. The TCDD-induced degradation of AhR involves ubiquitination of C-terminal half of AhR and its consequent degradation by the 26S proteasome (Hahn et al. 2009; Pollenz and Buggy 2006). Downregulation of AhR signalling is also mediated by the aryl hydrocarbon receptor repressor (AhRR) (Brauze et al. 2006). This protein is structurally similar to AhR, as it contains an amino-terminal bHLH motif, comprising both NLS and NES, and a PAS A domain; consequently it can bind ARNT as well as XREs. However, it localizes within the nucleus as an AhRR/ARNT dimer constitutively in a ligand-independent manner, which is consistent with the lack of PAS B domain. The COOH-terminal half of the sequence greatly differs from that of AhR and functions as transcriptional repressor of AhR activity. The way such repression occurs remains uncertain. Competition could arise between AhRR and AhR for association with ARNT, and/or between AhRR/ARNT and AhR/ARNT dimers to interact with consensus sequences. However, when both hypothesized mechanisms were precluded, through experimental manipulations, the ability of AhRR to repress was maintained (Hahn et al. 2009). That suggests the existence of another mechanisms, which could require AhRR establishing protein-protein interactions with corepressors without binding DNA directly. In any case, it has been demonstrated that AhRR expression can be induced by activated AhR, defining a negative feedback loop engaged in modulation of AhR activity, even though, tissue and context-dependent variability should be also taken into account. Conversely, signal amplification may occur by a positive feedback circuit involving AhR-induced expression of the receptor itself. Indeed, some studies on Sprague-Dawley rats report increased levels of AhR mRNA upon treatment of with TCDD, 3-methyl-colantrene (3-MC) and beta-naphthoflavone (BNF) (Nguyen and Bradfield 2008).

That said, further studies have revealed that AhR activity is far more complex. The receptor can ride alternative signaling routes intercepting other intracellular communication pathways with AhR exerting also co-regulatory functions.

1.2 Exogenous and Endogenous AhR Ligands

AhR has a promiscuous ligand binding site that can interact with a broad array of synthetic and natural ligands (Denison and Nagy 2003; Guyot et al. 2013). Exogenous and endogenous ligands critically shape AhR functions. The environmental pollutant TCDD dioxin is commonly used as surrogate model ligand for AhR, although its use is problematic in revealing true functions of AhR, because dioxin is not easily metabolized and eliminated from the body (Bunger et al. 2008). Indeed, a proper understanding of AhR biology must differentiate between effects triggered by toxic ligands, such as dioxin, and physiological effects triggered by endogenous or other exogenous AhR ligands. On considering the conformation-based model of agonism, it is likely that different AhR ligands preferentially bind distinct conformations of the AhR—each having a distinct set of crucial fingerprint residues—thus initiating different pathways of downstream signaling and transcriptional events (Nuti et al. 2014). This suggests that structurally and/or functionally distinct AhR ligands have distinct affinities for distinct conformations, each active conformation of the receptor recruiting specific downstream signaling molecules. Thus, in selected tissues, both ligand-intrinsic and cell-intrinsic factors may dictate the outcome of AhR activation. Recent data demonstrate that AhR is a convergence point of multiple signaling pathways that inform the cell of its external and internal environments (Tian et al. 2015). Within this scenario, three major factors appear to contribute to the outcome of gene transcriptional regulation by AhR, namely, nature of the ligand, local tissue microenvironment, and presence of coactivators in the cell. Altogether, these recent data suggest that simple extrapolation of the outcomes obtained for one ligand to all others, would not correctly reflect AhR biological activities. Although it is possible that AhR might function in the absence of a ligand (i.e. ligand independent AhR activation), overwhelming evidence suggests that ligands are required (Denison and Nagy 2003). The list of AhR ligands is very broad and striking in its diversity; however, it remains to be determined which of these are actually relevant in a physiological or immunotoxic sense. AhR has not yet been crystallized, so information on ligand-dependent structural changes is currently lacking. Ligand-protected protease digestion studies indicated that only one binding pocket for ligands exists (Kronenberg et al. 2000). Because different ligands result in different outcomes, it is generally assumed that the AhR signaling pathway uses ligands in a sort of adaptive fashion, depending on the tissue and specific environment. Heart and lung extracts (Chiaro et al. 2007) have been shown to contain AhR activators and several molecules have been identified as putative ligands such as, arachidonic acid and leukotrienes, heme metabolites and UV photoproducts of tryptophan (Nguyen and Bradfield 2008). The presence of AhR-activating compounds, such as oxidized low-density lipoprotein (LDL), has also been demonstrated in sheared human and bovine sera (McMillan and Bradfield 2007). Prototypical examples of endogenous AhR ligands are represented by microbiota-derived indole-3-aldehyde (IAld) active on gut innate lymphoid cells (Zelante et al. 2013); 6-formylindolo [3,2-b] carbazole (FICZ) important in skin keratinocytes (Di Meglio et al. 2014). Moreover, the tryptophan metabolite L-kynurenine was found to regulate dendritic cells and other cells in lymphoid tissues (Bessede et al. 2014), 2-(1'H-indole-3'-carbonyl)-thiazole-4-carboxylic acid methyl ester (ITE) identified in lung tissues (Song et al. 2002) and indoxyl sulfate (IS) in human produced by primary hepatocytes (Schroeder et al. 2010). Therefore, AhR not only elicits protection against environmental toxic molecules, but also serves as a sensor for invading pathogenic microbes (i.e. AhR-microbiome), suggesting a complex and diverse repertoire of AhR functions. Moreover, diet, particularly vegetable flavonoids, fruits and teas, are an important sources of AhR ligands (Quintana 2013). Flavonoids represent the largest group of naturally occurring dietary AhR ligands, which can have either agonist on antagonist activities on AhR activation (Ashida et al. 2008). Usually, dietary AhR ligands have low affinity for AhR, but are converted into high-affinity ligands by poorly characterized enzymatic reactions. For example, Bradfield and co-workers reported that the d-amino acid oxidase enzyme generates AhR ligands from tryptophan degradation (Nguyen et al. 2009). In addition, naturally derived AhR agonist tryptophan metabolite indole-3-carbinol (I3C) was reported to contribute to the maintenance of intestinal intraepithelial lymphocytes (IELs) protecting the gut epithelium from bacterial infection (Li et al. 2011). The pathophysiological roles of the natural ligands other than tryptophan-derived chemicals, need to be further elucidated. In addition to classical AhR ligands, new AhR xenobiotic activators, that are substantially different from polycyclic aromatic hydrocarbons have been identified and characterized (Ashida et al. 2008). These ligands have been described at the beginning as AhR antagonists because they usually antagonize CYP induction (Ashida et al. 2008). However, recent studies have demonstrated that these molecules activate the AhR, in a different fashion, through alternative AhR-mediated transcriptional responses.

1.3   AhR Plasticity: Alternative AhR Transcriptional Responses

The effect of a particular AhR ligand may depend as much on the adaptive interactions that it establishes with pathways and proteins expressed in a specific cell or tissue. Several lines of evidence now suggest that AhR is characterized by a high degree of signaling plasticity. In addition to the classical activation pathway, AhR is also involved in non-genomic and in various nuclear crosstalk pathways (Carlson and Perdew 2002). For instance, AhR can bind to the retinoblastoma protein (Puga et al. 2000), estrogen receptor (ER), the transcription factor E2F1 (Marlowe et al. 2008) and to the NFkB subunits RelA and RelB (Tian 2009). Crosstalk of AhR with other signaling pathways, such as via kinases (src, JNK, p38, MAPK) or competition for transcription cofactors, has also been reported (Esser and Rannug 2015). Bioinformatic analysis points to the existence of complex signal cross-talk between AhR and other transcription factors or transcription co-activators (Esser and Rannug 2015). The downstream targets of these pathways differ; for example, Cyp1A1 or specific cytokines, such as IL-2 can be induced via canonical signaling (Jeon and Esser 2000) while IL-8 can be induced via RelB (Vogel et al. 2007). In addition to the transcriptional activity, AhR non-genomic pathways does not involve translocation of the receptor to the nucleus, but instead contributes to a cascade of signaling events, including phosphorylation of mitogen activated protein kinases (MAPKs), activation of cytosolic phospholipase A2 (cPLA2), increase in intracellular calcium, release of arachidonic acid and induction of inflammatory mediators (Li and Matsumura 2008; Matsumura 2009; Sciullo et al. 2008). In nucleated cells, this pathway has been studied by mutating the AhR nuclear localization sequence thereby preventing AhR nuclear translocation (Sciullo et al. 2008), as well as by short interfering RNA technology, blocking the formation of AhR/ARNT heterodimers, required for initiation of transcription of xenobiotic response genes (Oesch-Bartlomowicz et al. 2005). One of these pathways can be activated by a dioxin-like compound and results in the release of c-SRC kinase from the AhR/HSP90/c-SRC complex, enabling c-SRC to target multiple cellular targets involved in cell proliferation and migration (Quintana and Sherr 2013). Moreover, AhR can function as an ubiquitin ligase, targeting specific interacting proteins, for proteasomal degradation (Ohtake et al. 2007). The proteasome system catalyzes the degradation of ubiquitinated proteins, an important process for cellular homeostasis (Hershko and Ciechanover 1998). Ubiquitin is covalently attached to target proteins through a sequence of reactions catalyzed by ubiquitin activating enzyme (E1), ubiquitin-conjugating enzyme (E2), and ubiquitin protein ligase (E3). E3 ubiquitin ligases control the specificity of the ubiquitination reaction (Hershko and Ciechanover 1998). Ohtake et al. (2007) demonstrated that the AhR is a ligand-dependent E3 ubiquitin ligase involved in the degradation of AhR-interacting proteins such as the estrogen receptor. This surprising finding suggests that, by means of its E3 ligase activity, AhR might promote the degradation of specific interacting partners. It is also clear that the phosphorylation state of AhR/ARNT heterodimer is a crucial determinant of biological responses to AhR ligands. Specifically, inhibition of serine/threonine phosphatases results in increased expression of AhR-controlled genes (Puga et al. 2009). Notably, phosphorylation of AhR can represent a ligand-independent activation mechanism. Additional evidence indicates that in a hepatocyte cell line, an increase in intracellular cAMP concentration, obtained by treatment with a membrane-permeating form of cAMP or with forskolin, a potent adenylate cyclase activator, addressed AhR into the nucleus in a PKA-dependent manner. Moreover, protein kinase A PKA-activated AhR was able to interact with canonical XRE, but was unable to transactivate CYP1A1 reporter gene, showing highest affinity for a slightly different consensus sequence (Oesch-Bartlomowicz et al. 2005).

All that suggests the existence of a novel signaling pathway which doesn't involve AhR/ARNT heterodimer formation, evoking unknown alternative responses. Another mechanism by which AhR can be activated is provided by p38, a member of MAPK family by a pathway that appear to be related to cell density. Specifically, in keratinocytes grown at low cell density or at confluence in a calcium-deficient medium, loss of cell-cell contact triggers signals promoting AhR translocation to the nucleus and expression of XRE-controlled reporter genes. In this case, AhR activation occurs through p38-mediated phosphorylation of serine 68 in the NES domain, which drives the receptor into the nuclear compartment (Ikuta et al. 2004). Numerous studies have identified co-regulators as key points of AhR signaling cross talk, and it is likely that additional interactions between AhR and signaling cascades through co-regulators will be uncovered.

As interactions between AhR and individual cell signaling pathways are further clarified, the interplay between multiple pathways should be deciphered and eventually allowing determination of the importance of the AhR in specific cell signaling.

2 AhR: A Multitasking Effector in Allergy Regulation

Recently identified as an immune regulator metabolic receptor, the ligand-activated transcription factor AhR mediates also toxicity to a variety of pollutant molecules, either by promoting gene expression or by physically interacting with other transcription factors. This multivariate response of AhR participates in the pathophysiology of allergy, in which a multitude of immune effector and regulatory cells are involved.

2.1 Ubiquitous AhR Expression and Activity in Multiple Cell Types Impacting on Allergic Responses

AhR Activation in DCs, T Helper Cells, Eosinophils and Mast Cells
Dendritic cells, the "pacemakers" of the immune system, were initially recognized as targets of the AhR ligand TCDD. Specifically, TCDD by binding to AhR was found to block the nuclear translocation of NF-kB/Rel (RelA and RelB) transcription factors to the nucleus reducing DC potential in T cell priming (Ruby et al. 2002). In the very recent years, other AhR ligands, either exogenous or endogenous, have been found to regulate specific DC functions by various mechanisms. In this regard, the AhR ligand, lipoxin A4, was shown to induce the suppressor of cytokine signaling protein 2 (SOCS-2), resulting in decreased DC activation (Machado et al. 2006). Moreover, it was reported that the major environmental pollutant benzo(a) pyrene (BP) suppressed growth and functional differentiation of mouse bone marrow (BM)-derived DCs in vitro (Hwang et al. 2007). Accordingly, the synthetic AhR ligand, VAF347, reduced the production of the pro-inflammatory cytokine IL-6 in DC subsets, leading to suppression of T helper (Th) cell differentiation, ultimately translating to an anti-allergic phenotype (Lawrence et al. 2008).

Inhibition of Th type 2 cell differentiation was also observed for another AhR agonist, M50367, an anti-inflammatory agent that may directly block the differentiation of naive T cells into Th2 cells by suppression of GATA-3 expression (Negishi et al. 2005). Oral treatment with this compound reduced Th2 response in vivo and suppressed disease progression in a murine model of atopic asthma (Negishi et al. 2005). In the contest of allergy, it has been proven that Th2 cells are inhibited in their differentiation thanks to AhR activation in lung DCs, which reduces Th2 expansion in an ovalbumin (OVA) mouse model (Thatcher et al. 2016). Thus, AhR when expressed in DCs and Th cells, appears to be a critical target for drugs and environmental chemicals initiating immunomodulatory effects. Among those, in addition to suppressing cytokine production, the AhR ligand TCDD was found to induce the immunoregulatory pathway of indoleamine 2,3-dioxygenase 1 (IDO1) and indoleamine 2,3-dioxygenase 2 (IDO2) in DCs (Vogel et al. 2008).

IDO1 metabolizes the essential amino acid tryptophan, producing the endogenous AhR ligand L-kynurenine in DCs, that down modulates inflammatory cytokine gene transcription, and induces transcription of genes encoding for the anti-inflammatory cytokines IL-10 and TGF-β. This mechanism involves the participation of non-canonical NF-κB family members and other molecules including phosphorylated IDO1, which acts as a substrate for AhR complex-associated Src kinase activity promoting Treg expansion (Bessede et al. 2014). Altogether, these studies highlight the importance of DCs in mediating immunomodulation operated by AhR ligands. Moreover, the outcome activation will be largely dictated by the nature of the ligand—which is able to initiate a specific sequence of downstream signaling events in a specific cell types.

Recent evidences, suggest that AhR fingerprint residues, required for activation by dioxin, are distinct from those necessary for activation by kynurenine, even when the response being measured was the same, namely, transcription of a gene (Cyp1a1), whose promoter contains AhR specific response elements. Notably, a mutated form of the receptor that does not bind kynurenine was able, instead, to bind dioxin with increased potency and affinity (Bessede et al. 2014). This suggests that structurally distinct AhR ligands have different affinities for distinct conformations, each active conformation of the receptor recruiting a unique set of downstream signaling molecules.

In addition to DCs, AhR is ubiquitously expressed in several immune cell types. Specific studies exploiting AhR functions in T and B cells have been performed in the last years. These studies, mostly using TCDD as prototypical AhR ligand, showed a clear function of AhR in regulating both T effector and humoral responses (Jeon and Esser 2000; Kerkvliet et al. 2002).

Specifically, complete opposite outcomes of AhR activation have been demonstrated in T cells. In CD4[+] T cells, TCDD binding to AhR, surprisingly suppresses experimental autoimmune encephalomyelitis (EAE) by promoting the development of Foxp3[+] Treg cells (Quintana et al. 2008), whereas FICZ enhanced EAE by inducing the differentiation of IL-17-producing T cells, but only when administered locally with the antigen (Marshall and Kerkvliet 2010). Interestingly when FICZ was administered systemically it was able to reduce signs of EAE (Duarte et al.

2013). In addition to Treg expansion, Wu et al. (2011) reported that the cytokine IL-27 could promote the development of Type 1 regulatory T (Tr1) cells mediating tolerogenic responses, by a mechanism requiring AhR activation. Understanding ligand-specific interference with AhR in immune responses will be pivotal in explain these different outcomes in a specific cell type.

In Th17 cells, activation of AhR is also a prerequisite for the production of the cytokine IL-22. IL-22 is linked to proinflammatory processes such as dermal inflammation, psoriasis, inflammatory bowel disease, and Crohn's disease. Moreover, IL-22 induces the production of antimicrobial proteins that are needed for the defense against pathogens in the skin and gut (Aujla et al. 2008; Wolk et al. 2006).

Innate lymphoid cells (ILCs), particularly type 3 innate lymphoid cells (ILC3) also express high levels of AhR, that protect against inflammatory bowel diseases, psoriasis, lung fibrosis *via* AhR-mediated IL-22 production (Stange and Veldhoen 2013).

Eosinophils play an essential role in the pathology of asthma because they contribute to tissue injury, vascular leakage, mucus secretion, and tissue remodeling, by releasing cytotoxic granule proteins, ROS, and lipid mediators. Peroxisome proliferator-activated receptor (PPARs) are among the important ligand-activated transcription factors that regulate the expression of genes involved in many cellular functions, including differentiation, immune responses, and inflammation (Issemann and Green 1990). It has been lately shown that PPARγ induction is suppressed during the activation of AhR by TCDD in eosinophils (Hanlon et al. 2003). Clarification of the interaction between AhR and PPARγ signals should broaden our understanding not only of the functional role of eosinophils but also of asthma regulation. Mast cells (MC) express high levels of AhR (Sibilano et al. 2012) and a recent study demonstrated that AhR knockout mice display a reduced number of MC and an impaired c-Kit expression (Zhou et al. 2013). MC are found in most tissues of the body, particularly in locations that are in close contact with the external environment, such as skin, airways, and intestine, representing a potential target cells for AhR ligands. Although MCs are best known for their role in mediating allergic diseases, for their stimulation of the high-affinity immunoglobulin E (IgE) receptor FcεRI (Gilfillan and Tkaczyk 2006), recent studies have highlighted the important role that these cells play in the protection against infection in a variety of organisms, suggesting a role in regulation of inflammatory responses mainly through IL-10 production (Hakim-Rad et al. 2009). Of interest, comparable levels of AhR expression between Th17 cells and MCs have been described (Sibilano et al. 2012). Moreover, it has been reported that short exposure to FICZ triggered MC degranulation, while long term exposure to the same AhR ligand reduced antigen-stimulated MC degranulation (Sibilano et al. 2012). These results highlight once again that the environment can profoundly influence the outcome of a specific AhR ligand in a selected cell.

AhR Activation in Tissue Barrier Cells

We are continuously exposed to pollutants, containing a variety of AhR ligands that may be inhaled, ingested or come into contact with skin, affecting host metabolism and immune responses (Neff-LaFord et al. 2007). Of interest, work from different groups has shown that AhR-deficient mice have an inherent weakness of the gut barrier (Lee et al. 2012; Li et al. 2011). The high availability of xenobiotics acting as AhR ligands such as dietary components, microbiome metabolites, and ubiquitous environmental pollutants makes the gut and lung epithelial barrier functions particularly dependent on AhR system. In this regards, AhR is highly expressed in intestinal epithelial cells (IECs) and in cells of gut-associated mucosal immune system both in mice and humans. Recently, we have demonstrated that IECs release a variety of antimicrobial peptides in response to IL-22, which is triggered by AhR$^+$ ILC3. The activation of AhR was mainly mediated by commensal metabolites and in particular by IALD, an indole-derivative of tryptophan (Zelante et al. 2013).

In addition to gut barrier cells, AhR is highly expressed in the lung, especially in ciliated and Clara cells, partly contributing to lung inflammation by promoting the release of IL-22 from invariant $\gamma\delta$ T cells (Vg6/Vd1) (Simonian et al. 2010). Specific studies reported both activation and exacerbation of inflammatory lung responses induced upon TCDD administration in vivo (Teske et al. 2005). Accordingly, it was found that TCDD increased expression of inflammatory cytokines, mucin 5AC, and a number of matrix metalloproteases in whole-lung samples in an in vivo C57BL/6 mouse model. Interestingly, these effects were not detected in mice in which AhR signaling was repressed (Wong et al. 2010). Moreover, AhR-mediated mucin increase was partially mediated by increased reactive oxygen species (Chiba et al. 2011).

In the skin, AhR is involved in multiple processes such as UV immunosuppression, pigmentation and proliferation of melanocytes (Bock and Kohle 2009; Esser et al. 2013). Dermal fibroblasts as well as keratinocytes express high levels of AhR. It has been shown that AhR signaling via the endogenous ligands contributes to decrease skin inflammation (Di Meglio et al. 2014).

All these data point to a crucial role of AhR in regulating tissue homeostasis acting on both immune and not immune cells. It remains to be clarified what are the direct target genes of AhR in different tissues and cells and how these can be linked to the regulation of the inflammatory networks. The postulated extensive interaction of AhR with other transcription factors, which may be cell type specific, presents a substantial challenge in identifying direct as well as indirect AhR-mediated effects in inflammatory responses.

2.2 AhR Role in Allergic Asthma

Allergic asthma is a quite common disease with a very complex immunological derangement. It has been considered for a long time as Th2-biased allergic airway disease, characterized by IgE reactivity to common inhaled allergens and airway remodeling. More recently considered as a syndrome, allergic asthma affects individuals with a heterogeneous presentation and distinct biological mechanisms, the latter better defined as endotypes.

Increased AhR expression has been found in asthmatic patients (Zhou et al. 2013). Moreover, animal models of allergic asthma have revealed specific impact of AhR in this disease. In a mouse model of OVA-induced allergic asthma (Jeong et al. 2012) FICZ intraperitoneal injection reduced Th2 differentiation and suppressed pulmonary eosinophilia, thus decreasing the pathophysiology of the disease (Jeong et al. 2012). Also, in a non-eosinophilic asthma model, gavage with 10 μg/kg TCDD resulted in reduced airway inflammation and hyper responsiveness, reduced Th17 cytokine expression and Th17 cell differentiation, and promoted Treg differentiation (Li et al. 2016). Similar outcomes were obtained by treating mice systemically with indoxyl 3-sulfate (Hwang et al. 2014). On the contrary, oral exposure to a common environmental AhR-binding molecule, belonging to the family of the alkylphenols (i.e., 4-nonylphenol), resulted in more severe OVA-induced allergic lung inflammation in mice (Suen et al. 2013).

Another interesting point is that the direct exposure of bronchial epithelial cells to AhR agonists strongly upregulated matrix metalloproteinase activity, which has been found to be increased in asthmatic patients (Tsai et al. 2014). In another study, TCDD induced AhR specific downstream genes *cyp1a1* and *cyp1b1*. Importantly, *in vitro* fibroblast exposure to TCDD triggered TGFβ1 release, affecting fibroblast differentiation, cell proliferation, and airway remodeling (Zhou et al. 2014). The direct impact of TCDD on bronchial epithelial cell function has been reported also recently. Indeed, TCDD-mediated AhR activation triggered the release of vascular endothelial growth factor, which in turn promoted angiogenesis *in vitro* (Tsai et al. 2015). AhR is also activated by particulates, released by vehicular traffic. Specifically, vehicular exhaust particulates activates Jagged-Notched pathway in AhR-dependent pathway, thus exacerbating lung inflammation (Xia et al. 2015).

Intriguingly, peripheral blood mononuclear cells (PBMCs) from patients with allergic asthma further increased their baseline IL-22, while IL-17A production decreased, after exposure to diesel exhaust particle (DEP)-polycyclic aromatic hydrocarbons (PAH) and BP. DEP-PAH-mediated IL-22 production was dependent on AhR activation, but not BP-mediated IL-22 release, the latter involving PI3K and JNK activation (Ple et al. 2015). The involvement of both AhR and IL-22 in the pathogenesis of allergic asthma and its severity has been revealed also by Zhu et al. (2011). These Authors, found that AhR mRNA expression in PBMCs and circulating IL-22 levels were higher in patients with allergic asthma than in healthy controls; moreover, plasma IL-22 levels were associated with impaired lung function.

Therefore, a direct activation of AhR at the epithelial respiratory layer seems to exacerbate the pathophysiology of allergic asthma whereas the systemic AhR recruitment may favor its immunomodulatory action, which is required for the regulation of lung inflammation.

2.3 AhR Role in Atopic Dermatitis

Atopic dermatitis (AD) is a chronically relapsing inflammatory skin disease of unknown origin. It is commonly diagnosed in childhood, and often associated to food allergy. When followed by allergic rhinitis and asthma later in childhood, the phenomenon is called 'atopic march'. The environment, diet and microbiome contribute to disease severity and may drive the therapeutic response. Interestingly, the exposure to air pollutants and allergens increases AD prevalence, exacerbates its progression and impairs clinical outcomes (Bieber 2008).

The disease is mainly caused by an imbalance of Th2 response and an extraordinary abundance of T cells producing IL-22 (T22) cells. These cells may be either CD4[+] T cells (Th22) or CD8[+] T cytotoxic cells (Tc22) infiltrating the skin layers and releasing high amount of IL-22. A positive correlation between Tc22 and AD severity has been recently found (Nograles et al. 2009). AhR is vastly studied in skin diseases, although the outcome of its activation is still poorly understood. Indeed, AhR may be either detrimental or protective in different skin disorders (Haarmann-Stemmann et al. 2015). Therefore, AhR may be considered to behave as a Janus-Faced molecule.

The extreme importance of AhR in the skin is mainly due to the great quantity of AhR ligands in the skin. Ligands may be of different origin. TCDD for example, together with PAH or BP are exogenous or synthetic. A second group of AhR ligands includes the exogenous/natural ligands produced by plants metabolism (e.g., flavonoids, indolcarbinols). Also, additional ligands may derive from skin microbial commensals. Thus, for instance, *Malassezia furfur* is able to metabolize tryptophan into multiple indole derivatives (e.g., malassezin, indolo[3,2-b] carbazole or ICZ, pityriacitrin, indirubin and 6-formyl-indolo[3,2-b]carbazole or 6-FICZ), which are potent AhR activators (Mexia et al. 2015). Lastly, endogenous AhR ligands, the kynurenines, are synthetized mainly from the amino acid tryptophan (Bessede et al. 2014). Kynurenines are generated by the activity of IDO in various skin immune cells. Importantly, UVB rays are absorbed by free tryptophan in the epidermal cells, forming the photoproduct FICZ (Denison and Nagy 2003; Esser et al. 2013).

The constitutive activation of AhR in keratinocytes recapitulated the AD phenotype (Tauchi et al. 2005) and induction of a number of AhR target genes such as *Il-33*, thymic stromal lymphopoietin (*Tslp*) and artemin (*Artn*), mimicking AD disease in mice (Hidaka et al. 2017). Artemin in particular, is important in triggering the intra-epidermal elongation of pruritus-related nerves in the skin (Hidaka et al. 2017). Another important observation is that chronic topic exposure of keratinocytes to an air pollutant AhR ligand 7,12-dimethylbenz[a]anthracene caused the acquisition of an AD-like phenotype, while FICZ did not. The impact of different ligands on the same receptor was mainly due to metabolic disruption of the ligand itself (i.e. FICZ metabolism in keratinocytes) preventing the upregulation of genes involved in the AD pathology (Hidaka et al. 2017). In a recent study it was reported that AhR mediates an anti-inflammatory feed-back mechanism in human Langerhans cells. Notably, AhR activation by FICZ reduced FcεRI and up-regulates IDO expression in human Langerhans cells in vitro (Koch et al. 2017). This AhR-mediated anti-inflammatory feedback mechanism may dampen the allergen-induced inflammation in AD. In addition, skin painting with BP stimulated Langerhans cell migration and Th17/Th2 response in a mouse model of AD.

These results highlight the detrimental effects of skin exposure to pollutants and their contribution to skin sensitization and AD progression (Hong et al. 2016). On the contrary, topical application of coal tar, which contains large amount of PAHs and has been traditionally used to cure AD, was found to trigger AhR, to upregulate the skin barrier protein filaggrin, and downregulate Th2 responses in the skin (van den Bogaard et al. 2013). Therefore, AhR may also help to create a sort of epithelial barrier, which avoids skin exposure to aberrant ligands, mediating AD disease. These results are supported by a recent study by Takei et al. showing that soybean tar Glyteer (i.e., a traditional Japanese alternative to coal tar) activated AhR in keratinocytes and upregulated filaggrin expression in an AhR-dependent manner.

2.4 AhR Role in Food Allergy

Food allergy (FA) is described as an 'adverse effect' to a given food. The incidence is increasing worldwide in both young and adult individuals, and clinical manifestations may range from mild allergic reactions to severe cases of anaphylaxis. Particularly, milk, nuts, fish, peanut, wheat and soy are particularly rich in allergens (Boyce et al. 2010; Sicherer and Sampson 2010). Gastric epithelium permeation of allergens triggers antigen presentation and polarization of Th2 cells, which release typical type 2 cytokines such as IL-4, IL-5, and IL-13. Allergen-mediated B cell activation promotes MC sensitization and eventually secondary responses with histamine release and local inflammation (Boyce et al. 2010). Occurrence of the disease is mainly due to oral tolerance disruption to food antigens (Sicherer and Sampson 2010).

Importantly, *Foxp3* deficiency strongly decreases Treg population and is accompanied by food allergy, as in polyendocrinopathy enteropathy X-linked syndrome (Torgerson et al. 2007). Accordingly, a possible treatment of FA is allergen-specific immunotherapy (SIT), which causes specific desensitization by inducing peripheral T cell tolerance and Treg differentiation (Fujita et al. 2012). In this regard, it has been shown that dioxin-like compounds suppress allergic sensitization by maintaining $CD4^+$ $CD25^+$ $Foxp3^+$ Treg cells and decreasing DC-mediated antigen presentation (Schulz et al. 2011, 2012a, 2013). Interestingly, it has been reported, that TCDD suppressed development of Th2-mediated food allergic responses. Specifically, TCDD treatment increased Treg cells in lymph nodes in mice sensitized to peanut, by an AhR dependent mechanism (Schulz et al. 2011). Similarly, in another study, TCDD, but not FICZ and b-naphthoflavone (b-NF) suppressed sensitization to peanut by increasing the percentage of Tregs (Schulz et al. 2012b).

AhR ligands normally present in food and dietary AhR ligands might influence the onset of intestinal diseases such as FA. Common AhR ligand-rich foods are cruciferous vegetables like broccoli, Brussels sprouts, cauliflower. These vegetables contain I3C, which is a secondary metabolite of glucobrassicin, that is converted in different AhR agonists such as ICZ, 3,3'-diindolylmethane (DIM), and 2-(indol-3-ylmethyl)-3,3'-diindolylmethane (LTr-1) by gastric acidity (Loub et al. 1975). In addition, fruits and vegetables are other source of AhR ligands, including flavonoids.

Other natural compounds such as baicalein (5,6,7-tritydroxyflavone), a natural flavone isolated from *Scutellaria baicalensis*, has been shown to improve desensitization in OVA-induced food allergy, through expansion of Treg cells and *via* AhR (Bae et al. 2016b). The herbal medicine Poria cocos bark (PCB) has been recently proposed to modulate AhR and FA. The orally administered PCB extract ameliorated symptoms of AD and FA. Moreover, PCB inhibited Th2-related cytokines and increased the population of Foxp3$^+$CD4$^+$Tregs in both AD and FA models. Intriguingly, the PCB-mediated differentiation of Foxp3$^+$CD4$^+$ Tregs resulted to be dependent on AhR activation (Bae et al. 2016a).

Future research should elucidate whether and how AhR activation can be used to interfere in food allergic responses in humans and in animals. This may lead to new prevention strategies and therapeutic possibilities for food allergy.

3 Conclusions

We reviewed studies on the relationship between AhR function in immunity and in specific allergic diseases (i.e. allergic asthma, AD, FA). Both environmental molecules and endogenous ligands continuously shape our immune responses. It seems increasingly apparent that AhR acts as a sensor of an extremely large number of endogenous and external molecules. This interaction results in several transcriptional responses that may depend on the specific cell type and microenvironment. The outcomes of AhR activation may impacts on T cells homeostasis, intraepithelial lymphocytes, lymphoid follicles, DC function, mast cell differentiation and growth and expansion of Treg cells. Taken together, these studies suggest that AhR may play a critical role in the regulation of allergic and inflammatory diseases (Fig. 1). This suggests that AhR pathway responses might provide potential therapeutic targets to treat allergic diseases.

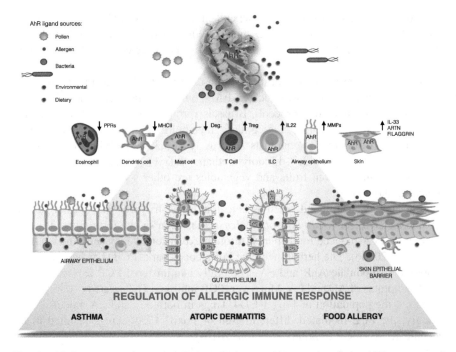

Fig. 1 AhR immune regulatory functions at the host-environment interfaces. AhR, a sensor of environmental or endogenous signals is highly expressed in immune cells such as eosinophils, DCs, MCs, T cells and ILCs and in the airways or skin epithelial cells. AhR specific ligands are derived from different sources: environmental (e.g. pollutants, pollen and allergens), dietary (e.g. food components), and bacterial (e.g. bacteria metabolites). Although the impact of AhR activation may diverge in selected cells (*upper panel*) and in specific tissues (*lower panel*); AhR-mediated immune regulation prevails, thereby preventing inflammatory type 2 allergic responses, as in the context of allergic asthma, atopic dermatitis or food allergy

References

Ashida H, Nishiumi S, Fukuda I (2008) An update on the dietary ligands of the AhR. Expert Opin Drug Metab Toxicol 4(11):1429–1447. https://doi.org/10.1517/17425255.4.11.1429

Aujla SJ, Chan YR, Zheng M, Fei M, Askew DJ, Pociask DA, Reinhart TA, McAllister F, Edeal J, Gaus K, Husain S, Kreindler JL, Dubin PJ, Pilewski JM, Myerburg MM, Mason CA, Iwakura Y, Kolls JK (2008) IL-22 mediates mucosal host defense against Gram-negative bacterial pneumonia. Nat Med 14(3):275–281. https://doi.org/10.1038/nm1710

Bae MJ, See HJ, Choi G, Kang CY (2016a) Regulatory T cell induced by Poria cocos bark exert therapeutic effects in murine models of atopic dermatitis and food allergy. Mediators Inflamm 2016:3472608. https://doi.org/10.1155/2016/3472608

Bae MJ, Shin HS, See HJ, Jung SY, Kwon DA, Shon DH (2016b) Baicalein induces CD4(+) Foxp3(+) T cells and enhances intestinal barrier function in a mouse model of food allergy. Sci Rep 6:32225. https://doi.org/10.1038/srep32225

Bessede A, Gargaro M, Pallotta MT, Matino D, Servillo G, Brunacci C, Bicciato S, Mazza EM, Macchiarulo A, Vacca C, Iannitti R, Tissi L, Volpi C, Belladonna ML, Orabona C, Bianchi R,

Lanz TV, Platten M, Della Fazia MA, Piobbico D, Zelante T, Funakoshi H, Nakamura T, Gilot D, Denison MS, Guillemin GJ, DuHadaway JB, Prendergast GC, Metz R, Geffard M, Boon L, Pirro M, Iorio A, Veyret B, Romani L, Grohmann U, Fallarino F, Puccetti P (2014) Aryl hydrocarbon receptor control of a disease tolerance defence pathway. Nature 511(7508): 184–190. https://doi.org/10.1038/nature13323

Bieber T (2008) Atopic dermatitis. N Engl J Med 358(14):1483–1494. https://doi.org/10.1056/NEJMra074081

Bock KW, Kohle C (2009) The mammalian aryl hydrocarbon (Ah) receptor: from mediator of dioxin toxicity toward physiological functions in skin and liver. Biol Chem 390(12): 1225–1235. https://doi.org/10.1515/BC.2009.138

Boyce JA, Assa'ad A, Burks AW, Jones SM, Sampson HA, Wood RA, Plaut M, Cooper SF, Fenton MJ, Arshad SH, Bahna SL, Beck LA, Byrd-Bredbenner C, Camargo CA Jr, Eichenfield L, Furuta GT, Hanifin JM, Jones C, Kraft M, Levy BD, Lieberman P, Luccioli S, McCall KM, Schneider LC, Simon RA, Simons FE, Teach SJ, Yawn BP, Schwaninger JM, NIAID-Sponsored Expert Panel (2010) Guidelines for the diagnosis and management of food allergy in the United States: summary of the NIAID-Sponsored Expert Panel report. J Allergy Clin Immunol 126(6):1105–1118. https://doi.org/10.1016/j.jaci.2010. 10.008

Brauze D, Widerak M, Cwykiel J, Szyfter K, Baer-Dubowska W (2006) The effect of aryl hydrocarbon receptor ligands on the expression of AhR, AhRR, ARNT, Hif1alpha, CYP1A1 and NQO1 genes in rat liver. Toxicol Lett 167(3):212–220. https://doi.org/10.1016/j.toxlet. 2006.09.010

Bunger MK, Glover E, Moran SM, Walisser JA, Lahvis GP, Hsu EL, Bradfield CA (2008) Abnormal liver development and resistance to 2,3,7,8-tetrachlorodibenzo-p-dioxin toxicity in mice carrying a mutation in the DNA-binding domain of the aryl hydrocarbon receptor. Toxicol Sci 106(1):83–92. https://doi.org/10.1093/toxsci/kfn149

Carlson DB, Perdew GH (2002) A dynamic role for the Ah receptor in cell signaling? Insights from a diverse group of Ah receptor interacting proteins. J Biochem Mol Toxicol 16(6): 317–325. https://doi.org/10.1002/jbt.10051

Chiaro CR, Patel RD, Marcus CB, Perdew GH (2007) Evidence for an aryl hydrocarbon receptor-mediated cytochrome p450 autoregulatory pathway. Mol Pharmacol 72(5):1369–1379. https://doi.org/10.1124/mol.107.038968

Chiba T, Uchi H, Tsuji G, Gondo H, Moroi Y, Furue M (2011) Arylhydrocarbon receptor (AhR) activation in airway epithelial cells induces MUC5AC via reactive oxygen species (ROS) production. Pulm Pharmacol Ther 24(1):133–140. https://doi.org/10.1016/j.pupt.2010.08.002

Denison MS, Nagy SR (2003) Activation of the aryl hydrocarbon receptor by structurally diverse exogenous and endogenous chemicals. Annu Rev Pharmacol Toxicol 43:309–334. https://doi.org/10.1146/annurev.pharmtox.43.100901.135828

Di Meglio P, Duarte JH, Ahlfors H, Owens ND, Li Y, Villanova F, Tosi I, Hirota K, Nestle FO, Mrowietz U, Gilchrist MJ, Stockinger B (2014) Activation of the aryl hydrocarbon receptor dampens the severity of inflammatory skin conditions. Immunity 40(6):989–1001. https://doi.org/10.1016/j.immuni.2014.04.019

Duarte JH, Di Meglio P, Hirota K, Ahlfors H, Stockinger B (2013) Differential influences of the aryl hydrocarbon receptor on Th17 mediated responses in vitro and in vivo. PLoS One 8(11): e79819. https://doi.org/10.1371/journal.pone.0079819

Esser C, Rannug A (2015) The aryl hydrocarbon receptor in barrier organ physiology, immunology, and toxicology. Pharmacol Rev 67(2):259–279. https://doi.org/10.1124/pr.114.009001

Esser C, Bargen I, Weighardt H, Haarmann-Stemmann T, Krutmann J (2013) Functions of the aryl hydrocarbon receptor in the skin. Semin Immunopathol 35(6):677–691. https://doi.org/10.1007/s00281-013-0394-4

Fernandez-Salguero PM, Ward JM, Sundberg JP, Gonzalez FJ (1997) Lesions of aryl-hydrocarbon receptor-deficient mice. Vet Pathol 34(6):605–614. https://doi.org/10.1177/030098589703400609

Fujii-Kuriyama Y, Mimura J (2005) Molecular mechanisms of AhR functions in the regulation of cytochrome P450 genes. Biochem Biophys Res Commun 338(1):311–317. https://doi.org/10.1016/j.bbrc.2005.08.162

Fujita H, Soyka MB, Akdis M, Akdis CA (2012) Mechanisms of allergen-specific immunotherapy. Clin Transl Allergy 2(1):2. https://doi.org/10.1186/2045-7022-2-2

Gilfillan AM, Tkaczyk C (2006) Integrated signalling pathways for mast-cell activation. Nat Rev Immunol 6(3):218–230. https://doi.org/10.1038/nri1782

Guyot E, Chevallier A, Barouki R, Coumoul X (2013) The AhR twist: ligand-dependent AhR signaling and pharmaco-toxicological implications. Drug Discov Today 18(9–10):479–486. https://doi.org/10.1016/j.drudis.2012.11.014

Haarmann-Stemmann T, Esser C, Krutmann J (2015) The janus-faced role of aryl hydrocarbon receptor signaling in the skin: consequences for prevention and treatment of skin disorders. J Invest Dermatol 135(11):2572–2576. https://doi.org/10.1038/jid.2015.285

Hahn ME (2002) Aryl hydrocarbon receptors: diversity and evolution. Chem Biol Interact 141(1–2):131–160

Hahn ME, Allan LL, Sherr DH (2009) Regulation of constitutive and inducible AHR signaling: complex interactions involving the AHR repressor. Biochem Pharmacol 77(4):485–497. https://doi.org/10.1016/j.bcp.2008.09.016

Hahn ME, Karchner SI, Merson RR (2017) Diversity as opportunity: insights from 600 million years of AHR evolution. Curr Opin Toxicol 2:58–71. https://doi.org/10.1016/j.cotox.2017.02.003

Hakim-Rad K, Metz M, Maurer M (2009) Mast cells: makers and breakers of allergic inflammation. Curr Opin Allergy Clin Immunol 9(5):427–430. https://doi.org/10.1097/ACI.0b013e32832e9af1

Hankinson O (1995) The aryl hydrocarbon receptor complex. Annu Rev Pharmacol Toxicol 35:307–340. https://doi.org/10.1146/annurev.pa.35.040195.001515

Hanlon PR, Ganem LG, Cho YC, Yamamoto M, Jefcoate CR (2003) AhR- and ERK-dependent pathways function synergistically to mediate 2,3,7,8-tetrachlorodibenzo-p-dioxin suppression of peroxisome proliferator-activated receptor-gamma1 expression and subsequent adipocyte differentiation. Toxicol Appl Pharmacol 189(1):11–27

Hershko A, Ciechanover A (1998) The ubiquitin system. Annu Rev Biochem 67:425–479. https://doi.org/10.1146/annurev.biochem.67.1.425

Hidaka T, Ogawa E, Kobayashi EH, Suzuki T, Funayama R, Nagashima T, Fujimura T, Aiba S, Nakayama K, Okuyama R, Yamamoto M (2017) The aryl hydrocarbon receptor AhR links atopic dermatitis and air pollution via induction of the neurotrophic factor artemin. Nat Immunol 18(1):64–73. https://doi.org/10.1038/ni.3614

Hoffer A, Chang CY, Puga A (1996) Dioxin induces transcription of fos and jun genes by Ah receptor-dependent and -independent pathways. Toxicol Appl Pharmacol 141(1):238–247. https://doi.org/10.1006/taap.1996.0280

Hong CH, Lee CH, Yu HS, Huang SK (2016) Benzopyrene, a major polyaromatic hydrocarbon in smoke fume, mobilizes Langerhans cells and polarizes Th2/17 responses in epicutaneous protein sensitization through the aryl hydrocarbon receptor. Int Immunopharmacol 36:111–117. https://doi.org/10.1016/j.intimp.2016.04.017

Hwang JA, Lee JA, Cheong SW, Youn HJ, Park JH (2007) Benzo(a)pyrene inhibits growth and functional differentiation of mouse bone marrow-derived dendritic cells. Downregulation of RelB and eIF3 p170 by benzo(a)pyrene. Toxicol Lett 169(1):82–90. https://doi.org/10.1016/j.toxlet.2007.01.001

Hwang YJ, Yun MO, Jeong KT, Park JH (2014) Uremic toxin indoxyl 3-sulfate regulates the differentiation of Th2 but not of Th1 cells to lessen allergic asthma. Toxicol Lett 225(1):130–138. https://doi.org/10.1016/j.toxlet.2013.11.027

Ikuta T, Kobayashi Y, Kawajiri K (2004) Cell density regulates intracellular localization of aryl hydrocarbon receptor. J Biol Chem 279(18):19209–19216. https://doi.org/10.1074/jbc.M310492200

Issemann I, Green S (1990) Activation of a member of the steroid hormone receptor superfamily by peroxisome proliferators. Nature 347(6294):645–650. https://doi.org/10.1038/347645a0

Jeon MS, Esser C (2000) The murine IL-2 promoter contains distal regulatory elements responsive to the Ah receptor, a member of the evolutionarily conserved bHLH-PAS transcription factor family. J Immunol 165(12):6975–6983

Jeong KT, Hwang SJ, Oh GS, Park JH (2012) FICZ, a tryptophan photoproduct, suppresses pulmonary eosinophilia and Th2-type cytokine production in a mouse model of ovalbumin-induced allergic asthma. Int Immunopharmacol 13(4):377–385. https://doi.org/10.1016/j.intimp.2012.04.014

Kagey MH, Newman JJ, Bilodeau S, Zhan Y, Orlando DA, van Berkum NL, Ebmeier CC, Goossens J, Rahl PB, Levine SS, Taatjes DJ, Dekker J, Young RA (2010) Mediator and cohesin connect gene expression and chromatin architecture. Nature 467(7314):430–435. https://doi.org/10.1038/nature09380

Karchner SI, Franks DG, Kennedy SW, Hahn ME (2006) The molecular basis for differential dioxin sensitivity in birds: role of the aryl hydrocarbon receptor. Proc Natl Acad Sci USA 103(16):6252–6257. https://doi.org/10.1073/pnas.0509950103

Kerkvliet NI, Shepherd DM, Baecher-Steppan L (2002) T lymphocytes are direct, aryl hydrocarbon receptor (AhR)-dependent targets of 2,3,7,8-tetrachlorodibenzo-p-dioxin (TCDD): AhR expression in both CD4+ and CD8+ T cells is necessary for full suppression of a cytotoxic T lymphocyte response by TCDD. Toxicol Appl Pharmacol 185(2):146–152

Kewley RJ, Whitelaw ML, Chapman-Smith A (2004) The mammalian basic helix-loop-helix/PAS family of transcriptional regulators. Int J Biochem Cell Biol 36(2):189–204

Kiss EA, Vonarbourg C, Kopfmann S, Hobeika E, Finke D, Esser C, Diefenbach A (2011) Natural aryl hydrocarbon receptor ligands control organogenesis of intestinal lymphoid follicles. Science 334(6062):1561–1565. https://doi.org/10.1126/science.1214914

Koch S, Stroisch TJ, Vorac J, Herrmann N, Leib N, Schnautz S, Kirins H, Forster I, Weighardt H, Bieber T (2017) AhR mediates an anti-inflammatory feed-back mechanism in human Langerhans cells involving FcepsilonRI and IDO. Allergy 72(11):1686–1693. https://doi.org/10.1111/all.13170

Kronenberg S, Esser C, Carlberg C (2000) An aryl hydrocarbon receptor conformation acts as the functional core of nuclear dioxin signaling. Nucleic Acids Res 28(12):2286–2291

Lawrence BP, Denison MS, Novak H, Vorderstrasse BA, Harrer N, Neruda W, Reichel C, Woisetschlager M (2008) Activation of the aryl hydrocarbon receptor is essential for mediating the anti-inflammatory effects of a novel low-molecular-weight compound. Blood 112(4):1158–1165. https://doi.org/10.1182/blood-2007-08-109645

Lee JS, Cella M, McDonald KG, Garlanda C, Kennedy GD, Nukaya M, Mantovani A, Kopan R, Bradfield CA, Newberry RD, Colonna M (2012) AHR drives the development of gut ILC22 cells and postnatal lymphoid tissues via pathways dependent on and independent of Notch. Nat Immunol 13(2):144–151. https://doi.org/10.1038/ni.2187

Li W, Matsumura F (2008) Significance of the nongenomic, inflammatory pathway in mediating the toxic action of TCDD to induce rapid and long-term cellular responses in 3T3-L1 adipocytes. Biochemistry 47(52):13997–14008. https://doi.org/10.1021/bi801913w

Li Y, Innocentin S, Withers DR, Roberts NA, Gallagher AR, Grigorieva EF, Wilhelm C, Veldhoen M (2011) Exogenous stimuli maintain intraepithelial lymphocytes via aryl hydrocarbon receptor activation. Cell 147(3):629–640. https://doi.org/10.1016/j.cell.2011.09.025

Li XM, Peng J, Gu W, Guo XJ (2016) TCDD-induced activation of aryl hydrocarbon receptor inhibits Th17 polarization and regulates non-eosinophilic airway inflammation in asthma. PLoS One 11(3):e0150551. https://doi.org/10.1371/journal.pone.0150551

Loub WD, Wattenberg LW, Davis DW (1975) Aryl hydrocarbon hydroxylase induction in rat tissues by naturally occurring indoles of cruciferous plants. J Natl Cancer Inst 54(4):985–988

Ma Q, Baldwin KT (2000) 2,3,7,8-tetrachlorodibenzo-p-dioxin-induced degradation of aryl hydrocarbon receptor (AhR) by the ubiquitin-proteasome pathway. Role of the transcription activaton and DNA binding of AhR. J Biol Chem 275(12):8432–8438

Machado FS, Johndrow JE, Esper L, Dias A, Bafica A, Serhan CN, Aliberti J (2006) Anti-inflammatory actions of lipoxin A4 and aspirin-triggered lipoxin are SOCS-2 dependent. Nat Med 12(3):330–334. https://doi.org/10.1038/nm1355

Marlowe JL, Fan Y, Chang X, Peng L, Knudsen ES, Xia Y, Puga A (2008) The aryl hydrocarbon receptor binds to E2F1 and inhibits E2F1-induced apoptosis. Mol Biol Cell 19(8):3263–3271. https://doi.org/10.1091/mbc.E08-04-0359

Marshall NB, Kerkvliet NI (2010) Dioxin and immune regulation: emerging role of aryl hydrocarbon receptor in the generation of regulatory T cells. Ann NY Acad Sci 1183:25–37. https://doi.org/10.1111/j.1749-6632.2009.05125.x

Matsumura F (2009) The significance of the nongenomic pathway in mediating inflammatory signaling of the dioxin-activated Ah receptor to cause toxic effects. Biochem Pharmacol 77(4):608–626. https://doi.org/10.1016/j.bcp.2008.10.013

McMillan BJ, Bradfield CA (2007) The aryl hydrocarbon receptor is activated by modified low-density lipoprotein. Proc Natl Acad Sci USA 104(4):1412–1417. https://doi.org/10.1073/pnas.0607296104

Mexia N, Gaitanis G, Velegraki A, Soshilov A, Denison MS, Magiatis P (2015) Pityriazepin and other potent AhR ligands isolated from Malassezia furfur yeast. Arch Biochem Biophys 571:16–20. https://doi.org/10.1016/j.abb.2015.02.023

Mimura J, Fujii-Kuriyama Y (2003) Functional role of AhR in the expression of toxic effects by TCDD. Biochim Biophys Acta 1619(3):263–268

Neff-LaFord H, Teske S, Bushnell TP, Lawrence BP (2007) Aryl hydrocarbon receptor activation during influenza virus infection unveils a novel pathway of IFN-gamma production by phagocytic cells. J Immunol 179(1):247–255

Negishi T, Kato Y, Ooneda O, Mimura J, Takada T, Mochizuki H, Yamamoto M, Fujii-Kuriyama-Y, Furusako S (2005) Effects of aryl hydrocarbon receptor signaling on the modulation of TH1/TH2 balance. J Immunol 175(11):7348–7356

Nguyen LP, Bradfield CA (2008) The search for endogenous activators of the aryl hydrocarbon receptor. Chem Res Toxicol 21(1):102–116. https://doi.org/10.1021/tx7001965

Nguyen LP, Hsu EL, Chowdhury G, Dostalek M, Guengerich FP, Bradfield CA (2009) D-amino acid oxidase generates agonists of the aryl hydrocarbon receptor from D-tryptophan. Chem Res Toxicol 22(12):1897–1904. https://doi.org/10.1021/tx900043s

Nograles KE, Zaba LC, Shemer A, Fuentes-Duculan J, Cardinale I, Kikuchi T, Ramon M, Bergman R, Krueger JG, Guttman-Yassky E (2009) IL-22-producing "T22" T cells account for upregulated IL-22 in atopic dermatitis despite reduced IL-17-producing TH17 T cells. J Allergy Clin Immunol 123(6):1244–1252.e1242. https://doi.org/10.1016/j.jaci.2009.03.041

Nuti R, Gargaro M, Matino D, Dolciami D, Grohmann U, Puccetti P, Fallarino F, Macchiarulo A (2014) Ligand binding and functional selectivity of L-tryptophan metabolites at the mouse aryl hydrocarbon receptor (mAhR). J Chem Inf Model 54(12):3373–3383. https://doi.org/10.1021/ci5005459

Oesch-Bartlomowicz B, Huelster A, Wiss O, Antoniou-Lipfert P, Dietrich C, Arand M, Weiss C, Bockamp E, Oesch F (2005) Aryl hydrocarbon receptor activation by cAMP vs. dioxin: divergent signaling pathways. Proc Natl Acad Sci USA 102(26):9218–9223. https://doi.org/10.1073/pnas.0503488102

Ohtake F, Baba A, Takada I, Okada M, Iwasaki K, Miki H, Takahashi S, Kouzmenko A, Nohara K, Chiba T, Fujii-Kuriyama Y, Kato S (2007) Dioxin receptor is a ligand-dependent E3 ubiquitin ligase. Nature 446(7135):562–566. https://doi.org/10.1038/nature05683

Okino ST, Whitlock JP Jr (1995) Dioxin induces localized, graded changes in chromatin structure: implications for Cyp1A1 gene transcription. Mol Cell Biol 15(7):3714–3721

Ple C, Fan Y, Ait Yahia S, Vorng H, Everaere L, Chenivesse C, Balsamelli J, Azzaoui I, de Nadai P, Wallaert B, Lazennec G, Tsicopoulos A (2015) Polycyclic aromatic hydrocarbons reciprocally regulate IL-22 and IL-17 cytokines in peripheral blood mononuclear cells from both healthy and asthmatic subjects. PLoS One 10(4):e0122372. https://doi.org/10.1371/journal.pone.0122372

Pollenz RS, Buggy C (2006) Ligand-dependent and -independent degradation of the human aryl hydrocarbon receptor (hAHR) in cell culture models. Chem Biol Interact 164(1–2):49–59. https://doi.org/10.1016/j.cbi.2006.08.014

Puga A, Barnes SJ, Dalton TP, Chang C, Knudsen ES, Maier MA (2000) Aromatic hydrocarbon receptor interaction with the retinoblastoma protein potentiates repression of E2F-dependent transcription and cell cycle arrest. J Biol Chem 275(4):2943–2950

Puga A, Ma C, Marlowe JL (2009) The aryl hydrocarbon receptor cross-talks with multiple signal transduction pathways. Biochem Pharmacol 77(4):713–722. https://doi.org/10.1016/j.bcp.2008.08.031

Quintana FJ (2013) The aryl hydrocarbon receptor: a molecular pathway for the environmental control of the immune response. Immunology 138(3):183–189. https://doi.org/10.1111/imm.12046

Quintana FJ, Sherr DH (2013) Aryl hydrocarbon receptor control of adaptive immunity. Pharmacol Rev 65(4):1148–1161. https://doi.org/10.1124/pr.113.007823

Quintana FJ, Basso AS, Iglesias AH, Korn T, Farez MF, Bettelli E, Caccamo M, Oukka M, Weiner HL (2008) Control of T(reg) and T(H)17 cell differentiation by the aryl hydrocarbon receptor. Nature 453(7191):65–71. https://doi.org/10.1038/nature06880

Ruby CE, Leid M, Kerkvliet NI (2002) 2,3,7,8-Tetrachlorodibenzo-p-dioxin suppresses tumor necrosis factor-alpha and anti-CD40-induced activation of NF-kappaB/Rel in dendritic cells: p50 homodimer activation is not affected. Mol Pharmacol 62(3):722–728

Schroeder JC, Dinatale BC, Murray IA, Flaveny CA, Liu Q, Laurenzana EM, Lin JM, Strom SC, Omiecinski CJ, Amin S, Perdew GH (2010) The uremic toxin 3-indoxyl sulfate is a potent endogenous agonist for the human aryl hydrocarbon receptor. Biochemistry 49(2):393–400. https://doi.org/10.1021/bi901786x

Schulz VJ, Smit JJ, Willemsen KJ, Fiechter D, Hassing I, Bleumink R, Boon L, van den Berg M, van Duursen MB, Pieters RH (2011) Activation of the aryl hydrocarbon receptor suppresses sensitization in a mouse peanut allergy model. Toxicol Sci 123(2):491–500. https://doi.org/10.1093/toxsci/kfr175

Schulz VJ, Smit JJ, Bol-Schoenmakers M, van Duursen MB, van den Berg M, Pieters RH (2012a) Activation of the aryl hydrocarbon receptor reduces the number of precursor and effector T cells, but preserves thymic CD4+CD25+Foxp3+ regulatory T cells. Toxicol Lett 215(2): 100–109. https://doi.org/10.1016/j.toxlet.2012.09.024

Schulz VJ, Smit JJ, Huijgen V, Bol-Schoenmakers M, van Roest M, Kruijssen LJ, Fiechter D, Hassing I, Bleumink R, Safe S, van Duursen MB, van den Berg M, Pieters RH (2012b) Non-dioxin-like AhR ligands in a mouse peanut allergy model. Toxicol Sci 128(1):92–102. https://doi.org/10.1093/toxsci/kfs131

Schulz VJ, van Roest M, Bol-Schoenmakers M, van Duursen MB, van den Berg M, Pieters RH, Smit JJ (2013) Aryl hydrocarbon receptor activation affects the dendritic cell phenotype and function during allergic sensitization. Immunobiology 218(8):1055–1062. https://doi.org/10.1016/j.imbio.2013.01.004

Sciullo EM, Vogel CF, Li W, Matsumura F (2008) Initial and extended inflammatory messages of the nongenomic signaling pathway of the TCDD-activated Ah receptor in U937 macrophages. Arch Biochem Biophys 480(2):143–155. https://doi.org/10.1016/j.abb.2008.09.017

Sibilano R, Frossi B, Calvaruso M, Danelli L, Betto E, Dall'Agnese A, Tripodo C, Colombo MP, Pucillo CE, Gri G (2012) The aryl hydrocarbon receptor modulates acute and late mast cell responses. J Immunol 189(1):120–127. https://doi.org/10.4049/jimmunol.1200009

Sicherer SH, Sampson HA (2010) Food allergy. J Allergy Clin Immunol 125(2 Suppl 2): S116–S125. https://doi.org/10.1016/j.jaci.2009.08.028

Simonian PL, Wehrmann F, Roark CL, Born WK, O'Brien RL, Fontenot AP (2010) Gammadelta T cells protect against lung fibrosis via IL-22. J Exp Med 207(10):2239–2253. https://doi.org/10.1084/jem.20100061

Sogawa K, Fujii-Kuriyama Y (1997) Ah receptor, a novel ligand-activated transcription factor. J Biochem 122(6):1075–1079

Song J, Clagett-Dame M, Peterson RE, Hahn ME, Westler WM, Sicinski RR, DeLuca HF (2002) A ligand for the aryl hydrocarbon receptor isolated from lung. Proc Natl Acad Sci USA 99(23):14694–14699. https://doi.org/10.1073/pnas.232562899

Stange J, Veldhoen M (2013) The aryl hydrocarbon receptor in innate T cell immunity. Semin Immunopathol 35(6):645–655. https://doi.org/10.1007/s00281-013-0389-1

Suen JL, Hsu SH, Hung CH, Chao YS, Lee CL, Lin CY, Weng TH, Yu HS, Huang SK (2013) A common environmental pollutant, 4-nonylphenol, promotes allergic lung inflammation in a murine model of asthma. Allergy 68(6):780–787. https://doi.org/10.1111/all.12156

Tauchi M, Hida A, Negishi T, Katsuoka F, Noda S, Mimura J, Hosoya T, Yanaka A, Aburatani H, Fujii-Kuriyama Y, Motohashi H, Yamamoto M (2005) Constitutive expression of aryl hydrocarbon receptor in keratinocytes causes inflammatory skin lesions. Mol Cell Biol 25(21): 9360–9368. https://doi.org/10.1128/MCB.25.21.9360-9368.2005

Teske S, Bohn AA, Regal JF, Neumiller JJ, Lawrence BP (2005) Activation of the aryl hydrocarbon receptor increases pulmonary neutrophilia and diminishes host resistance to influenza A virus. Am J Physiol Lung Cell Mol Physiol 289(1):L111–L124. https://doi.org/10.1152/ajplung. 00318.2004

Thatcher TH, Williams MA, Pollock SJ, McCarthy CE, Lacy SH, Phipps RP, Sime PJ (2016) Endogenous ligands of the aryl hydrocarbon receptor regulate lung dendritic cell function. Immunology 147(1):41–54. https://doi.org/10.1111/imm.12540

Tian Y (2009) Ah receptor and NF-kappaB interplay on the stage of epigenome. Biochem Pharmacol 77(4):670–680. https://doi.org/10.1016/j.bcp.2008.10.023

Tian Y, Ke S, Chen M, Sheng T (2003) Interactions between the aryl hydrocarbon receptor and P-TEFb. Sequential recruitment of transcription factors and differential phosphorylation of C-terminal domain of RNA polymerase II at cyp1a1 promoter. J Biol Chem 278(45): 44041–44048. https://doi.org/10.1074/jbc.M306443200

Tian J, Feng Y, Fu H, Xie HQ, Jiang JX, Zhao B (2015) The aryl hydrocarbon receptor: a key bridging molecule of external and internal chemical signals. Environ Sci Technol 49(16): 9518–9531. https://doi.org/10.1021/acs.est.5b00385

Torgerson TR, Linane A, Moes N, Anover S, Mateo V, Rieux-Laucat F, Hermine O, Vijay S, Gambineri E, Cerf-Bensussan N, Fischer A, Ochs HD, Goulet O, Ruemmele FM (2007) Severe food allergy as a variant of IPEX syndrome caused by a deletion in a noncoding region of the FOXP3 gene. Gastroenterology 132(5):1705–1717. https://doi.org/10.1053/j.gastro.2007.02. 044

Tsai MJ, Hsu YL, Wang TN, Wu LY, Lien CT, Hung CH, Kuo PL, Huang MS (2014) Aryl hydrocarbon receptor (AhR) agonists increase airway epithelial matrix metalloproteinase activity. J Mol Med 92(6):615–628. https://doi.org/10.1007/s00109-014-1121-x

Tsai MJ, Wang TN, Lin YS, Kuo PL, Hsu YL, Huang MS (2015) Aryl hydrocarbon receptor agonists upregulate VEGF secretion from bronchial epithelial cells. J Mol Med 93(11): 1257–1269. https://doi.org/10.1007/s00109-015-1304-0

van den Bogaard EH, Bergboer JG, Vonk-Bergers M, van Vlijmen-Willems IM, Hato SV, van der Valk PG, Schroder JM, Joosten I, Zeeuwen PL, Schalkwijk J (2013) Coal tar induces AHR-dependent skin barrier repair in atopic dermatitis. J Clin Invest 123(2):917–927. https://doi.org/10.1172/JCI65642

Vogel CF, Sciullo E, Li W, Wong P, Lazennec G, Matsumura F (2007) RelB, a new partner of aryl hydrocarbon receptor-mediated transcription. Mol Endocrinol 21(12):2941–2955. https://doi. org/10.1210/me.2007-0211

Vogel CF, Goth SR, Dong B, Pessah IN, Matsumura F (2008) Aryl hydrocarbon receptor signaling mediates expression of indoleamine 2,3-dioxygenase. Biochem Biophys Res Commun 375(3): 331–335. https://doi.org/10.1016/j.bbrc.2008.07.156

Wang S, Hankinson O (2002) Functional involvement of the Brahma/SWI2-related gene 1 protein in cytochrome P4501A1 transcription mediated by the aryl hydrocarbon receptor complex. J Biol Chem 277(14):11821–11827. https://doi.org/10.1074/jbc.M110122200

Wang S, Ge K, Roeder RG, Hankinson O (2004) Role of mediator in transcriptional activation by the aryl hydrocarbon receptor. J Biol Chem 279(14):13593–13600. https://doi.org/10.1074/jbc. M312274200

Wolk K, Witte E, Wallace E, Docke WD, Kunz S, Asadullah K, Volk HD, Sterry W, Sabat R (2006) IL-22 regulates the expression of genes responsible for antimicrobial defense, cellular differentiation, and mobility in keratinocytes: a potential role in psoriasis. Eur J Immunol 36(5):1309–1323. https://doi.org/10.1002/eji.200535503

Wong PS, Vogel CF, Kokosinski K, Matsumura F (2010) Arylhydrocarbon receptor activation in NCI-H441 cells and C57BL/6 mice: possible mechanisms for lung dysfunction. Am J Respir Cell Mol Biol 42(2):210–217. https://doi.org/10.1165/rcmb.2008-0228OC

Wu HY, Quintana FJ, da Cunha AP, Dake BT, Koeglsperger T, Starossom SC, Weiner HL (2011) In vivo induction of Tr1 cells via mucosal dendritic cells and AHR signaling. PLoS One 6(8): e23618. https://doi.org/10.1371/journal.pone.0023618

Xia M, Viera-Hutchins L, Garcia-Lloret M, Noval Rivas M, Wise P, McGhee SA, Chatila ZK, Daher N, Sioutas C, Chatila TA (2015) Vehicular exhaust particles promote allergic airway inflammation through an aryl hydrocarbon receptor-notch signaling cascade. J Allergy Clin Immunol 136(2):441–453. https://doi.org/10.1016/j.jaci.2015.02.014

Yin H, Li Y, Sutter TR (1994) Dioxin-enhanced expression of interleukin-1 beta in human epidermal keratinocytes: potential role in the modulation of immune and inflammatory responses. Exp Clin Immunogenet 11(2–3):128–135

Zelante T, Iannitti RG, Cunha C, De Luca A, Giovannini G, Pieraccini G, Zecchi R, D'Angelo C, Massi-Benedetti C, Fallarino F, Carvalho A, Puccetti P, Romani L (2013) Tryptophan catabolites from microbiota engage aryl hydrocarbon receptor and balance mucosal reactivity via interleukin-22. Immunity 39(2):372–385. https://doi.org/10.1016/j.immuni.2013.08.003

Zhou Y, Tung HY, Tsai YM, Hsu SC, Chang HW, Kawasaki H, Tseng HC, Plunkett B, Gao P, Hung CH, Vonakis BM, Huang SK (2013) Aryl hydrocarbon receptor controls murine mast cell homeostasis. Blood 121(16):3195–3204. https://doi.org/10.1182/blood-2012-08-453597

Zhou Y, Mirza S, Xu T, Tripathi P, Plunkett B, Myers A, Gao P (2014) Aryl hydrocarbon receptor (AhR) modulates cockroach allergen-induced immune responses through active TGFbeta1 release. Mediat Inflamm 2014:591479. https://doi.org/10.1155/2014/591479

Zhu J, Cao Y, Li K, Wang Z, Zuo P, Xiong W, Xu Y, Xiong S (2011) Increased expression of aryl hydrocarbon receptor and interleukin 22 in patients with allergic asthma. Asian Pac J Allergy Immunol 29(3):266–272

The Gut Microbiome and Its Marriage to the Immune System: Can We Change It All?

Eva Untersmayr

Abstract Major research efforts during the past decades have revealed the close interaction between the host immune system and the symbionts colonizing the surfaces of the human body. Due to the high number of microbes found in this environment correlating with its role as an essential immune organ, the gastrointestinal tract has been a major focus of research. This chapter reviews current knowledge regarding the role of microbial colonization with beneficial microbes and the influence of released mediators on innate as well as adaptive immune cells. Several life style factors seems to substantially influence the composition of the microbiome especially during the first period of life. Nevertheless, also later in life factors such as dietary habits, medication intake or life style might alter the gut microbiota. Thus, strategies to beneficially modulate the overall microbial composition are essential to treat dysbiosis, which has been described to contribute to the onset or progression of various immune mediated intestinal diseases.

1 Introduction: The Gastrointestinal Microbial Colonization

For a long time microbes colonizing the human body were considered a health hazard being associated with infections and disease. Research efforts mainly focused on advances in defense strategies against microbes developing disinfectants and novel antibiotic, antiviral and antifungal therapies to kill bacteria, viruses and fungi. Only during the last decades the symbiotic relationship with the microbes on the surfaces of the human body were increasingly recognized and the dual role of e.g. bacteria was accepted. While some bacteria cause disease upon infections or autoimmunity (see chapter "Microbial Triggers in Autoimmunity, Severe Allergy, and Autoallergy"), others fulfill essential tasks to ensure human

E. Untersmayr
Institute of Pathophysiology and Allergy Research, Center of Pathophysiology, Infectiology and Immunology, Medical University of Vienna, Vienna, Austria
e-mail: eva.untersmayr@meduniwien.ac.at

© Springer International Publishing AG 2017   191
C.B. Schmidt-Weber (ed.), *Allergy Prevention and Exacerbation*, Birkhäuser
Advances in Infectious Diseases, https://doi.org/10.1007/978-3-319-69968-4_10

health by contributing to nutrient digestion, production of bioactive substances, essential micronutrients and vitamins such as B vitamins and Vitamin K.

Representing the logical continuation of the human genome project, the human microbiome project (Turnbaugh et al. 2007) resulted in a boost of knowledge and understanding regarding composition, function and diversity of the human microbiota. The inner and outer surfaces of the human body are colonized by an enormous number of microorganisms (bacteria, fungi, archaea, protozoa and viruses), which are collectively termed microbiota (Scarpellini et al. 2015). The number of these microbes exceeds human cell number by factor 10. They are found in highest density in the distal human colon (Backhed et al. 2005). The biomass of the intestinal microbial symbionts may weight up to 1.5 kg and the combined genome of the gut microbiota—termed microbiome—contains 150 times more genes than the human genome with a total number of 23,000 protein-coding genes of human origin (Gill et al. 2006).

Data from the human microbiome project and other studies revealed the important influence of diversity being defined by number and abundance of microbes within a certain habitat on human surfaces regarding health and disease (Human Microbiome Project Consortium 2012). Within the gastrointestinal tract the concentration of microbiota ranges from 10^1–10^3 cfu/mL in the stomach to 10^{11}–10^{12} cfu/mL in the colon. When correlating microbial diversity with disease, variations between the different body compartments have to be taken into consideration. Low diversity in the gut is associated with obesity and intestinal inflammation, while high vaginal diversity is attributed to bacterial vaginosis (Fredricks et al. 2005; Turnbaugh et al. 2009).

The diversity of bacteria found within a healthy gut was suggested to go beyond 1000 different species (Mandal et al. 2015; Marchesi 2010). The overall composition is influenced by factors such as e.g. diet, age, medication and health status (Claesson et al. 2012; Zhernakova et al. 2016) resulting in a highly individual, unique "fingerprint" of microbiota (Franzosa et al. 2015). Nevertheless, there are common pattern shared within the population. The dominant bacteria phyla in the human gut are Firmicutes, Bacterioidetes, Actinobacteria and Proteobacteria while on genus level *Bacterioides* species dominate representing 30% of all intestinal bacteria (Linares et al. 2016).

2 Early Microbial Colonization and Immune Programming

The initiation and progression of microbial colonization seems to be a crucial period shaping the immune response (see chapter "The Role of the Gut in Type 2 Immunity") and the overall health of the host by extensive interactions between host and microbiota not only in the gut but also on other outer surfaces (Bosch et al. 2016; Fouhy et al. 2012). Despite early evidence coming from culturing of bacteria contained in the first meconium passed after delivery (Burrage 1927; Hall and O'Toole 1934; Snyder 1936), microbial colonization was considered to be initiated during delivery until a decade ago. Only in 2005 first evidence was published that even before birth beneficial interactions between microbes and the fetal immune system might take place. Jimenez and co-workers reported the presence of commensal bacteria from the genus Enterococcus, Streptococcus, Staphylococcus, or Propionibacterium in umbilical cord blood collected after elective Cesarean section of healthy mothers (Jimenez et al. 2005). In the same study detection of labeled commensals orally fed during pregnancy was detected in the amniotic fluid after Cesarean section in a mouse model. Microbes were additionally found in fetal membranes after preterm and term delivery and in placental tissue by 16S-based operational taxonomic unit analyses (Aagaard et al. 2014; Jones et al. 2009).

The intrauterine exposure of the fetal immune system to microbes and microbial metabolites might substantially contribute to shaping a healthy immune system (Koleva et al. 2015). Since expression and signaling of Toll-like receptors (TLRs) is well established for newborns (van den Berk et al. 2009), it has been speculated that in utero interaction between the fetal immune system and microbes might involve TLRs (Koleva et al. 2015). Furthermore, maternal gut microbiota might influence the fetal immune system by bacterial metabolites. Short chain fatty acids (SCFA) are transported through gut epithelial cells by monocarboxylate transporters, which are found in murine placenta (Nagai et al. 2010). Nevertheless, the detailed impact of intrauterine microbial exposure on the developing fetal immune system remains to be elucidated by further research.

The main boost of microbial colonization takes place in the perinatal period. In the gut the pioneer colonizers are aerobe microbes, followed by facultative anaerobes creating an anaerobic luminal environment. The earliest colonizers include E. coli and enterococci, while consumption of oxygen allows further colonization by the later predominating strict anaerobes such as bifidobacteria, Bacteroides and Clostridium spp. (Edwards and Parrett 2002; Vaishampayan et al. 2010). The composition of early intestinal microbiota is shaped by various factors such as mode of delivery, infant diet, weaning and introduction of solid diet, use of medication and infant environment (Fouhy et al. 2012). While vaginally born infants are first colonized by vaginal and fecal microbes of the mother resulting in a domination of lactobacilli within hours after birth, Cesarean section delivery is associated with colonization by skin and environmental microbes. After Cesarean sections a 100-fold lower bifidobacteria and Bacteroides fragilis number together with a 100-fold increased colonization with Clostridium difficile was detected (Penders et al. 2006). In a recent study microbiota were partially restored by transfer of vaginal microbes after Cesarean section delivery (Dominguez-Bello et al. 2016). Exclusive breastfeeding in early life is known to be associated with higher intestinal bifidobacteria and lactobacilli levels and lower E. coli and clostridia counts compared to formula-fed children (Penders et al. 2006; Vaishampayan et al. 2010; Bezirtzoglou et al. 2011; Le Huerou-Luron et al. 2010). The profound influence of breastmilk on microbiota composition is due to its neutralizing antibody content and due to modulating factors such as prebiotic carbohydrates and antimicrobials (Verhasselt 2010). Moreover, infant diet provides specific microbial substrates in this early phase of life. The introduction of solid food together with or without partial breastfeeding represents an additional event of substantial alteration of microbiota composition. A shift towards a more adult-type microbiota with an enrichment of Bacteroidetes und Firmicutes is associated with introduction of solid food (Koenig et al. 2011). Besides these events with evident impact on microbiota composition less well established factors such as family structure or maternal weight and diet might also influence microbial colonization as discussed by Fouhy and co-authors (Fouhy et al. 2012). Here, the impact might partially be ascribed to environmental shaping of the gut microbiota. By the age of 2–3 years the microbiota composition reaches a steady state and is considered to remain relatively stable throughout the rest of the life with continuous reconfiguration of its metagenomics layout based on life style variations (Quercia et al. 2014).

In early life intestinal microbial colonization is paralleled by the development of a functional immune response (Kelly et al. 2007). Studies in germ-free (GF) mice underline the essential role of microbiota colonization for maturation of the intestinal tract and the mucosal immune system. For investigating the essential crosstalk between commensals and the gastrointestinal compartment including its immune cells GF mouse models have been indispensable. In these experimental models an overall changed intestinal architecture associated with a reduced epithelia barrier formation was demonstrated under GF conditions (Kozakova et al. 2016). Early studies reported different structure and composition of intestinal lymphatic tissue in association with changed pattern in gut homing of MLN cells in GF mice compared to conventionally housed animals (Vetvicka et al. 1983; Tlaskalova-Hogenova et al. 1983). The same and another group reported a dependence of antibacterial antibody formation on the presence of intestinal microbiota (Tlaskalova-Hogenova et al. 1983; Macpherson et al. 2000). Moreover, peripheral T cell expansion was microbiota depend with expansion after colonization in previously GF mice (Umesaki et al. 1993; Dobber et al. 1992), however, with variations in different intraepithelial lymphocyte compartments of the gut (Kawaguchi et al. 1993). Additionally, GF housing is associated with incomplete development of mucosal tolerance resulting in higher allergy susceptibility (Rodriguez et al. 2011). In addition to the above reviewed knowledge on deviated immune maturation and immune response in GF mice, a lack of an acute local inflammatory response and an impaired defense against viral and parasite infections was reported in these animals (Oliveira et al. 2005; Lima et al. 2016; Souza et al. 2004).

3 Intestinal Microbiota and Immune System Interactions

The reciprocal communication and influence between the gut symbionts and the intestinal epithelial cells (IECs) and mucosal immune cells is well established (Maynard et al. 2012) with major impact of direct microbial contact and released microbial mediators on the innate as well as the adaptive immune system

(Smolinska et al. 2017). However, the immune response also limits microbial contact with intestinal epithelial surfaces.

To prevent systemic exposure to potentially pathogenic bacteria there is a clear separation between intraluminal microbes and immune cells minimizing direct interactions. Secretory antimicrobial proteins like α-defenins and RegIIIγ shape luminal microbiota composition and reduce epithelial contact especially in the small intestine where the mucus overlying the intestinal epithelium shows a different composition compared to the large bowel (Vaishnava et al. 2011; Johansson et al. 2011). In the colon a two-layered mucus is formed by glycoproteins secreted by intestinal goblet cells with the inner layer showing special resistance against bacterial penetration (Johansson et al. 2008). Moreover, secretory immunoglobulin A (sIgA) antibodies specific for intestinal microbes and toxins prevent bacterial translocation through the epithelium, neutralize toxins and contribute to immunological tolerance against microbiota (Fig. 1) (Macpherson et al. 2000; Macpherson and Uhr 2004).

IECs are able to sense conserved microbial associated molecular pattern (MAMPS; e.g. wall components) by innate pattern recognition receptors (PRR) such as TLRs and Nod-like receptors. Interaction with bacteria was shown to control epithelial cell proliferation and was suggested to regulate intestinal epithelial barrier function (Lin and Zhang 2017). Furthermore, microbial metabolites such as SCFAs and bacterial quorum sensing (QS) signals have direct effects on IECs. SCFAs activate the inflammasome pathway by binding to G-Protein coupled receptors (GPRs) on the IEC surface. By enhancing IL-18 secretion a direct influence on the epithelium integrity is well recognized and more antimicrobial peptides are produced (Chinthrajah et al. 2016; Levy et al. 2015). Butyrate has an anti-inflammatory action mediated by inhibiting NFκB transcription and induces mucin synthesis (Segain et al. 2000; Burger-van Paassen et al. 2009). QS signals from Gram positive bacteria such as e.g. *Bacillus subtilis* have immunomodulatory potential, stimulate survival pathways and contributing to intestinal homeostasis by downregulation of pro-inflammatory cytokines and upregulation of anti-inflammatory markers such as IL-10 (Fujiya et al. 2007; Okamoto et al. 2012). Despite these defined measures to separate luminal microbes and the mucosal immune system also direct interaction takes place.

3.1 Microbial Effects on Innate Immune Cells

Dendritic cells have a large variety of interaction partners with microbial ligands such as TLRs and c-type lectin receptors. Upon binding of bacterial components a phenotypic and cytokine production profile change of dendritic cells occurs, which substantially influences the adaptive immune response (Fig. 1) (Smolinska et al. 2017). Moreover, also CX3CR1$^+$ macrophages are well accepted to sample antigens within the intestinal lumen (Berin and Shreffler 2016). With substantial impact on oral tolerance induction, TLR binding of commensal bacteria on macrophages is

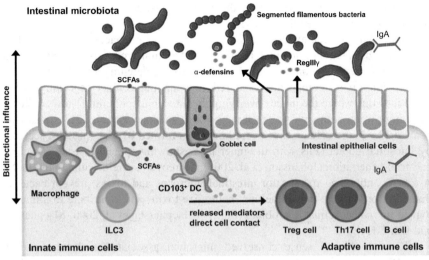

Fig. 1 Reciprocal interaction of intestinal microbes and microbial metabolites with the local and systemic immune system. Due to the close interaction between luminal symbionts and pathogens, microbial metabolites, intestinal epithelial cells and mucosal immune cells, the luminal content as well as the local and systemic immune response is shaped

associated with MyD88 signaling pathway activation. Bacteria-stimulated MyD88 signaling in CD103$^+$ myeloid cells leads to IL-23 secretion, which activates innate lymphoid cells 3 (ILC3) to release IL-22 (Kinnebrew et al. 2012). IL-22 promotes epithelial RegIIIγ release as well as glycosylation of IEC surface proteins, which is required for pathogen defense (Zheng et al. 2008; Goto et al. 2014). Moreover, MyD88 signaling in bacteria stimulated macrophages induces IL-1β secretion, which triggers the secretion of Csf2 by ILC3s. Even though generally considered to be a proinflammatory mediator Csf2 stimulates expansion of CD103$^+$ dendritic cells, which drive the induction and expansion of regulatory T cells (Tregs) (Mortha et al. 2014).

3.2 Cross-Talk Between the Gut Microbes and Adaptive Immune Cells

Activated by microbial antigens, innate immune cells are able to polarize adaptive immune cells by direct cellular contact and by cytokine secretion (Fig. 1) (Smolinska et al. 2017). It is well established that formation of Tregs is microbiota-dependent, as GF mice have abrogated oral tolerance development (Garn et al. 2013). Mouse models showed a special effect of Clostridia strains on Treg induction and presence of specific Clostridia bacteria was even reported to be associated with protection against food allergy development (Atarashi et al. 2011, 2013; Diesner et al. 2016; Narushima et al. 2014; Stefka et al. 2014; Lyons et al. 2010). In the context of Treg induction special relevance has the intake of dietary fiber, which is fermented to SCFAs by bacteria. Of interest, Clostridia strains being recognized for their potential as Treg inducers express a number of genes involved in SCFA biosynthesis (Narushima et al. 2014). Nevertheless, several studies indicate a broad range of bacteria including bifidobacteria and lactobacilli to beneficially modulate the immune response, with distinct immunomodulatory capacity of microbes related to animal age in experimental models using newborn and adult animals (Faith et al. 2014; Lyons et al. 2010; Karimi et al. 2009; Tang et al. 2015).

Besides the beneficial induction of Tregs, intestinal bacteria might also be associated with T-cell responses contributing to the development of autoimmune reactions. Epithelial adherence of segmented filamentous bacteria was suggested to contribute to Th17 driven autoimmune diseases in hosts with genetic predisposition (Honda and Littman 2016). Again dietary habits and especially lipid uptake seems to play an essential role as long chain fatty acids were reported to promote Th17 responses (Haghikia et al. 2015).

In the past few years a bidirectional influence of microbiota and natural killer T (NKT) cells have been revealed (Dowds et al. 2015), however, the precise functional role and consequences of microbe induced NKT cell activation still remains to be elucidated.

Besides this impact on T-cells, also B-cell activation and production of microbiota-specific, high-affinity sIgA are suggested to be initiated by microbial contact (Fig. 1). In Peyer's patches B-cell may get activated via T-cells after uptake of bacterial antigens via M cells. In contrast, T-cell independent B-cell activation takes place in the lamina propria and in isolated lymphoid follicles involving antigen passage through epithelial cells, antigen presentation by dendritic cells, as well as mediator release by ILCs (Honda and Littman 2016). In both cases an IgA secreting B-cell pool is generated, which contributes to control of luminal microbiota composition (Moon et al. 2015).

4 Factors Associated with Changes of the Gut Microbiota

With this knowledge about the close interaction between microbes and the immune system, it becomes obvious that a changed microbiome might substantially influence the development and phenotype of immunologically driven intestinal diseases. The intestinal microbial composition is considered to be relatively stable after 2 years of age (Quercia et al. 2014). However, several factors associated with changes in microbial composition have been identified.

Dietary habits influence to a great extent gut microbiota. Children eating Western diet have a different composition of the gut microflora compared to rural African children (De Filippo et al. 2010). A ground-breaking study by David and colleagues reported that only 5 days of dietary intervention eating either plant-based or animal-based diet was associated with a shift in the abundance of fecal microbial communities (David et al. 2014b). In healthy subjects high fiber intake by whole grain barley was revealed to be associated with increases in lactobacilli, while other bacterial strains such as *Enterobacteriaceae* and *Bacteroides* decreased (De Angelis et al. 2015). Even 1 day of high fiber intake was associated with changed abundance of 15% of total bacterial communities on the consecutive day (David et al. 2014a). Of interest, gluten-free diet was reported to change counts of different bacterial strains in healthy adults as well as in patients with celiac disease (De Palma et al. 2009; Nistal et al. 2016) underlining the important interaction between diet, microbiota and host metabolism (Barcik et al. 2016). Thus, protection against allergy development by food diversity early in life, which has been reported by a large cohort study (Roduit et al. 2014), might also be seen in association with the impact of diet on the intestinal microflora. However, nutrition might also detrimentally influence intestinal homeostasis. Recent work demonstrated commonly used emulsifiers to induce low-grade inflammation, metabolic syndrome and colitis associated with microbiota shifts in predisposed animals. Transfer of this pro-inflammatory condition to GF mice was possible by fecal transplantation (Chassaing et al. 2015).

Another factor influencing the bacterial composition in the gut is medication intake. It is well established that antibiotic treatment is associated with dysbiosis of the gut especially when given in early life (Vangay et al. 2015; Cotter et al. 2012). An associated between antibiotic use in childhood and food sensitization or the development of other allergies later in life was reported (Hirsch et al. 2017). However, further studies evaluating the direct correlations between antibiotic use, microbial change and allergy development are needed to in depth evaluate under-lying mechanisms. Gastric acid suppression, which has previously been linked with food allergies in pediatric as well as adult patients (Trikha et al. 2013; Untersmayr et al. 2005), was found to induce a substantial changes in gut microbiome with a shift towards bacterial strains predisposing patients to *Clostridium difficile* infec-tions (Imhann et al. 2016). In an experimental model, protection from food allergy development under gastric acid suppression was associated with presence of dis-tinct bacterial strains in fecal microbiota (Diesner et al. 2016).

Also other life style factors such as chronic psychological stress are associated with a changed microbiome and are linked to the onset of intestinal inflammation (Gur and Bailey 2016; Watanabe et al. 2016).

As microbial changes are reported for a large variety of metabolic and immu-nologically mediated diseases (Lin and Zhang 2017), the here summarized factors might substantially influence disease onset and outcome. With regards to intestinal allergic diseases changed commensal composition were reported in infants as well as adult patients (Chen et al. 2016; Hua et al. 2016), and regulation of food allergen sensitization by gut bacteria is well recognized (Cao et al. 2014).

5 Clinical Implications: Microbiota Modulation Potentially Influencing Immunological Diseases

The qualitative and quantitave change of gut microbes associated with metabolic alterations and changed local distribution might be of pathophysiological relevance for various intestinal diseases. Due to the close interaction between intestinal microbes and the local as well as systemic immune system, microbiota modulating strategies are a promising field modulating or treating immune mediated disorders. Several strategies are currently available or matter of further research (Table 1).

Prebiotics are dietary components specifically acting as substrates for single strains or a small number of beneficial bacteria in the colon as they are neither degraded by host enzymes nor absorbed in the upper gastrointestinal (Brahe et al. 2016). The application of prebiotics and their effect on the immune response with focus on allergy prevention has been extensively studied in non-breastfed or non-exclusively breastfed infants using prebiotic-supplemented milk formula with transient beneficial effects. Still, more studies are needed to enable clear conclusions and recommendations for usage of prebiotics (Forsberg et al. 2016). The World health organization defined probiotics as live microbes confering a health benefit to the host when administered in an adequate amount (Pineiro and Stanton 2007). Most studies evaluating allergy preventive effects focus on single strains or combinations of lactobacilli and bifidobacteria in experimental as well as human studies (West et al. 2015). However, also SCFA released by bacteria such as *Bacteroides* or Clostridia strains might exert immunomodulatory properties and

Table 1 Currently available microbiota modulation strategies

Modulation strategies	Examples and explanation	References
Prebiotics	Inulin, fructo-oligosaccharides (FOS), galacto-oligosaccharides (GOS), resistant starch, xylo-oligosaccharides, arabinoxylan-oligosaccharides	Brahe et al. (2016)
Probiotics	lactobacilli and bifidobacteria strains; unconventional probiotics such as *Bacteroides* or Clostridia strains	West et al. (2015)
Synbiotics	Mixtures of prebiotics and probiotics in a single product	Perez Martinez et al. (2014)
Postbiotics	Microbial metabolites such as SCFA produced by fermentation of dietary fiber	Tsilingiri and Rescigno (2013)
Antimicrobials	Narrow-spectrum antibiotics, sub-therapeutic antibiotic therapy, Bacteriocin-producing bacteria	Walsh et al. (2014)
Dietary intervention	Increase of dietary fiber uptake, Plant-based diet, SCFA uptake e.g. via butter consumption	Smolinska et al. (2017), David et al. (2014b)
Fecal microbiota transplantation	Transfer of microbiota from a healthy individual to a patient with changed microbiota composition associated with disease	Kump and Hogenauer (2016), Wu et al. (2011)

are focus of future research (Stefka et al. 2014). The application of bacterial metabolites such as the above mentioned SCFAs butyrate or propionate, being associated with a modulation of the host's immune response, is matter of experimental studies and has been suggested to represent a safe alternative to probiotic strategies (Tsilingiri and Rescigno 2013). Despite the negative effect of broad-spectrum antibiotics on intestinal microbial populations and overall host's health (Cotter et al. 2012), the use of narrow-spectrum antibiotics, sub-therapeutic antibiotic levels or the application of bacteria-produced antimicrobials might represent an intriguing approach for gut microbiota modulate, which warrants further research efforts (Walsh et al. 2014). Moreover, dietary habits represent an environmental factor of major relevance shaping microbial communities (Claesson et al. 2012; David et al. 2014b). A balanced intake of the three main nutritional components, i.e. proteins, carbohydrates and fat, is essential for gut health as these nutrients are partially degraded by microbes with the metabolites acting back on the host. Repeatedly, changes of dietary habits were reported to be associated with substantial alterations of microbiota composition (Wu et al. 2011). Last but not least transfer of bacterial community from healthy donors to recipients with diseases associated with a changed microbiome might represent a potent strategy to restore microbial composition (van Nood et al. 2013). This approach is mainly used for treatment of recurrent *Clostridium difficile* infections after antibiotic therapy, for which substantial efficacy and safety has been proven (Aroniadis et al. 2016). However, it is increasingly also applied for a larger variety of gastrointestinal disorders, such as inflammatory bowel disease, for which effectiveness still needs to be proven due to variations in study design and influencing factors on both donor and recipient side (Kump and Hogenauer 2016).

Besides these different currently available strategies for microbiota modulation also the time-point of intervention might influence the outcome. Several crucial time-periods with major impact on overall microbiota composition and with influence of microbiota on immune maturation might be taken into consideration: pregnancy, first years of life and major lifestyle changes, e.g. medication intake. As in these situations host-microbiota interactions might be especially active and/or vulnerable for development of dysbiosis, interventions might be of special relevance. To give an example also intrauterine microbial exposure might influence the development of atopic disease. However there is a single study on meconium composition and atopy-related diseases (Gosalbes et al. 2013) underlining the urgent need for further research studies evaluating the impact of timing regarding microbial interventions.

6   Conclusion and Future Outlook

Based on the here reviewed data, the marriage between the gut microbiome and the host's immune system becomes evident. Despite the enormous boost of knowledge generated by scientific studies during the past decades, we are still far from understanding in detail the complex interaction between our symbionts, our immune system and environmental factors such as diet. One of the difficulties is the highly individual composition of the microbiota, which is based on various influencing factors throughout our entire life. Thinking about clinical applications of microbiota interventions a holistic approach has to be chosen as targeting the human microbiome can only be successful if the concept of personalized medicine is applied.

Acknowledgements The research of the author is supported by grants KLI284-B00 and WKP39 of the Austrian Science Fund FWF, by grant P1621673 of the Austrian Ministry of Science, Research and Economy and a research project supported by Nordmark GmbH.

References

Aagaard K, Ma J, Antony KM, Ganu R, Petrosino J, Versalovic J (2014) The placenta harbors a unique microbiome. Sci Transl Med 6(237):237–265. https://doi.org/10.1126/scitranslmed. 3008599

Aroniadis OC, Brandt LJ, Greenberg A, Borody T, Kelly CR, Mellow M, Surawicz C, Cagle L, Neshatian L, Stollman N, Giovanelli A, Ray A, Smith R (2016) Long-term follow-up study of fecal microbiota transplantation for severe and/or complicated clostridium difficile infection: a multicenter experience. J Clin Gastroenterol 50(5):398–402. https://doi.org/10.1097/MCG. 0000000000000374

Atarashi K, Tanoue T, Shima T, Imaoka A, Kuwahara T, Momose Y, Cheng G, Yamasaki S, Saito T, Ohba Y, Taniguchi T, Takeda K, Hori S, Ivanov II, Umesaki Y, Itoh K, Honda K (2011) Induction of colonic regulatory T cells by indigenous Clostridium species. Science 331 (6015):337–341. https://doi.org/10.1126/science.1198469

Atarashi K, Tanoue T, Oshima K, Suda W, Nagano Y, Nishikawa H, Fukuda S, Saito T, Narushima S, Hase K, Kim S, Fritz JV, Wilmes P, Ueha S, Matsushima K, Ohno H, Olle B, Sakaguchi S, Taniguchi T, Morita H, Hattori M, Honda K (2013) Treg induction by a rationally selected mixture of Clostridia strains from the human microbiota. Nature 500(7461):232–236. https://doi.org/10.1038/nature12331

Backhed F, Ley RE, Sonnenburg JL, Peterson DA, Gordon JI (2005) Host-bacterial mutualism in the human intestine. Science 307(5717):1915–1920. https://doi.org/10.1126/science.1104816

Barcik W, Untersmayr E, Pali-Schöll I, O'Mahony L, Frei R (2016) Influence of microbiome and diet on immune responses in food allergy models. Drug Discov Today Dis Model 17:71–82

Berin MC, Shreffler WG (2016) Mechanisms underlying induction of tolerance to foods. Immunol Allergy Clin N Am 36(1):87–102. https://doi.org/10.1016/j.iac.2015.08.002

Bezirtzoglou E, Tsiotsias A, Welling GW (2011) Microbiota profile in feces of breast- and formula-fed newborns by using fluorescence in situ hybridization (FISH). Anaerobe 17(6): 478–482. https://doi.org/10.1016/j.anaerobe.2011.03.009

Bosch AA, Levin E, van Houten MA, Hasrat R, Kalkman G, Biesbroek G, de Steenhuijsen Piters WA, de Groot PK, Pernet P, Keijser BJ, Sanders EA, Bogaert D (2016) Development of upper respiratory tract microbiota in infancy is affected by mode of delivery. EBioMedicine 9:336–345. https://doi.org/10.1016/j.ebiom.2016.05.031

Brahe LK, Astrup A, Larsen LH (2016) Can we prevent obesity-related metabolic diseases by dietary modulation of the gut microbiota? Adv Nutr 7(1):90–101. https://doi.org/10.3945/an. 115.010587

Burger-van Paassen N, Vincent A, Puiman PJ, van der Sluis M, Bouma J, Boehm G, van Goudoever JB, van Seuningen I, Renes IB (2009) The regulation of intestinal mucin MUC2 expression by short-chain fatty acids: implications for epithelial protection. Biochem J 420 (2):211–219. https://doi.org/10.1042/BJ20082222

Burrage S (1927) Bacteria in the supposedly sterile meconium. J Bacteriol 13(1):47–48

Cao S, Feehley TJ, Nagler CR (2014) The role of commensal bacteria in the regulation of sensitization to food allergens. FEBS Lett 588(22):4258–4266. https://doi.org/10.1016/j. febslet.2014.04.026

Chassaing B, Koren O, Goodrich JK, Poole AC, Srinivasan S, Ley RE, Gewirtz AT (2015) Dietary emulsifiers impact the mouse gut microbiota promoting colitis and metabolic syndrome. Nature 519(7541):92–96. https://doi.org/10.1038/nature14232

Chen CC, Chen KJ, Kong MS, Chang HJ, Huang JL (2016) Alterations in the gut microbiotas of children with food sensitization in early life. Pediatr Allergy Immunol 27(3):254–262. https:// doi.org/10.1111/pai.12522

Chinthrajah RS, Hernandez JD, Boyd SD, Galli SJ, Nadeau KC (2016) Molecular and cellular mechanisms of food allergy and food tolerance. J Allergy Clin Immunol 137(4):984–997. https://doi.org/10.1016/j.jaci.2016.02.004

Claesson MJ, Jeffery IB, Conde S, Power SE, O'Connor EM, Cusack S, Harris HM, Coakley M, Lakshminarayanan B, O'Sullivan O, Fitzgerald GF, Deane J, O'Connor M, Harnedy N, O'Connor K, O'Mahony D, van Sinderen D, Wallace M, Brennan L, Stanton C, Marchesi JR, Fitzgerald AP, Shanahan F, Hill C, Ross RP, O'Toole PW (2012) Gut microbiota composition correlates with diet and health in the elderly. Nature 488(7410):178–184. https://doi.org/10.1038/nature11319

Cotter PD, Stanton C, Ross RP, Hill C (2012) The impact of antibiotics on the gut microbiota as revealed by high throughput DNA sequencing. Discov Med 13(70):193–199

David LA, Materna AC, Friedman J, Campos-Baptista MI, Blackburn MC, Perrotta A, Erdman SE, Alm EJ (2014a) Host lifestyle affects human microbiota on daily timescales. Genome Biol 15(7):R89. https://doi.org/10.1186/gb-2014-15-7-r89

David LA, Maurice CF, Carmody RN, Gootenberg DB, Button JE, Wolfe BE, Ling AV, Devlin AS, Varma Y, Fischbach MA, Biddinger SB, Dutton RJ, Turnbaugh PJ (2014b) Diet rapidly and reproducibly alters the human gut microbiome. Nature 505(7484):559–563. https://doi.org/10.1038/nature12820

De Angelis M, Montemurno E, Vannini L, Cosola C, Cavallo N, Gozzi G, Maranzano V, Di Cagno R, Gobbetti M, Gesualdo L (2015) Effect of whole-grain barley on the human fecal microbiota and metabolome. Appl Environ Microbiol 81(22):7945–7956. https://doi.org/10.1128/AEM.02507-15

De Filippo C, Cavalieri D, Di Paola M, Ramazzotti M, Poullet JB, Massart S, Collini S, Pieraccini G, Lionetti P (2010) Impact of diet in shaping gut microbiota revealed by a comparative study in children from Europe and rural Africa. Proc Natl Acad Sci USA 107(33):14691–14696. https://doi.org/10.1073/pnas.1005963107

De Palma G, Nadal I, Collado MC, Sanz Y (2009) Effects of a gluten-free diet on gut microbiota and immune function in healthy adult human subjects. Br J Nutr 102(8):1154–1160. https://doi.org/10.1017/S0007114509371767

Diesner SC, Bergmayr C, Pfitzner B, Assmann V, Krishnamurthy D, Starkl P, Endesfelder D, Rothballer M, Welzl G, Rattei T, Eiwegger T, Szepfalusi Z, Fehrenbach H, Jensen-Jarolim E, Hartmann A, Pali-Scholl I, Untersmayr E (2016) A distinct microbiota composition is associated with protection from food allergy in an oral mouse immunization model. Clin Immunol 173:10–18. https://doi.org/10.1016/j.clim.2016.10.009

Dobber R, Hertogh-Huijbregts A, Rozing J, Bottomly K, Nagelkerken L (1992) The involvement of the intestinal microflora in the expansion of CD4+ T cells with a naive phenotype in the periphery. Dev Immunol 2(2):141–150

Dominguez-Bello MG, De Jesus-Laboy KM, Shen N, Cox LM, Amir A, Gonzalez A, Bokulich NA, Song SJ, Hoashi M, Rivera-Vinas JI, Mendez K, Knight R, Clemente JC (2016) Partial restoration of the microbiota of cesarean-born infants via vaginal microbial transfer. Nat Med 22(3):250–253. https://doi.org/10.1038/nm.4039

Dowds CM, Blumberg RS, Zeissig S (2015) Control of intestinal homeostasis through crosstalk between natural killer T cells and the intestinal microbiota. Clin Immunol 159(2):128–133. https://doi.org/10.1016/j.clim.2015.05.008

Edwards CA, Parrett AM (2002) Intestinal flora during the first months of life: new perspectives. Br J Nutr 88(Suppl 1):S11–S18. https://doi.org/10.1079/BJN2002625

Faith JJ, Ahern PP, Ridaura VK, Cheng J, Gordon JI (2014) Identifying gut microbe-host phenotype relationships using combinatorial communities in gnotobiotic mice. Sci Transl Med 6(220):220ra211. https://doi.org/10.1126/scitranslmed.3008051

Forsberg A, West CE, Prescott SL, Jenmalm MC (2016) Pre- and probiotics for allergy prevention: time to revisit recommendations? Clin Exp Allergy 46(12):1506–1521. https://doi.org/10.1111/cea.12838

Fouhy F, Ross RP, Fitzgerald GF, Stanton C, Cotter PD (2012) Composition of the early intestinal microbiota: knowledge, knowledge gaps and the use of high-throughput sequencing to address these gaps. Gut Microbes 3(3):203–220. https://doi.org/10.4161/gmic.20169

Franzosa EA, Huang K, Meadow JF, Gevers D, Lemon KP, Bohannan BJ, Huttenhower C (2015) Identifying personal microbiomes using metagenomic codes. Proc Natl Acad Sci USA 112 (22):E2930–E2938. https://doi.org/10.1073/pnas.1423854112

Fredricks DN, Fiedler TL, Marrazzo JM (2005) Molecular identification of bacteria associated with bacterial vaginosis. N Engl J Med 353(18):1899–1911. https://doi.org/10.1056/NEJMoa043802

Fujiya M, Musch MW, Nakagawa Y, Hu S, Alverdy J, Kohgo Y, Schneewind O, Jabri B, Chang EB (2007) The Bacillus subtilis quorum-sensing molecule CSF contributes to intestinal homeostasis via OCTN2, a host cell membrane transporter. Cell Host Microbe 1(4):299–308. https://doi.org/10.1016/j.chom.2007.05.004

Garn H, Neves JF, Blumberg RS, Renz H (2013) Effect of barrier microbes on organ-based inflammation. J Allergy Clin Immunol 131(6):1465–1478. https://doi.org/10.1016/j.jaci.2013.04.031

Gill SR, Pop M, Deboy RT, Eckburg PB, Turnbaugh PJ, Samuel BS, Gordon JI, Relman DA, Fraser-Liggett CM, Nelson KE (2006) Metagenomic analysis of the human distal gut microbiome. Science 312(5778):1355–1359. https://doi.org/10.1126/science.1124234

Gosalbes MJ, Llop S, Valles Y, Moya A, Ballester F, Francino MP (2013) Meconium microbiota types dominated by lactic acid or enteric bacteria are differentially associated with maternal eczema and respiratory problems in infants. Clin Exp Allergy 43(2):198–211. https://doi.org/10.1111/cea.12063

Goto Y, Obata T, Kunisawa J, Sato S, Ivanov II, Lamichhane A, Takeyama N, Kamioka M, Sakamoto M, Matsuki T, Setoyama H, Imaoka A, Uematsu S, Akira S, Domino SE, Kulig P, Becher B, Renauld JC, Sasakawa C, Umesaki Y, Benno Y, Kiyono H (2014) Innate lymphoid cells regulate intestinal epithelial cell glycosylation. Science 345(6202):1254009. https://doi.org/10.1126/science.1254009

Gur TL, Bailey MT (2016) Effects of stress on commensal microbes and immune system activity. Adv Exp Med Biol 874:289–300. https://doi.org/10.1007/978-3-319-20215-0_14

Haghikia A, Jorg S, Duscha A, Berg J, Manzel A, Waschbisch A, Hammer A, Lee DH, May C, Wilck N, Balogh A, Ostermann AI, Schebb NH, Akkad DA, Grohme DA, Kleinewietfeld M, Kempa S, Thone J, Demir S, Muller DN, Gold R, Linker RA (2015) Dietary fatty acids directly impact central nervous system autoimmunity via the small intestine. Immunity 43(4):817–829. https://doi.org/10.1016/j.immuni.2015.09.007

Hall IC, O'Toole E (1934) Bacterial flora of first specimens of meconium passaed by fifty new-born infants. Am J Dis Child 47(6):1279–1285

Hirsch AG, Pollak J, Glass TA, Poulsen MN, Bailey-Davis L, Mowery J, Schwartz BS (2017) Early-life antibiotic use and subsequent diagnosis of food allergy and allergic diseases. Clin Exp Allergy 47(2):236–244. https://doi.org/10.1111/cea.12807

Honda K, Littman DR (2016) The microbiota in adaptive immune homeostasis and disease. Nature 535(7610):75–84. https://doi.org/10.1038/nature18848

Hua X, Goedert JJ, Pu A, Yu G, Shi J (2016) Allergy associations with the adult fecal microbiota: analysis of the American Gut Project. EBioMedicine 3:172–179. https://doi.org/10.1016/j. ebiom.2015.11.038

Human Microbiome Project Consortium (2012) Structure, function and diversity of the healthy human microbiome. Nature 486(7402):207–214. https://doi.org/10.1038/nature11234

Imhann F, Bonder MJ, Vich Vila A, Fu J, Mujagic Z, Vork L, Tigchelaar EF, Jankipersadsing SA, Cenit MC, Harmsen HJ, Dijkstra G, Franke L, Xavier RJ, Jonkers D, Wijmenga C, Weersma RK, Zhernakova A (2016) Proton pump inhibitors affect the gut microbiome. Gut 65(5): 740–748. https://doi.org/10.1136/gutjnl-2015-310376

Jimenez E, Fernandez L, Marin ML, Martin R, Odriozola JM, Nueno-Palop C, Narbad A, Olivares M, Xaus J, Rodriguez JM (2005) Isolation of commensal bacteria from umbilical cord blood of healthy neonates born by cesarean section. Curr Microbiol 51(4):270–274. https://doi.org/10.1007/s00284-005-0020-3

Johansson ME, Phillipson M, Petersson J, Velcich A, Holm L, Hansson GC (2008) The inner of the two Muc2 mucin-dependent mucus layers in colon is devoid of bacteria. Proc Natl Acad Sci USA 105(39):15064–15069. https://doi.org/10.1073/pnas.0803124105

Johansson ME, Ambort D, Pelaseyed T, Schutte A, Gustafsson JK, Ermund A, Subramani DB, Holmen-Larsson JM, Thomsson KA, Bergstrom JH, van der Post S, Rodriguez-Pineiro AM, Sjovall H, Backstrom M, Hansson GC (2011) Composition and functional role of the mucus layers in the intestine. Cell Mol Life Sci 68(22):3635–3641. https://doi.org/10.1007/s00018-011-0822-3

Jones HE, Harris KA, Azizia M, Bank L, Carpenter B, Hartley JC, Klein N, Peebles D (2009) Differing prevalence and diversity of bacterial species in fetal membranes from very preterm and term labor. PLoS One 4(12):e8205. https://doi.org/10.1371/journal.pone.0008205

Karimi K, Inman MD, Bienenstock J, Forsythe P (2009) Lactobacillus reuteri-induced regulatory T cells protect against an allergic airway response in mice. Am J Respir Crit Care Med 179(3):186–193. https://doi.org/10.1164/rccm.200806-951OC

Kawaguchi M, Nanno M, Umesaki Y, Matsumoto S, Okada Y, Cai Z, Shimamura T, Matsuoka Y, Ohwaki M, Ishikawa H (1993) Cytolytic activity of intestinal intraepithelial lymphocytes in germ-free mice is strain dependent and determined by T cells expressing gamma delta T-cell antigen receptors. Proc Natl Acad Sci USA 90(18):8591–8594

Kelly D, King T, Aminov R (2007) Importance of microbial colonization of the gut in early life to the development of immunity. Mutat Res 622(1–2):58–69. https://doi.org/10.1016/j.mrfmmm. 2007.03.011

Kinnebrew MA, Buffie CG, Diehl GE, Zenewicz LA, Leiner I, Hohl TM, Flavell RA, Littman DR, Pamer EG (2012) Interleukin 23 production by intestinal CD103(+)CD11b(+) dendritic cells in response to bacterial flagellin enhances mucosal innate immune defense. Immunity 36(2): 276–287. https://doi.org/10.1016/j.immuni.2011.12.011

Koenig JE, Spor A, Scalfone N, Fricker AD, Stombaugh J, Knight R, Angenent LT, Ley RE (2011) Succession of microbial consortia in the developing infant gut microbiome. Proc Natl Acad Sci USA 108(Suppl 1):4578–4585. https://doi.org/10.1073/pnas.1000081107

Koleva PT, Kim JS, Scott JA, Kozyrskyj AL (2015) Microbial programming of health and disease starts during fetal life. Birth Defects Res C Embryo Today 105(4):265–277. https://doi.org/10. 1002/bdrc.21117

Kozakova H, Schwarzer M, Tuckova L, Srutkova D, Czarnowska E, Rosiak I, Hudcovic T, Schabussova I, Hermanova P, Zakostelska Z, Aleksandrzak-Piekarczyk T, Koryszewska-Baginska A, Tlaskalova-Hogenova H, Cukrowska B (2016) Colonization of germ-free mice with a mixture of three lactobacillus strains enhances the integrity of gut mucosa and amelio-rates allergic sensitization. Cell Mol Immunol 13(2):251–262. https://doi.org/10.1038/cmi. 2015.09

Kump P, Hogenauer C (2016) Any future for fecal microbiota transplantation as treatment strategy for inflammatory bowel diseases? Dig Dis 34(Suppl 1):74–81. https://doi.org/10.1159/ 000447379

Le Huerou-Luron I, Blat S, Boudry G (2010) Breast- v. formula-feeding: impacts on the digestive tract and immediate and long-term health effects. Nutr Res Rev 23(1):23–36. https://doi.org/10.1017/S0954422410000065

Levy M, Thaiss CA, Zeevi D, Dohnalova L, Zilberman-Schapira G, Mahdi JA, David E, Savidor A, Korem T, Herzig Y, Pevsner-Fischer M, Shapiro H, Christ A, Harmelin A, Halpern Z, Latz E, Flavell RA, Amit I, Segal E, Elinav E (2015) Microbiota-modulated metabolites shape the intestinal microenvironment by regulating NLRP6 inflammasome signaling. Cell 163(6):1428–1443. https://doi.org/10.1016/j.cell.2015.10.048

Lima MT, Andrade AC, Oliveira GP, Calixto RS, Oliveira DB, Souza EL, Trindade GS, Nicoli JR, Kroon EG, Martins FS, Abrahao JS (2016) Microbiota is an essential element for mice to initiate a protective immunity against Vaccinia virus. FEMS Microbiol Ecol 92(2). https://doi.org/10.1093/femsec/fiv147

Lin L, Zhang J (2017) Role of intestinal microbiota and metabolites on gut homeostasis and human diseases. BMC Immunol 18(1):2. https://doi.org/10.1186/s12865-016-0187-3

Linares DM, Ross P, Stanton C (2016) Beneficial microbes: the pharmacy in the gut. Bioengineered 7(1):11–20. https://doi.org/10.1080/21655979.2015.1126015

Lyons A, O'Mahony D, O'Brien F, MacSharry J, Sheil B, Ceddia M, Russell WM, Forsythe P, Bienenstock J, Kiely B, Shanahan F, O'Mahony L (2010) Bacterial strain-specific induction of Foxp3+ T regulatory cells is protective in murine allergy models. Clin Exp Allergy 40 (5):811–819. https://doi.org/10.1111/j.1365-2222.2009.03437.x

Macpherson AJ, Uhr T (2004) Induction of protective IgA by intestinal dendritic cells carrying commensal bacteria. Science 303(5664):1662–1665. https://doi.org/10.1126/science.1091334

Macpherson AJ, Gatto D, Sainsbury E, Harriman GR, Hengartner H, Zinkernagel RM (2000) A primitive T cell-independent mechanism of intestinal mucosal IgA responses to commensal bacteria. Science 288(5474):2222–2226

Mandal RS, Saha S, Das S (2015) Metagenomic surveys of gut microbiota. Genomics Proteomics Bioinformatics 13(3):148–158. https://doi.org/10.1016/j.gpb.2015.02.005

Marchesi JR (2010) Prokaryotic and eukaryotic diversity of the human gut. Adv Appl Microbiol 72:43–62. https://doi.org/10.1016/S0065-2164(10)72002-5

Maynard CL, Elson CO, Hatton RD, Weaver CT (2012) Reciprocal interactions of the intestinal microbiota and immune system. Nature 489(7415):231–241. https://doi.org/10.1038/nature11551

Moon C, Baldridge MT, Wallace MA, Burnham CA, Virgin HW, Stappenbeck TS (2015) Vertically transmitted faecal IgA levels determine extra-chromosomal phenotypic variation. Nature 521(7550):90–93. https://doi.org/10.1038/nature14139

Mortha A, Chudnovskiy A, Hashimoto D, Bogunovic M, Spencer SP, Belkaid Y, Merad M (2014) Microbiota-dependent crosstalk between macrophages and ILC3 promotes intestinal homeostasis. Science 343(6178):1249288. https://doi.org/10.1126/science.1249288

Nagai A, Takebe K, Nio-Kobayashi J, Takahashi-Iwanaga H, Iwanaga T (2010) Cellular expression of the monocarboxylate transporter (MCT) family in the placenta of mice. Placenta 31(2):126–133. https://doi.org/10.1016/j.placenta.2009.11.013

Narushima S, Sugiura Y, Oshima K, Atarashi K, Hattori M, Suematsu M, Honda K (2014) Characterization of the 17 strains of regulatory T cell-inducing human-derived Clostridia. Gut Microbes 5(3):333–339. https://doi.org/10.4161/gmic.28572

Nistal E, Caminero A, Herran AR, Perez-Andres J, Vivas S, Ruiz de Morales JM, Saenz de Miera LE, Casqueiro J (2016) Study of duodenal bacterial communities by 16S rRNA gene analysis in adults with active celiac disease vs non-celiac disease controls. J Appl Microbiol 120 (6):1691–1700. https://doi.org/10.1111/jam.13111

Okamoto K, Fujiya M, Nata T, Ueno N, Inaba Y, Ishikawa C, Ito T, Moriichi K, Tanabe H, Mizukami Y, Chang EB, Kohgo Y (2012) Competence and sporulation factor derived from Bacillus subtilis improves epithelial cell injury in intestinal inflammation via immunomodulation and cytoprotection. Int J Color Dis 27(8):1039–1046. https://doi.org/10.1007/s00384-012-1416-8

Oliveira MR, Tafuri WL, Afonso LC, Oliveira MA, Nicoli JR, Vieira EC, Scott P, Melo MN, Vieira LQ (2005) Germ-free mice produce high levels of interferon-gamma in response to infection with Leishmania major but fail to heal lesions. Parasitology 131(Pt 4):477–488. https://doi.org/10.1017/S0031182005008073

Penders J, Thijs C, Vink C, Stelma FF, Snijders B, Kummeling I, van den Brandt PA, Stobberingh EE (2006) Factors influencing the composition of the intestinal microbiota in early infancy. Pediatrics 118(2):511–521. https://doi.org/10.1542/peds.2005-2824

Perez Martinez G, Bauerl C, Collado MC (2014) Understanding gut microbiota in elderly's health will enable intervention through probiotics. Benef Microbes 5(3):235–246. https://doi.org/10.3920/BM2013.0079

Pineiro M, Stanton C (2007) Probiotic bacteria: legislative framework—requirements to evidence basis. J Nutr 137(3 Suppl 2):850S–853S

Quercia S, Candela M, Giuliani C, Turroni S, Luiselli D, Rampelli S, Brigidi P, Franceschi C, Bacalini MG, Garagnani P, Pirazzini C (2014) From lifetime to evolution: timescales of human gut microbiota adaptation. Front Microbiol 5(587). https://doi.org/10.3389/fmicb.2014.00587

Rodriguez B, Prioult G, Bibiloni R, Nicolis I, Mercenier A, Butel MJ, Waligora-Dupriet AJ (2011) Germ-free status and altered caecal subdominant microbiota are associated with a high susceptibility to cow's milk allergy in mice. FEMS Microbiol Ecol 76(1):133–144. https://doi.org/10.1111/j.1574-6941.2010.01035.x

Roduit C, Frei R, Depner M, Schaub B, Loss G, Genuneit J, Pfefferle P, Hyvarinen A, Karvonen AM, Riedler J, Dalphin JC, Pekkanen J, von Mutius E, Braun-Fahrlander C, Lauener R (2014) Increased food diversity in the first year of life is inversely associated with allergic diseases. J Allergy Clin Immunol 133(4):1056–1064. https://doi.org/10.1016/j.jaci.2013.12.1044

Scarpellini E, Ianiro G, Attili F, Bassanelli C, De Santis A, Gasbarrini A (2015) The human gut microbiota and virome: potential therapeutic implications. Dig Liver Dis 47(12):1007–1012. https://doi.org/10.1016/j.dld.2015.07.008

Segain JP, Raingeard de la Bletiere D, Bourreille A, Leray V, Gervois N, Rosales C, Ferrier L, Bonnet C, Blottiere HM, Galmiche JP (2000) Butyrate inhibits inflammatory responses through NFkappaB inhibition: implications for Crohn's disease. Gut 47(3):397–403

Smolinska S, Groeger D, O'Mahony L (2017) Biology of the Microbiome 1. Interactions with the Host Immune Response. Gastroenterol Clin N Am 46(1):19–35. https://doi.org/10.1016/j.gtc.2016.09.004

Snyder ML (1936) The bacterial flora of meconium specimens collected from sixty-four infants within four hours after delivery. J Pediatr 9:624–632

Souza DG, Vieira AT, Soares AC, Pinho V, Nicoli JR, Vieira LQ, Teixeira MM (2004) The essential role of the intestinal microbiota in facilitating acute inflammatory responses. J Immunol 173(6):4137–4146

Stefka AT, Feehley T, Tripathi P, Qiu J, McCoy K, Mazmanian SK, Tjota MY, Seo GY, Cao S, Theriault BR, Antonopoulos DA, Zhou L, Chang EB, YX F, Nagler CR (2014) Commensal bacteria protect against food allergen sensitization. Proc Natl Acad Sci USA 111(36):13145–13150. https://doi.org/10.1073/pnas.1412008111

Tang C, Kamiya T, Liu Y, Kadoki M, Kakuta S, Oshima K, Hattori M, Takeshita K, Kanai T, Saijo S, Ohno N, Iwakura Y (2015) Inhibition of dectin-1 signaling ameliorates colitis by inducing lactobacillus-mediated regulatory T cell expansion in the intestine. Cell Host Microbe 18(2):183–197. https://doi.org/10.1016/j.chom.2015.07.003

Tlaskalova-Hogenova H, Sterzl J, Stepankova R, Dlabac V, Veticka V, Rossmann P, Mandel L, Rejnek J (1983) Development of immunological capacity under germfree and conventional conditions. Ann N Y Acad Sci 409:96–113

Trikha A, Baillargeon JG, Kuo YF, Tan A, Pierson K, Sharma G, Wilkinson G, Bonds RS (2013) Development of food allergies in patients with gastroesophageal reflux disease treated with gastric acid suppressive medications. Pediatr Allergy Immunol 24(6):582–588. https://doi.org/10.1111/pai.12103

Tsilingiri K, Rescigno M (2013) Postbiotics: what else? Benef Microbes 4(1):101–107. https://doi. org/10.3920/BM2012.0046

Turnbaugh PJ, Ley RE, Hamady M, Fraser-Liggett CM, Knight R, Gordon JI (2007) The human microbiome project. Nature 449(7164):804–810. https://doi.org/10.1038/nature06244

Turnbaugh PJ, Hamady M, Yatsunenko T, Cantarel BL, Duncan A, Ley RE, Sogin ML, Jones WJ, Roe BA, Affourtit JP, Egholm M, Henrissat B, Heath AC, Knight R, Gordon JI (2009) A core gut microbiome in obese and lean twins. Nature 457(7228):480–484. https://doi.org/10.1038/nature07540

Umesaki Y, Setoyama H, Matsumoto S, Okada Y (1993) Expansion of alpha beta T-cell receptor-bearing intestinal intraepithelial lymphocytes after microbial colonization in germ-free mice and its independence from thymus. Immunology 79(1):32–37

Untersmayr E, Bakos N, Scholl I, Kundi M, Roth-Walter F, Szalai K, Riemer AB, Ankersmit HJ, Scheiner O, Boltz-Nitulescu G, Jensen-Jarolim E (2005) Anti-ulcer drugs promote IgE formation toward dietary antigens in adult patients. FASEB J 19(6):656–658. https://doi.org/10.1096/fj.04-3170fje

Vaishampayan PA, Kuehl JV, Froula JL, Morgan JL, Ochman H, Francino MP (2010) Comparative metagenomics and population dynamics of the gut microbiota in mother and infant. Genome Biol Evol 2:53–66. https://doi.org/10.1093/gbe/evp057

Vaishnava S, Yamamoto M, Severson KM, Ruhn KA, Yu X, Koren O, Ley R, Wakeland EK, Hooper LV (2011) The antibacterial lectin RegIIIgamma promotes the spatial segregation of microbiota and host in the intestine. Science 334(6053):255–258. https://doi.org/10.1126/science.1209791

van den Berk LC, Jansen BJ, Siebers-Vermeulen KG, Netea MG, Latuhihin T, Bergevoet S, Raymakers RA, Kogler G, Figdor CC, Adema GJ, Torensma R (2009) Toll-like receptor triggering in cord blood mesenchymal stem cells. J Cell Mol Med 13(9B):3415–3426. https://doi.org/10.1111/j.1582-4934.2009.00653.x

van Nood E, Vrieze A, Nieuwdorp M, Fuentes S, Zoetendal EG, de Vos WM, Visser CE, Kuijper EJ, Bartelsman JF, Tijssen JG, Speelman P, Dijkgraaf MG, Keller JJ (2013) Duodenal infusion of donor feces for recurrent Clostridium difficile. N Engl J Med 368(5):407–415. https://doi.org/10.1056/NEJMoa1205037

Vangay P, Ward T, Gerber JS, Knights D (2015) Antibiotics, pediatric dysbiosis, and disease. Cell Host Microbe 17(5):553–564. https://doi.org/10.1016/j.chom.2015.04.006

Verhasselt V (2010) Oral tolerance in neonates: from basics to potential prevention of allergic disease. Mucosal Immunol 3(4):326–333. https://doi.org/10.1038/mi.2010.25

Vetvicka V, Tlaskalova-Hogenova H, Stepankova R (1983) Effects of microflora antigens on lymphocyte migration patterns in germfree and conventional rats. Folia Biol (Praha) 29(6):412–418

Walsh CJ, Guinane CM, O'Toole PW, Cotter PD (2014) Beneficial modulation of the gut microbiota. FEBS Lett 588(22):4120–4130. https://doi.org/10.1016/j.febslet.2014.03.035

Watanabe Y, Arase S, Nagaoka N, Kawai M, Matsumoto S (2016) Chronic psychological stress disrupted the composition of the murine colonic microbiota and accelerated a murine model of inflammatory bowel disease. PLoS One 11(3):e0150559. https://doi.org/10.1371/journal.pone.0150559

West CE, Renz H, Jenmalm MC, Kozyrskyj AL, Allen KJ, Vuillermin P, Prescott SL (2015) The gut microbiota and inflammatory noncommunicable diseases: associations and potentials for gut microbiota therapies. J Allergy Clin Immunol 135(1):3–13. https://doi.org/10.1016/j.jaci.2014.11.012

Wu GD, Chen J, Hoffmann C, Bittinger K, Chen YY, Keilbaugh SA, Bewtra M, Knights D, Walters WA, Knight R, Sinha R, Gilroy E, Gupta K, Baldassano R, Nessel L, Li H, Bushman FD, Lewis JD (2011) Linking long-term dietary patterns with gut microbial enterotypes. Science 334(6052):105–108. https://doi.org/10.1126/science.1208344

Zheng Y, Valdez PA, Danilenko DM, Hu Y, Sa SM, Gong Q, Abbas AR, Modrusan Z, Ghilardi N, de Sauvage FJ, Ouyang W (2008) Interleukin-22 mediates early host defense against attaching and effacing bacterial pathogens. Nat Med 14(3):282–289. https://doi.org/10.1038/nm1720

Zhernakova A, Kurilshikov A, Bonder MJ, Tigchelaar EF, Schirmer M, Vatanen T, Mujagic Z, Vila AV, Falony G, Vieira-Silva S, Wang J, Imhann F, Brandsma E, Jankipersadsing SA, Joossens M, Cenit MC, Deelen P, Swertz MA, Weersma RK, Feskens EJ, Netea MG, Gevers D, Jonkers D, Franke L, Aulchenko YS, Huttenhower C, Raes J, Hofker MH, Xavier RJ, Wijmenga C, Fu J (2016) Population-based metagenomics analysis reveals markers for gut microbiome composition and diversity. Science 352(6285):565–569. https://doi.org/10.1126/science.aad3369

Specific Therapies for Asthma Endotypes: A New Twist in Drug Development

Ulrich M. Zissler

Abstract Asthma is a heterogeneous disease mainly affecting the lower airways and hallmarked by a chronic lung tissue inflammation that is strongly influenced by environmental and immunological factors. In a subgroup of asthmatics, respiratory infections are associated with the development of chronic disease and more frequent inflammatory exacerbations. Nevertheless, patients suffering from different asthma subgroups are treated with mainly non-specific and symptom-related using anti-inflammatory drugs and bronchodilators. Therefore, it is of great importance to define molecular patterns of asthma and allergy as basis for endotype definitions, rather than using clinical disease phenotypes. In addition, a number of factors can contribute to poor responses, while underlying patho-biological differences are increasingly recognized to play a role. These factors include novel biomarkers as well as well-known pathways but also consideration of the influence of the airway microbiome. With respect to this, specific drugs are developed targeting cellular components (e.g. eosinophils, neutrophils, Th9 cells) or specific mediators (e.g. IgE, FceR, IL5, TSLP) but also key regulators such as GATA3. In addition, also viral-induced asthma is attracting notice and development of anti-viral therapy strategies are promoted.

Abbreviations

ACQ   Asthma control questionnaire
BHR   Bronchial hyperresponsiveness
CCL   CC-chemokine ligand

U.M. Zissler
Center of Allergy & Environment (ZAUM), Technical University of Munich, Munich, Germany

Helmholtz Center Munich, German Research Center for Environmental Health, Munich, Germany

German Center for Lung Research (DZL), CPC-M, Munich, Germany
e-mail: ulrich.zissler@tum.de

© Springer International Publishing AG 2017
C.B. Schmidt-Weber (ed.), *Allergy Prevention and Exacerbation*, Birkhäuser
Advances in Infectious Diseases, https://doi.org/10.1007/978-3-319-69968-4_11

CD   Cluster of differentiation
CLCA1   Calcium-activated chloride channel protein 1
FcεRI   High-affinity IgE receptor I
FDA   US Food and Drug Administration
FeNO   Fraction of exhaled nitric oxide
FEV1   Volume of air expelled in the first second of a forced expiration
huMAb   Humanized monoclonal antibody
ICS   Inhaled corticosteroids
Ig   Immunoglobulin
IL   Interleukin
iNOS   Inducible nitric oxide synthase
LABA   Long acting beta2 agonist
LPS   Lipopolysaccharide
mAb   Monoclonal antibody
OCS   Oral corticosteroid
POSTN   Periostin
RSV   Respiratory syncytial virus
Th   T helper
TSLP   Thymic stromal lymphopoietin

1   Key Mechanisms of Asthma

Diagnosis and long-term management of asthma were remarkably advanced within the last years, however it remains a serious problem for public health worldwide. It is now one of the most common chronic diseases in developed countries with an ascending tendency in newly industrialized countries (Braman 2006). The key pathological features of asthma are highly complex with multiple features including infiltration of the airways by activated lymphocytes, eosinophils, and neutrophils. Further, mast cell degranulation and mucous gland hyperplasia characterize the asthma pathology. Asthmatic epithelium exhibits matrix remodeling as well as cilia dysfunction accompanied by collagen deposition in the sub-epithelial area. A strong association is existing between asthma pathology and the release of pro-inflammatory mediators including lipid mediators, chemokines, cytokines. In addition to infiltrating leukocytes, the structural cells in the airways, including airway epithelial cells, smooth muscle cells, endothelial cells, and fibroblasts, are important sources of asthma-related or enhancing mediators (Zissler et al. 2016a, b). Allergies and Asthma are common but at once heterogeneous chronic diseases. While the definition of asthma is rather general and non-specific, multiple clinical phenotypes meet these criteria (National Asthma Education and Prevention Program 2007). The current treatment of asthma is mainly non-specific and symptom-related using anti-inflammatory drugs and bronchodilators, which are successfully used in most patients (Sheehan et al. 2016; Billington et al. 2016). However, even responses to these treatments vary. In addition, intense research of respiratory diseases has rewarded us with improved pharmacological treatment strategies such as inhaled corticosteroids, leukotriene receptor antagonists, β-agonists and more recently targeted monoclonal therapies which alleviated the burden by asthma. Of particular note is the recent confirmation by research that the environment of the lungs are not sterile as had previously been believed (Marsland 2013). Previously it was generally assumed that a positive detection of bacteria in the lower respiratory tract represented an abnormal health condition and that the healthy lung was sterile (Aho et al. 2015). This was mainly due to poor, ineffective microbial cultivation techniques, which could not represent the environmental conditions in the lung (Marsland and Gollwitzer 2014). However it is clear that although the lung contains a smaller bacterial burden than in the upper respiratory tract there is a distinct microbiome present in the lower airways of healthy individuals (Segal and Blaser 2014). This finding that the lungs show a unique composition of the respiratory microbiome independent of health status has led to the increase in research now taking place in order to categorize not only the clear microbiota of the diseased lung but also the healthy lung.

Further, a number of factors can contribute to poor responses, while underlying patho-biological differences are increasingly recognized to play a role. In contrast to asthmatic disease, allergic reactions are triggered when preformed IgE is cross-link allergens bound to the high-affinity receptor FcεRI on mast cells. This antigen-dependent activation of tissue mast cells that have specific IgE bound to their surface is the central event in acute allergic reactions (Galli and Tsai 2012).

Type I immediate hypersensitivity allergic reactions, characterized by short-lived mediators such as histamine, include allergic rhinitis as well as allergic asthma, while the later events involve leukotrienes, cytokines, and chemokines, which recruit and activate eosinophils and basophils. This late phase response can evolve into chronic inflammation with a characteristic presence of effector T cells and eosinophils, most clearly seen in chronified allergic asthma. Despite these characteristics, there is some heterogeneity in patients with allergic disease impeding a phenotypic characterization, which includes a complex interaction of many genetic and environmental factors in conjunction with observable characteristics, such as lung function in asthma or allergen-specific IgE responses. It is of great importance to define molecular patterns of asthma and allergy as basis for endotype definitions, rather than using clinical disease phenotypes following the definition of Wenzel (Wenzel 2013; Zissler et al. 2016b). Thus, the definition of an endotype is a molecular phenotype that supports a clinical outcome Asthma phenotyping has involved biased and unbiased approaches to grouping clinical, physiologic, and hereditary characteristics (Wenzel 2006; Fitzpatrick et al. 2011). Studies have supported the importance of age of onset, eosinophils, and lung function, but definitive clustering of these characteristics or their relation to pathobiology remains uncertain (Moore et al. 2010; Haldar et al. 2008; Wu et al. 2014; Miranda et al. 2004).

2 Type 2 Biomarkers in Asthma

Biomarkers are of importance to target type 2 therapies to influence the disease outcome of asthma patients (Fig. 1). Recent type 2 biomarkers include periostin, fraction of exhaled nitric oxide (FeNO), and systemic or local, sputum-assessed eosinophils. Since years, sputum eosinophil numbers were used to predict response to systemic corticosteroids, while later studies directly correlated their absence with poor corticosteroid responses (Brown 1958; Pavord et al. 1999). However, several studies suggested a relationship between Th2 cytokines and eosinophil counts, but this result was not confirmed until antibodies against IL-5 almost totally depleted blood eosinophils (Leckie et al. 2000; Flood-Page et al. 2003; Haldar et al. 2009; Nair et al. 2009; Pavord et al. 2012; Leung et al. 2017). Blood eosinophil counts correlated reasonably well with IL-13 mRNA levels in sputum,however these counts were shown not to be dependent on IL-4 or IL-13 (Corren et al. 2011a; Wenzel et al. 2013). Serum periostin (POSTN) was identified in airway epithelial brushings as a downstream signature of IL-13, a classifier for eosinophilic airway inflammation (Woodruff et al. 2009; Jia et al. 2012). In addition, FeNO primarily

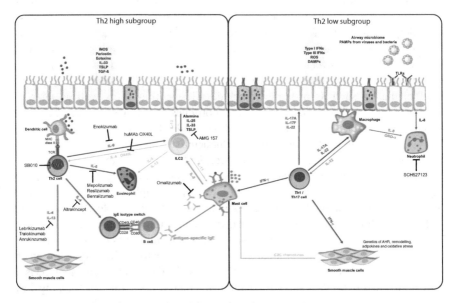

Fig. 1 Th2 "high" and Th2 "low" subgroups and related drug targets. The onset of asthma begins with the development of Th2 cells, which are activated by antigen-presenting cells (e.g. Dendritic cells) and their following production of the Th2 key mediators IL-4, IL-5 and IL-13. These cytokines promote eosinophilic inflammation, while they can be inhibited by several small-molecule drugs such as Altrakincept targeting IL-4. Mepolizumab, Reslizumab, and Benralizumab act on IL-5, while Lebrikizumab, Tralokinumab, and Anrukinzumab block IL-13. Soluble IgE can be blocked by Omalizumab, but also transcription factors such as GATA-3 can be inhibited by specific approaches such as DNAzyme SB010. In addition, changes in epithelial and smooth-muscle cells further contribute to asthma pathobiology

generated by inducible nitric oxide synthase (iNOS), can be strongly induced in airway epithelial cells by IL-13 (Chibana et al. 2008; Dweik et al. 2010). While FeNO levels correlate with lung eosinophilia (Dweik et al. 2010), however FeNO levels are independent of IL-5 (Haldar et al. 2009). These results indicated that Th2-related biomarkers and their relation to the downstream signaling pathways, however, their specific benefit for tailor-made *"Th2 therapies"* is not yet clear.

3   Approaches to Target Asthma Subgroups by Biologicals

3.1   Th2 Pattern and Related Biomarkers

Identification of Th1 and Th2 immunity and widespread efficacy of inhaled corti-costeroids (ICSs) suggested that asthma is primarily driven by Th2 immunity involving the key cytokines IL-4, IL-5, and IL-13 (Robinson et al. 1992; Brusselle et al. 1995; Grunig et al. 1998; Wills-Karp et al. 1998). Numerous Th2-specific biologics target on immunoglobulin E (IgE), the central Th2 key cytokines IL-4 and IL-13, further on IL-5, thymic stromal lymphopoietin (TSLP), and IL-4 receptor α (IL-4Rα) , but also on the Th2/type-2 key transcription factor GATA-3. A recent study showed that a polymorphism in the gene encoding the IL-4Rα chain is associated with severe asthma and promotes conversion of induced Treg (iTreg) cells toward a T helper 17 (Th17) cell fate (Massoud et al. 2016). Interleukin-4Rα, the common receptor subchain for both IL-4 and IL-13, is present in both the type-1 (dimerized with γc) and type-2 (dimerized with IL-13Ra1) receptors, while on airway epithelial cells only type 2 is expressed (Zissler et al. 2016a). Interleukin-4 is able to bind to two receptor (type I and type II) heterodimers consisting of either gc/IL-13Ra (type I) or IL-4Ra/IL-13Ra (type II), whereas binding of IL-13 is restricted to type II (Zissler et al. 2016a). Following this, IL-13 is not able to activate T cells, but both cytokines IL-4 and IL-13 can promote an IgE isotype switching in B cells (Grunig et al. 1998; Wills-Karp et al. 1998; Zhu 2015). In turn, IgE binds to the high-affinity IgE receptor, which activates mast cells by cross-linking with an allergen. This mast cell development is further promoted by type 2 innate lymphoid cell-derived IL-13 (Martinez-Gonzalez et al. 2015). Further, IL-5 as a pro-eosinophilic cytokine promotes development and survival of eosinophils. Interestingly, like IL-13, IL-5 is generated in high amounts also by ILC-2 cells (Martinez-Gonzalez et al. 2015). However, the Th2 key cytokine IL-4 is the main factor for Th2 priming and maturation, while Th2 differentiation is supported by cytokines, such as TSLP (Allakhverdi et al. 2007).

Despite this, studies investigating Th2-related asthma pathobiology suggested differences in immune processes across asthmatic patients, while early studies of Th2-targeted therapies were not efficacious (Wenzel et al. 1999; Leckie et al. 2000; Flood-Page et al. 2003). Asthma subgroups were rarely considered for therapeutic approaches, despite corticosteroid responses were reported to be dependent on lung eosinophils for years (Leckie et al. 2000; Haldar et al. 2008; Wu et al. 2014).

Therapeutic approaches considering specific endotypes began to evolve when a mAb to IL-5 (mepolizumab) was specifically developed for eosinophilic asthma. This anti-eosinophilic therapeutic approach efficiently reduced exacerbations and systemic corticosteroid use in patients with eosinophilic asthma, which was however not successful in patients with refractory eosinophilic asthma and recurrent severe exacerbations (Haldar et al. 2009; Nair et al. 2009). By using a different approach, epithelial biomarkers for Th2-driven inflammation, mainly characterized by IL-4 and IL-13 were first identified in vitro and then identified ex vivo in a subset of patients (Woodruff et al. 2007, 2009; Zissler et al. 2016a). In contrast to smokers also investigated in this study, three genes including POSTN up-regulated in response to IL-13 in vitro were identified in human airway epithelial cells from approximately 50% of corticosteroid-naive patients with mild asthma, while healthy subjects and the other asthmatic patients showed only low expression of this Th2 signature (Woodruff et al. 2007). Those in the "type-2-high" cluster showed increased atopy, higher tissue eosinophil counts, and more bronchial hyperresponsiveness. Interestingly, inhaled fluticasone was more beneficial for these patients expressing this "Th2-high" cluster than those without a Th2 signature(Woodruff et al. 2009). These approaches showed the importance of targeted and untargeted therapies directly related to patients characteristics. The contribution of molecular pathways to clinical outcomes could further lead to identification of disease endotypes, which could be directly targeted, however are only partially understood in asthma (Anderson 2008; Lotvall et al. 2011).

3.2 Approaches to Target Immunoglobulin E

For many years, a strong correlation was suggested for asthma as an allergic disease with allergen-specific IgE. A recombinant humanized anti-IgE mAb (omalizumab) was developed to inhibit interactions of IgE with high-affinity IgE receptors (FcεRI). This anti-IgE approach suppressed the early immediate hypersensitivity response to nebulized inhaled allergen in corticosteroid-naive patients with mild asthma (Boulet et al. 1997; Table 1). Further studies showed decreased asthmatic response, early as well as late, accompanied by reduced sputum and tissue eosinophil counts, submucosal IgE and FcεRI1 (Fahy et al. 1997; van Rensen et al. 2009). Omalizumab also improved rhinitis symptoms and increased lung function parameters after allergen exposure in asthmatic patients (Corren et al. 2011b; Vignola et al. 2004). Therefore, anti-IgE seems to be consistently effective in allergic reactions of rhinitis and asthmatic patients. Further, early studies on omalizumab in chronic asthma patients took aim on an allergic subgroups, which however was poorly defined by atopy and a total serum IgE level of between 30 and 700 IU/mL. Previous studies of patients with moderate allergic asthma and use of

Table 1 Molecular targets for novel endotype-related therapies

Target	Biological	Phase of the trial	Endotype	Clinical result	References
IgE	Omalizumab/rhuMAb-E25 (Monoclonal anti-IgE antibody)	FDA, EMA	Mild-to moderate allergic asthma	↓ Early and late asthmatic response, ↓ serum free IgE	Milgrom et al. (1999), Fahy et al. (1997), van Rensen et al. (2009), and Corren et al. (2011b)
			Chronic moderate-to-severe allergic asthma	↓ Asthma exacerbations, ↓ serum free IgE	Milgrom et al. (1999, 2001), Busse et al. (2001), Soler et al. (2001), Holgate et al. (2004), Humbert et al. (2005), Hanania et al. (2011), Busse et al. (2011), Lanier et al. (2009), and Vignola et al. (2004)
			Chronic severe allergic asthma	↓ asthma exacerbations for type 2-high subgroup, patients with ↑ FENO levels, ↑ blood eosinophil counts, and ↑ serum periostin levels	Hanania et al. (2013)
			Moderate-severe allergic asthma, on ICS (6–12 years)	↓ rescue medication use, ↓ exacerbations	Vignola et al. (2004)
	QGE031	II	Mild allergic asthma	↑ FcεRI expression on basophils, ↓ surface IgE levels on basophils	Gauvreau et al. (2016)
	Quilizumab (humanized IgG1 monoclonal antibody)	II	Moderate- to severe allergic asthma, on ICS	↓ serum IgE	Harris et al. (2016)
IL4R	Pitrakinra (Mutant IL-4 molecule)	II	Chronic moderate-to-severe asthma	No effect on pre-specified clinical asthma	Wenzel et al. (2007)

(continued)

Table 1 (continued)

Target	Biological	Phase of the trial	Endotype	Clinical result	References
	AMG317 (Humanized monoclonal antibody to IL4Rα)	II	Moderate, ICS treated asthma	↓ serum IgE, Subgroup effect: worst asthma control modest benefit	Corren et al. (2010)
	Dupilumab (Humanized monoclonal antibody to IL4Rα)	II	Moderate-severe asthma, mid-high dose ICS/LABA with Type-2 High Phenotype	↓ β-agonist use, ↓ upper airway symptoms, ↓ FeNO, ↑ FEV1, ↑ asthma control	Wenzel et al. (2016)
		II	Chronic moderate-to-severe asthma, with Type-2 High Phenotype	↓ Asthma exacerbations, ↓ FENO, ↓ β-agonist use, ↑ FEV1	Wenzel et al. (2013)
IL-4	Altrakincept (recombinant human IL-4 receptor)	II	Moderate atopic asthma	↑ FEV1	Borish et al. (1999)
IL-13	IMA-638 (Humanized IL13 antibody)	II	Mild atopic asthma	No effect on airway hyperresponsiveness or sputum eosinophils	Gauvreau et al. (2011)
	Lebrikizumab (monoclonal IL13 antibody)	II/III	Moderate-severe, not controlled by mid-high dose ICS/LABA; Type-2 "*High*" Phenotype (Periostin)	Periostin-high group: ↑ FEV1, ↓ FeNO, ↓ exacerbation in trend	Corren et al. (2011a) and Hanania et al. (2016)
	Tralokinumab (monoclonal IL13 antibody)	II	Moderate-severe uncontrolled asthma, Subgroup of Type-2 "High" Phenotype (IL-13 in sputum)	↓ β-agonist use in IL-13+ group, trend to ↑ FEV1	Piper et al. (2013)
		II	Uncontrolled asthma	inconsistent reduction of exacerbations in Type-2 high phenotype	Hanania et al. (2016)
	GSK679586 (monoclonal IL13 antibody)	II	Severe Asthma	No effect on pre-specified clinical asthma outcomes	De Boever et al. (2014)
	Anrukinzumab (monoclonal IL13 antibody)	II	Mild atopic asthma (allergen challenge)	↓ early-/late-phase asthmatic FEV1 response No effect on AHR, eosinophils, and total IgE	Gauvreau et al. (2011)

		Phase	Asthma type/phenotype	Effects	Reference
IL-5	SB-240563 (anti-IL5)		Mild, allergic asthmatic males, no ICS use (allergen challenge)	↓ blood and sputum eosinophils	Leckie et al. (2000)
	Mepolizumab (anti-IL5)	II/III	Mild, stable, atopic asthma, no ICS	↓ airway, bone marrow, and blood eosinophils	Flood-Page et al. (2003)
			Mild-to-moderate allergic asthma	↓ airway, bone marrow, and blood eosinophils	Flood-Page et al. (2007)
			Moderate, persistent asthma on ICS	↓ blood/sputum eosinophils, trend to ↓ exacerbations	Fahy et al. (1997)
			Severe refractory asthma, Type-2 "High" Phenotype	↓ blood and sputum eosinophils, ↓ airway wall thickening, improved quality of life	Haldar et al. (2009)
			Severe asthma, prednisone-dependent, Type-2 "High" Phenotype	↓ sputum/blood eosinophils, ↑ asthma control and FEV1	Nair et al. (2009)
			Severe chronic refractory asthma, Type-2-high phenotype	↓ Asthma exacerbations, ↓ blood/sputum eosinophils, ↑ FEV1	van Rensen et al. (2009)
	Reslizumab (anti-IL5)	III	Poorly controlled asthma on high dose ICS + additional controller and Type-2 "High" Phenotype	↓ sputum eosinophils, trended to ↑ FEV1 and asthma control, results best in nasal polyp subgroup	Castro et al. (2011)
	Benralizumabα (anti-IL-5Rα)	III	Chronic eosinophilic asthma, Type-2 "High" Phenotype	↓ sputum, blood, and bone marrow eosinophils	Tan et al. (2016)
		III	Severe asthma uncontrolled ICS/LABA	↑ exacerbation, ↑ pre-bronchodilator FEV1 and ↑ total asthma symptom score	FitzGerald et al. (2016)
		II	Uncontrolled eosinophilic asthma	↓ exacerbation rates (in group 100mg)	Castro et al. (2014)
TSLP	AMG 157 (human anti-TSLP monoclonal antibody)		Mild allergic asthma (allergen challenge)	↓ Late asthmatic response, ↓ blood/sputum eosinophils, ↓ FENO levels	Gauvreau et al. (2014b)

(continued)

Table 1 (continued)

Target	Biological	Phase of the trial	Endotype	Clinical result	References
CD25	Daclizumab (huMAb anti-IL-2)	II	Moderate- to severe chronic asthma, poorly controlled on ICS	↑ FEV1(% predicted), ↑ FEV1/FVC, ↑ asthma control, ↓ SABA use, ↓ blood eosinophils	Busse et al. (2008)
OX40L	huMAb OX40L (anti-OX40L monoclonal antibody)	II	Mild allergic asthma	↓ serum total IgE, ↓ airway eosinophils	Gauvreau et al. (2014a)
GATA3	SB010 (DNAzyme-based GATA-3 antagonist)	II	Allergic asthma with sputum eosinophilia	↓ asthmatic responses after allergen provocation	Krug et al. (2015)
IL9	Enokizumab (monoclonal IL9 antibody)	II	Chronic eosinophilic asthma and Th2/Type-2 "High" Phenotype	↓ Eosinophils in airway mucosa/submucosa, sputum, bone marrow, and blood	Antoniu (2010)
TNF-α	Etanercept (soluble TNF-α inhibitor)	II	Severe refractory asthma on ICS	↑ post-bronchodilator FEV1, ↓CRP	Morjaria et al. (2008)
Neutro-phils	CXCR2 receptor antagonist (SCH527123)		Severe asthma with sputum neutrophilia (>40%)	trend to improvement in ACQ, no effect on FEV1	Nair et al. (2012)
IL-17	Brodalumab (anti-IL17 receptor antibody)	II	Moderate-severe asthma on ICS	Marginal ACQ change in high-reversibility group (>20% bronchodilator response)	Busse et al. (2013)
RSV	Palivizumab (anti-RSV monoclonal antibody)		recurrent wheezing	no suppression of the onset of atopic asthma, ↓ recurrent wheezing during first 6 years of life	Mochizuki et al. (2017)

ICSs demonstrated reductions in asthma exacerbation rates as well as corticosteroid requirements (Milgrom et al. 1999, 2001).

Subcutaneous injection of omalizumab reduced asthma exacerbations in moderate allergic asthma patients, while symptoms and lung function were only minimal affected (Soler et al. 2001; Busse et al. 2001). In patients with severe atopic asthma taking ICSs and long-acting ß-agonists (LABAs), omalizumab reduced ICS requirements compared with placebo controls (Holgate et al. 2004). In contrast, in a trial on severe asthma patients, add-on omalizumab did not significantly reduce asthma exacerbation rates (Humbert et al. 2005). In patients with the most severe asthma studied, omalizumab reduced asthma exacerbations over 48 weeks compared with placebo. While clinical scores of these patients showed changes, no improvement could be stated for lung function parameters (Hanania et al. 2011). Furthermore, trials were able to show efficacy for omalizumab in reducing asthma exacerbation rates (Milgrom et al. 2001; Lanier et al. 2009; Busse et al. 2011). Although some effects on allergic reactions could be related to allergic exacerbations, a study in children suffering from persistent asthma showed that omalizumab significantly decreased viral exacerbations, while the underlying mechanisms are still unclear (Busse et al. 2011). Primarily, anti-IgE was developed for the treatment of allergic asthma, however it was not effective in all patients. In addition, treatment with omalizumab is limited by costs and compliance, thus to improve the ability to identify responsive patients is of high importance. However, QGE031 a further investigational anti-IgE antibody that binds IgE with higher affinity than omalizumab showed greater efficacy than omalizumab on inhaled allergen responses in patients with mild allergic asthma (Gauvreau et al. 2016; Table 1). Furthermore, a humanized IgG1 monoclonal antibody Quilizumab, targets the M1-prime segment of membrane-expressed IgE, wich leads to a depletion of IgE-switched and memory B cells (Harris et al. 2016; Table 1). However, the results of this study showed no benefits for uncontrolled asthma patients, when targeting the IgE pathway via depletion of IgE-switched and memory B cells. Based on median splits of blood eosinophils, POSTN, and FeNO Hanania et al. observed stronger reductions of exacerbation rates in patients above the median compared to patients below (Hanania et al. 2011, 2013). This approach could be helpful to categorize patients even if other outcome parameters are not altered.

3.3   Approaches to Target Th2 Key Cytokines IL-4 and IL-13

Interleukin-4 and IL-13 are type 2 cytokines known to play an important role in allergic airway inflammation (Brusselle et al. 1995; Grunig et al. 1998; Wills-Karp et al. 1998; Zissler et al. 2016a). However, studies on molecules targeting these pathways showed no success until they were linked with type 2 biomarkers beyond atopy or IgE. Given the existing evidence for microbial activation of T-cell subsets in the gastrointestinal tract, the current hypothesis is that the overall heterogeneity in asthma-associated immune dysfunction is related to the presence of distinct microbial populations in the airways. Pitrakinra, a recombinant human interleukin-4 variant blocks the ability of human IL-4 or IL-13 to bind to IL-4Rα. Nebulized pitrakinra reduced the late asthmatic response in patients with mild atopic asthma compared with a placebo group (Wenzel et al. 2007; Table 1). Similarly, an IL-13 antibody (IMA-638) decreased both early and late asthmatic responses (Gauvreau et al. 2011; Table 1). Although both approaches seemed to improve lung function parameters after challenge, they did not influence bronchial hyperresponsiveness or eosinophil counts in sputum. However, FeNO levels were reduced implicating a pivotal role of IL-4/IL-13 pathways and suggesting its use as a type-2 biomarker (Wenzel et al. 2007). While earlier studies targeting IL-4 in patients with chronic asthma did not target specific asthma subgroups, only one pilot study supported the efficacy of blocking IL-4 in patients with moderate asthma (Borish et al. 2001). Many years passed before the type 2 key cytokines IL-4 and IL-13 were again targeted in patients with chronic asthma. In adults with moderate-to-severe asthma and inadequately controlled symptoms, subcutaneous administration of the humanized IgG4 mAb to IL-13, lebrikizumab, improved FEV1 was observed (Corren et al. 2011a; Table 1). In the LAVOLTA studies of adult patients with uncontrolled asthma, lebrikizumab did not consistently show reduced asthma exacerbation rates in biomarker-high patients (Hanania et al. 2016). However, lebrikizumab blocked interleukin-13 as evidenced by the effect on interleukin-13-related pharmacodynamic biomarkers, and clinically relevant changes could not be ruled out (Hanania et al. 2016). Another humanized IgG4 mAb to IL-13, tralokinumab only brought moderate improvement in lung function and some decrease in ß-agonist use (Piper et al. 2013). However, pitrakinra did not decrease induced asthma exacerbation rates when background medication was withdrawn in a population with moderate-to-severe asthma (Slager et al. 2012). Similarly, a study of a humanized mAb to IL-4Rα (AMG317) reduced serum IgE levels, suggesting some biologic activity (Corren et al. 2010; Table 1). Thus in patients with non-subtyped asthma, blockade of IL-13 alone or in combination with IL-4 demonstrated only moderate clinically improvements.

Other studies pre-specified additional analysis in relation to type-2 biomarkers (Corren et al. 2011a; Piper et al. 2013). In a lebrikizumab study, analyses identified patients with "Th2-high" asthma, which was identified by using serum IgE and blood eosinophil levels, while this only moderately improved the efficacy of lebrikizumab on exacerbations (Corren et al. 2011a). However, serum periostin levels were also used to pre-define the "Th2-high" and "Th2-low" subgroups. Patients with higher periostin levels treated with lebrikizumab showed an improvement in FEV1, whereas those with low periostin levels showed no improvement (Corren et al. 2011a). High FeNO levels predicted also a better response to lebrikizumab, whereas high levels of both markers in combination, FeNO and periostin, were even more predictive for therapy success (Corren et al. 2011a). Similar to pitrakinra, lebrikizumab also decreased FeNO levels, which is supportive of biologic effect on active type 2-associated pathways.

Tralokinumab (anti-IL-13) showed better results in asthmatic patients with measurable type-2 signatures, as defined by detectable IL-13 levels in induced sputum compared with IL-13-negative groups treated with tralokinumab or placebo (Piper et al. 2013). However, a study on GSK679586, a humanized mAb, which inhibits IL-13 binding to both IL-13 receptor α1 and α2 showed no effects on pre-specified clinical outcomes in patients suffering from severe asthma (De Boever et al. 2014). The first study to prospectively target a type-2-high subgroup, as defined by systemic or local, sputum-assessed eosinophilia, used dupilumab, a humanized mAb to IL-4Ra (Wenzel et al. 2013), while in a study with uncontrolled moderate-to-severe asthmatics, dupilumab was associated with decreased asthma exacerbation rates of 87% compared with placebo (Wenzel et al. 2013). In this Th2-high population dupilumab improved asthma control, symptoms, and lung function on top of combined ICS/LABA treatment when background therapy was withdrawn or even when no background therapy remained. In addition, symptoms of the upper airways were shown to be significantly improved in this study (Wenzel et al. 2013). Furthermore, some investigated type-2 biomarkers, including FeNO, CCL17, CCL26, and IgE levels, were also decreased. In fact, the change in FeNO levels inversely correlated with changes in lung function, which was highly supportive of a clinically meaningful effect of dupilumab on its target biologic pathway (Wenzel et al. 2013). Similar to lebrikizumab, however, dupilumab did not decrease blood eosinophil counts.

While Polymorphisms in type-2 cytokine pathway genes have been shown to be linked to asthma (Ober and Yao 2011), single nucleotide polymorphisms in IL4RA are associated with both asthma and severe asthma (Massoud et al. 2016).

Also on allergen challenge, greater improvement in response to pitrakinra was seen in patients expressing certain IL4RA alleles (Slager et al. 2010). More importantly, in a pitrakinra study, asthmatic patients homozygous for the common alleles at rs8832, rs1029489, or rs8832 showed a dose-related decrease in asthma exacerbation rates and activity limitation compared with those receiving placebo (Slager et al. 2010, 2012). These studies suggest that the type-2 genetic background could play a pivotal role in predicting response to type-2-directed interventions.

3.4 Approaches to Target IL-5

Interleukin-5, also a type-2 effector cytokine produced by Th2- and ILC2-cells, is the most potent eosinophilic cytokine known, with its receptor IL-5Ra expressed on eosinophils and some basophils (Clutterbuck et al. 1989; Aron and Akbari 2017). Blockage of IL-5 selectively affects the eosinophils and not other type-2-like inflammatory elements. However, eosinophilia can be present in the early-onset allergic asthma but is not mandatory to be associated with IgE-mediated allergy (Miranda et al. 2004; Haldar et al. 2008; Moore et al. 2010; Amelink et al. 2013). This allergic asthma subtype, although not associated with traditional allergic biomarkers or symptoms, often involves upper airways disease, such as sinus disease, nasal polyps, but also higher urinary leukotrienes, and more aspirin-sensitive disease (Miranda et al. 2004; Haldar et al. 2008; Moore et al. 2010;

Amelink et al. 2013; Wu et al. 2014). With respect to the association of eosinophils and Th2 inflammation with allergen challenge, it is not surprising that anti-IL-5 was initially investigated in an allergen challenge study. However, anti-IL-5 therapy did not inhibit asthmatic responses or airway hyperreactivity, despite strongly decreased eosinophil counts (Leckie et al. 2000; Table 1). This study implicates eosinophils and IL-5 not as the most critical key players to allergic asthma endotypes and their related allergic reactions. Two studies with three intravenous doses of mepolizumab in patients with mild atopic or persistent asthma showed negative results for clinical parameters (Flood-Page et al. 2003, 2007; Table 1). Eosinophil granulocytes were decreased systemically and in the airways, while no effects on lung function or other clinical parameters were observed (Flood-Page et al. 2003, 2007). Thus anti-IL-5, similar to approaches targeting the IL-4/IL-13 pathway, was not effective in non-subgrouped asthmatic patients. In contrast to IL-4/IL-13, mepolizumab was not successful in an allergic model of asthma. Furthermore, two studies were crucial to an overall subgrouping approach targeting patients with eosinophilic asthma. Both studies defined eosinophilic asthma as persistent sputum eosinophilia higher than 3% compared to previous year (Haldar et al. 2009; Nair et al. 2009). One study investigated patients suffering from moderate-to-severe asthma (Haldar et al. 2009), while the other targeted on OCS-dependent patients (Nair et al. 2009). Intravenous administration of mepolizumab for one year reduced asthma exacerbations by approximately 48%, however only with a moderate effect on life quality of asthmatic patients, going along with decreased eosinophil counts in periphery and lower airways (Haldar et al. 2009). While lung function parameters, symptoms, and FeNO levels were not affected, decreased airway wall thickening became visible on computed tomographic (CT) imaging (Haldar et al. 2009). In a trial, studying corticosteroid-dependent patients, mepolizumab treatment was accompanied by significant reductions in steroid dose, and decreased sputum eosinophil counts (Nair et al. 2009). Because of the complexity of sputum induction, local sputum eosinophil counts are not generally available, while subsequent studies identified an "eosinophilic type-2-high asthma" based on systemic eosinophil counts. A study on mepolizumab in severe asthma patients receiving high-dose ICS/LABA treatment and evidence for this eosinophilic type-2-high asthma identified blood eosinophil counts higher than 300 per milliliter as biomarker or reliable predictor of treatment response (Pavord et al. 2012). Further, three different doses of mepolizumab demonstrated appropriate effects in decreasing asthma exacerbation rates compared with placebo controls. The most prominent effect was seen in those asthma patients with the highest blood eosinophil counts and exacerbation history (Pavord et al. 2012). Mepolizumab appeared to have the same efficiency in patients with eosinophilic asthma independent of treatment with systemic corticosteroids (Prazma et al. 2014). In a recent study of asthmatic patients hallmarked by systemic eosinophilia accompanied by recurrent exacerbation rates despite high-dose ICSs, monthly intravenous or subcutaneous mepolizumab administration reduced the exacerbation rate with simultaneously increased lung function and moderately improved symptom scores (Ortega et al. 2016). Mepolizumab was also studied in patients with systemic corticosteroid-dependent severe asthma, demonstrating subcutaneous mepolizumab administered for 20 weeks to be highly effective in decreasing daily corticosteroid doses with simulanteous improvement of clinical scores (Bel et al. 2014). In this study, 40% of mepolizumab-treated patients were able to reduce their OCS dosage more than 75% (Bel et al. 2014). In addition, reslizumab, also an anti-IL-5 antibody, was studied in patients with poorly controlled asthma and compared to a control group with persistent sputum eosinophils over 3% (Castro et al. 2011). Like mepolizumab, intravenous administration of reslizumab decreased systemic and local, sputum-assessed eosinophil counts (Castro et al. 2011; Table 1). While mepolizumab and reslizumab directly target to IL-5, benralizumab is a humanized monoclonal antibody directed against IL-5Ra (Table 1). When benralizumab binds the IL-5Ra, it activates opsonization to destroy eosinophils and basophils expressing the receptor on their surface (Kolbeck et al. 2010). In a study of adult asthmatic patients, benralizumab decreased systemic and local eosinophil counts (Iida et al. 2006; Laviolette et al. 2013), while in adults with uncontrolled eosinophilic asthma, benralizumab significantly reduced asthma exacerbation rates (Castro et al. 2014). Taken together, approaches specifically targeting the IL-5 pathway have been demonstrated to be efficacious, with representative effects on asthma exacerbation rates. However, it remains unclear whether treatment with benralizumab shows differences in efficacy or target endotype from those targeting IL-5. Thus, the benefit of approaches blocking IL-5/IL-5Ra in patients with allergic type-2 disorders remains unclear.

3.5 Approaches to Target TSLP

Recently described, the innate "alarmin" thymic stromal lymphopoietin (TSLP) is a cytokine influencing Th2 responses by suppression of IL-12 production, thus skewing Th0 cells toward Th2 pattern. TSLP is reported to be increased on mRNA and protein levels in human asthmatic airways in relation to healthy control subjects, highlighting TSLP as a potential type-2-target (Ying et al. 2005, 2008). In a Th2-associated allergen challenge model, AMG157, a human anti-TSLP monoclonal IgG2λ, decreased allergen-induced late asthmatic responses (Gauvreau et al. 2014b; Table 1). These findings were accompanied by decreased systemic and local airway eosinophil counts before and after allergen challenge and simultaneously decreased FeNO levels (Gauvreau et al. 2014b). According to these findings, TSLP blockade could be suggested to affect inflammatory pathways more powerful than IL-4/IL-13 or IL-5. However, it is still unclear whether these effects will be visible in severe corticosteroid-treated asthma patients, probably expressing an allergic Th2 subtype. Interestingly, recently published data showed that TSLP and IL33 induction was reduced by LPS treatment which is also produced by microbes within the airways (Lin et al. 2016). Recent treatment studies suggested type-2 asthma itself representing several different molecular endotypes. Data on late-onset asthma associated with an accompanying eosinophilia showed improved response to IL-5-targeted therapies than in early-onset asthma patients, whereas targeting anti-IL-4/IL-13 appeared to be more effective in allergic models of asthma (Haldar et al. 2009; Pavord et al. 2012; Wenzel et al. 2013). However, the rate of non-responders to these treatments was about 50% of mepolizumab treated asthma patients (Bel et al. 2014). This heterogeneity of type-2 inflammation patterns is supported by biomarker studies including eosinophil counts and FeNO levels, which seem to be present in a wide range of asthma severities expressing distinct gene signatures (Wu et al. 2014; Modena et al. 2014). Moreover, a distinct biomarker can be present while another is not illustrating the type 2 biomarker expression. A study, highlighting the epithelial gene expression patterns of asthmatic patients and healthy control subjects identified more than five hundered genes, which are correlating with the FeNO (Modena et al. 2014). These FeNO-associated genes classified five distinct patient clusters, including three of these clusters with high FeNO levels. Although all of these patterns were identified by the expression of a type-2 gene signature, these three FeNO-related pattern showed different relations to other genes expressed in the respective pattern. (Modena et al. 2014). Whether Th2-targeted therapies will work equally well in all three Th2-high clusters is not yet clear but could help to understand why a humanized mAb to IL-13 (GSK679586; Table 1) inhibiting the binding of IL-13 to IL-13Rα1 and IL-13Rα2 was not effective in severe asthma patients, even not in asthma patients with increased systemic eosinophil counts (De Boever et al. 2014). Although the missing efficacy might be based on the molecule itself, it is also possible that inhibition of type-2 immunity in refractory asthma patients might only affect a low number of complex inflammatory processes.

3.6   Approaches to Target CD25

A humanized monoclonal antibody Daclizumab, which binds specifically to the alpha-subunit (CD25) of the high-affinity IL-2 receptor, inhibits IL-2 binding and its biological activity (Queen et al. 1989). In patients with moderate-to-severe persistent asthma, Daclizumab showed slightly improved FEV1, reduced asthma symptoms and exacerbations, and reduced use of rescue medications (Busse et al. 2008; Table 1). Further studies are needed to assess the potential of daclizumab in patients with asthma.

3.7   Approaches to Target OX40L

Recently, a phase II randomized clinical trial in patients with mild asthma investigated efficacy, safety, and tolerability of huMAb OX40L administered intravenous over 3 months, with allergen challenges on days 56 and 113 (Gauvreau et al. 2014a; Table 1). The huMAb OX40L did not attenuate the late-phase asthmatic response or early-phase asthmatic response, bronchial hyperresponsiveness, serum IgE, and blood and sputum eosinophils. A significant reduction was only seen for serum IgE levels after the second allergen challenge. However, huMAb OX40L significantly reduced sputum eosinophils before the allergen challenge, but no difference was found following the challenge. Although these results did not support its use in mild asthma, further studies are needed to assess the efficacy of huMAb OX40L in moderate-to-severe asthma.

3.8   Approaches to Target GATA-3

A new strategy in asthma treatment could be the strategy to target on the type-2 master transcription factor GATA-3, which binds on the DNA and activates downstream Th2 cytokines (Wang et al. 2013). It is suggested that lacking the activation by GATA-3 impedes generation and maintenance of Th2 cells. As GATA-3 is also expressed in granulocytes, mast cells, and epithelial cells, it takes a key role in the general development of allergic immunoreactions. Facing this central role of GATA-3, an inhaled DNAzyme (SB010) targeting GATA3 mRNA was tested in an experimental model of chronic asthma (Sel et al. 2008; Homburg et al. 2015). In a randomized, double-blind, placebo-controlled, multi-center clinical trial of SB010 showed dampening of the late asthmatic response by 34%, whereas the early asthmatic response was attenuated by 11% (Krug et al. 2015; Table 1). In summary, the successful evolution of type-2-targeted therapies has been strongly influenced by the recognition that only a portion of asthmatic patients manifest Th2 inflammatory processes. To date, these biologic approaches have been reported to be safe and well tolerated, although with greater numbers of patients exposed for more prolonged periods of time, adverse aspects to their use are likely to emerge. Remaining questions include those related to long-term safety, whether one of these approaches will be more successful than others or will be selectively more effective in some type-2 subgroups than others, whether certain biomarkers will better identify treatment-responsive patients, and, importantly, whether any of these molecules will produce a disease-modifying effect.

4 Biologic Approaches in Non-Th2 Asthma

Approximately 50% of asthmatic patients do not express type-2 inflammation and seem to be characterized by normal eosinophil numbers or even (Woodruff et al. 2009; Dweik et al. 2010; McGrath et al. 2012). This non-Th2 entity is represented by the absence of characteristic Th2 cytokines and their downstream signatures, without specific biomarkers, making molecular phenotyping and targeted therapy more complicated. Some of these patients could lack type-2 biomarkers simply because corticosteroid use substantially reducing this pathway. Non-type 2 patients generally show adult-onset disease, often associated with obesity, post-infectious, neutrophilia, and smoking-related factors, and show no atopic or allergic characteristics (Moore et al. 2010; Wu et al. 2014; Miranda et al. 2004; Dixon et al. 2011).

4.1 Approaches to Target TNF-α

TNF-α is a prototypic member of the TNF superfamily identified in innate, type 1, and type 2 immunity (Howarth et al. 2005). Blocking TNF-α and its downstream pathways seems to be highly effective in patients mainly type 1-driven inflammatory diseases. Further, TNF-α is known to contribute to neutrophilic inflammation and bronchial hyperresponsiveness (Kips et al. 1992). A study in severe refractory asthma patients reported effects for etanercept, a soluble TNF-α receptor inhibitor, which showed decreased bronchial hyperresponsiveness with simultaneously increase of post bronchodilator lung function parameters (Berry et al. 2006; Table 1). Nevertheless, others showed only slightly effects on CRP levels in patients with severe refractory asthma (Morjaria et al. 2008). Infliximab, a mAb directly targeting against TNF-α, seemed to decrease the frequency of asthma exacerbations (Erin et al. 2006). Furthermore, golimumab, an antagonistic TNF-α mAb, showed no improvements on lung function or exacerbation rate in patients with severe uncontrolled asthma, while patients with adult-onset disease showed lower risk of exacerbations (Wenzel et al. 2009). Strikingly, treatment with golimumab seems to be associated with an increased rate of severe adverse events such as systemic infections and cancer (Wenzel et al. 2009). Thus, the controversial efficacy of golimumab accompanied by severe side effects stopped further development of this antagonistic TNF-α mAb in asthma.

4.2 Approaches to Target Th9 Cells

Before uncovering the existence of Th9 cells, it was reported that CD4+ T cells are able to produce IL-9 depending on IL-2 (Schmitt et al. 1994). These IL-9 producing cells can be promoted by IL-4 and TGF-ß, while IFN-γ seemed to be a potent inhibitor of IL-9 expression (Schmitt et al. 1994). The effects of monoclonal antibody against IL-9 enokizumab on asthma have been evaluated in four clinical trials. In phase II trials, no significant improvement was found in the asthma symptom score following enokizumab, although there were some indications that enokizumab was able to reduce exacerbation rates and improve asthma symptoms (White et al. 2009; Parker et al. 2011; Table 1). A limitation of these trials consisted in the low number of patients enrolled. A recent phase-IIb clinical trial including more than 300 subjects with uncontrolled asthma demonstrated that enokizumab failed to show any improvement in asthma symptoms, lung function, or reduced asthma exacerbation rates compared to placebo (Oh et al. 2013). However, completed trials provided negative results and further studies are needed to evaluate whether a particular sub-type of asthma could benefit from treatment with enokizumab.

4.3   Targeting Neutrophils and Related Microbial Influence

Neutrophilia, a predominance of neutrophilic granulocytes in circulation as well as in the lower airways mirrored by induced sputum in asthmatic patients has been associated with decreased lung function and increased need for ICS (Wenzel et al. 1997; Jatakanon et al. 1999). Furthermore, two sputum-based studies identified two groups with lower airways neutrophilia, one mixed with additional eosinophilia and one without, each showing different clinical outcomes (Hastie et al. 2010; Moore et al. 2014). The only one study directly targeting on lower airways neutrophilia investigated CXCR2 the receptor for IL-8 mediates neutrophil migration to the tissue sites of inflammation. Although systemic and local neutrophil counts were decreased by a CXCR2 antagonist SCH527123, there was no additional clinical benefit (Nair et al. 2012; Table 1). In addition, the airway microbiome can enhance neutrophil killing, which is essential for microbe host defense. Pulmonary Th2 responses induced e.g. by *Aspergillus ssp.* are influenced by the composition of the gut microbiome (Kolwijck and van de Veerdonk 2014). Thus, the local microbiome seems to play an important role in modulating systemic Treg cell functions, which can inhibit these local Th2 responses in the lower airways. Members of the IL-17 familiy, namely IL-17A-D and IL-17F, are also associated with the Th17 pathway and the related downstream signaling, which play a pivotal role in bacterial host defense and subsequent infiltration by neutrophils (Ye et al. 2001; Schmidt-Weber et al. 2007). Brodalumab, an anti-IL-17R antibody, blocks receptor binding of IL-17A and IL-17F but additionally blocks binding of the type 2-associated IL17 family member IL-25, also known as IL-17E (Rickel et al. 2008; Table 1). Patients with severe asthma showed no benefit from brodalumab treatment, while also sub-types determined by the presence of systemic neutrophils or eosinophils did not improve the prediction of treatment success or the identification of responding patients (Busse et al. 2013). Only a subgroup with lung function improvement showed slightly improvement of asthma symptoms, while it is questionably whether this difference in asthma subtypes is useful to identify Th17-associated asthma responsive to targeted therapy (Busse et al. 2013).

4.4 Targeting RSV and Dysbalanced Microbiome

It is increasingly recognized that human asthma is a heterogeneous disease, including endotypes such as non-atopic and viral asthma. A standard treatment option for a viral-induced asthma may include bronchodilators for mild symptoms and increased steroids for more severe or prolonged attacks. Inhaled steroids have been shown to be effective in adults without asthma who have asthma-like symptoms after a viral infection (Tan 2005). Respiratory syncytial virus (RSV) infection presents a significant health challenge in children, the elderly and immunocompromised patients and Dysbalanced Microbiome. Hence, factors that influence the development and/or function of the immune system may critically impact the host response to RSV. It is recognized that the composition and diversity of the microbiota, which stabilizes around 12 months of age in humans, fundamentally affects host physiology and immunity (Hooper et al. 2012). Several recent studies have demonstrated changes in the respiratory microbiome in subjects with RSV-induced asthma (Depner et al. 2017; Mansbach et al. 2016; Teo et al. 2015). However, whether microbial dysbiosis in the composition of the lung microbiota predisposes to RSV-induced asthma or is simply a consequence of disease remains unclear. There is still a high unmet medical need among patients with RSV infection, and no specific antiviral therapy is available. The only approved agents are palivizumab, which has to be given prophylactically, mainly in high-risk infants, and ribavirin, which is rarely used due to toxicity concerns and questionable benefits (Shook and Lin 2017). Palivizumab in patients with recurrent wheezing showed no suppression of the onset of atopic asthma but a trend to reduction of recurrent wheezing during first 6 years of life (Shook and Lin 2017). However, an alternative treatment could be the nasal administration of immunobiotics like *Lactobacillus ssp.*, which induced protection against respiratory syncytial virus infection in mice (Tomosada et al. 2013).

Although the testing of biologic medications is mandatory to start in animal models, the response of the human immune system in asthma is more complex. Moving forward, it will be necessary to uncover optimized systemic and local biomarkers to identify patients, who receive the maximum therapeutic benefit from specific designed and targeted drugs. Identification of biomarkers on protein and transcriptomic levels might increase the chance to identify these patients who get most benefit of specifically developed biologics. This biologic approach targeting type-2 inflammation is increasing the options for new therapies for most of the asthmatic patients. However, regarding long-term efficacy, safety, and the cost-effectiveness, these specific therapeutic approaches have to be more investigated with a simultaneous identification of patients receiving most benefit. Further integration of standard treatment schemes as related to type-2 directed therapeutic approaches has the potential to make these among the first successful individualized medicines.

References

Aho VT, Pereira PA, Haahtela T, Pawankar R, Auvinen P, Koskinen K (2015) The microbiome of the human lower airways: a next generation sequencing perspective. World Allergy Organ J 8 (1):23. https://doi.org/10.1186/s40413-015-0074-z

Allakhverdi Z, Comeau MR, Jessup HK, Yoon BR, Brewer A, Chartier S, Paquette N, Ziegler SF, Sarfati M, Delespesse G (2007) Thymic stromal lymphopoietin is released by human epithelial cells in response to microbes, trauma, or inflammation and potently activates mast cells. J Exp Med 204(2):253–258. https://doi.org/10.1084/jem.20062211

Amelink M, de Groot JC, de Nijs SB, Lutter R, Zwinderman AH, Sterk PJ, ten Brinke A, Bel EH (2013) Severe adult-onset asthma: a distinct phenotype. J Allergy Clin Immunol 132 (2):336–341. https://doi.org/10.1016/j.jaci.2013.04.052

Anderson GP (2008) Endotyping asthma: new insights into key pathogenic mechanisms in a complex, heterogeneous disease. Lancet 372(9643):1107–1119. https://doi.org/10.1016/ S0140-6736(08)61452-X

Antoniu SA (2010) MEDI-528, an anti-IL-9 humanized antibody for the treatment of asthma. Curr Opin Mol Ther 12(2):233–239

Aron JL, Akbari O (2017) Regulatory T cells and type 2 innate lymphoid cell dependent asthma. Allergy 72(8):1148–1155. https://doi.org/10.1111/all.13139

Bel EH, Wenzel SE, Thompson PJ, Prazma CM, Keene ON, Yancey SW, Ortega HG, Pavord ID, Investigators S (2014) Oral glucocorticoid-sparing effect of mepolizumab in eosinophilic asthma. N Engl J Med 371(13):1189–1197. https://doi.org/10.1056/NEJMoa1403291

Berry MA, Hargadon B, Shelley M, Parker D, Shaw DE, Green RH, Bradding P, Brightling CE, Wardlaw AJ, Pavord ID (2006) Evidence of a role of tumor necrosis factor alpha in refractory asthma. N Engl J Med 354(7):697–708. https://doi.org/10.1056/NEJMoa050580

Billington CK, Penn RB, Hall IP (2016) beta2 agonists. Handb Exp Pharmacol 237:23–40. https:// doi.org/10.1007/164_2016_64

De Boever EH, Ashman C, Cahn AP, Locantore NW, Overend P, Pouliquen IJ, Serone AP, Wright TJ, Jenkins MM, Panesar IS, Thiagarajah SS, Wenzel SE (2014) Efficacy and safety of an anti-IL-13 mAb in patients with severe asthma: a randomized trial. J Allergy Clin Immunol 133 (4):989–996. doi:https://doi.org/10.1016/j.jaci.2014.01.002

Borish LC, Nelson HS, Lanz MJ, Claussen L, Whitmore JB, Agosti JM, Garrison L (1999) Interleukin-4 receptor in moderate atopic asthma. A phase I/II randomized, placebo-controlled trial. Am J Respir Crit Care Med 160(6):1816–1823. https://doi.org/10.1164/ajrccm.160.6. 9808146

Borish LC, Nelson HS, Corren J, Bensch G, Busse WW, Whitmore JB, Agosti JM, IL-4R Asthma Study Group (2001) Efficacy of soluble IL-4 receptor for the treatment of adults with asthma. J Allergy Clin Immunol 107(6):963–970. https://doi.org/10.1067/mai.2001.115624

Boulet LP, Chapman KR, Cote J, Kalra S, Bhagat R, Swystun VA, Laviolette M, Cleland LD, Deschesnes F, Su JQ, DeVault A, Fick RB Jr, Cockcroft DW (1997) Inhibitory effects of an anti-IgE antibody E25 on allergen-induced early asthmatic response. Am J Respir Crit Care Med 155(6):1835–1840. https://doi.org/10.1164/ajrccm.155.6.9196083

Braman SS (2006) The global burden of asthma. Chest 130(1 Suppl):4S–12S. https://doi.org/10. 1378/chest.130.1_suppl.4S

Brown HM (1958) Treatment of chronic asthma with prednisolone; significance of eosinophils in the sputum. Lancet 2(7059):1245–1247

Brusselle G, Kips J, Joos G, Bluethmann H, Pauwels R (1995) Allergen-induced airway inflammation and bronchial responsiveness in wild-type and interleukin-4-deficient mice. Am J Respir Cell Mol Biol 12(3):254–259. https://doi.org/10.1165/ajrcmb.12.3.7873190

Busse W, Corren J, Lanier BQ, McAlary M, Fowler-Taylor A, Cioppa GD, van As A, Gupta N (2001) Omalizumab, anti-IgE recombinant humanized monoclonal antibody, for the treatment of severe allergic asthma. J Allergy Clin Immunol 108(2):184–190. https://doi.org/10.1067/mai.2001.117880

Busse WW, Israel E, Nelson HS, Baker JW, Charous BL, Young DY, Vexler V, Shames RS, Daclizumab Asthma Study Group (2008) Daclizumab improves asthma control in patients with moderate to severe persistent asthma: a randomized, controlled trial. Am J Respir Crit Care Med 178(10):1002–1008. https://doi.org/10.1164/rccm.200708-1200OC

Busse WW, Morgan WJ, Gergen PJ, Mitchell HE, Gern JE, Liu AH, Gruchalla RS, Kattan M, Teach SJ, Pongracic JA, Chmiel JF, Steinbach SF, Calatroni A, Togias A, Thompson KM, Szefler SJ, Sorkness CA (2011) Randomized trial of omalizumab (anti-IgE) for asthma in inner-city children. N Engl J Med 364(11):1005–1015. https://doi.org/10.1056/NEJMoa1009705

Busse WW, Holgate S, Kerwin E, Chon Y, Feng J, Lin J, Lin SL (2013) Randomized, double-blind, placebo-controlled study of brodalumab, a human anti-IL-17 receptor monoclonal antibody, in moderate to severe asthma. Am J Respir Crit Care Med 188(11):1294–1302. https://doi.org/10.1164/rccm.201212-2318OC

Castro M, Mathur S, Hargreave F, Boulet LP, Xie F, Young J, Wilkins HJ, Henkel T, Nair P, Res-5-0010 Study Group (2011) Reslizumab for poorly controlled, eosinophilic asthma: a randomized, placebo-controlled study. Am J Respir Crit Care Med 184(10):1125–1132. https://doi.org/10.1164/rccm.201103-0396OC

Castro M, Wenzel SE, Bleecker ER, Pizzichini E, Kuna P, Busse WW, Gossage DL, Ward CK, Wu Y, Wang B, Khatry DB, van der Merwe R, Kolbeck R, Molfino NA, Raible DG (2014) Benralizumab, an anti-interleukin 5 receptor alpha monoclonal antibody, versus placebo for uncontrolled eosinophilic asthma: a phase 2b randomised dose-ranging study. Lancet Respir Med 2(11):879–890. https://doi.org/10.1016/S2213-2600(14)70201-2

Chibana K, Trudeau JB, Mustovich AT, Hu H, Zhao J, Balzar S, Chu HW, Wenzel SE (2008) IL-13 induced increases in nitrite levels are primarily driven by increases in inducible nitric oxide synthase as compared with effects on arginases in human primary bronchial epithelial cells. Clin Exp Allergy 38(6):936–946. https://doi.org/10.1111/j.1365-2222.2008.02969.x

Clutterbuck EJ, Hirst EM, Sanderson CJ (1989) Human interleukin-5 (IL-5) regulates the production of eosinophils in human bone marrow cultures: comparison and interaction with IL-1, IL-3, IL-6, and GMCSF. Blood 73(6):1504–1512

Corren J, Busse W, Meltzer EO, Mansfield L, Bensch G, Fahrenholz J, Wenzel SE, Chon Y, Dunn M, Weng HH, Lin SL (2010) A randomized, controlled, phase 2 study of AMG 317, an IL-4Ralpha antagonist, in patients with asthma. Am J Respir Crit Care Med 181(8):788–796. https://doi.org/10.1164/rccm.200909-1448OC

Corren J, Lemanske RF, Hanania NA, Korenblat PE, Parsey MV, Arron JR, Harris JM, Scheerens H, Wu LC, Su Z, Mosesova S, Eisner MD, Bohen SP, Matthews JG (2011a) Lebrikizumab treatment in adults with asthma. N Engl J Med 365(12):1088–1098. https://doi.org/10.1056/NEJMoa1106469

Corren J, Wood RA, Patel D, Zhu J, Yegin A, Dhillon G, Fish JE (2011b) Effects of omalizumab on changes in pulmonary function induced by controlled cat room challenge. J Allergy Clin Immunol 127(2):398–405. https://doi.org/10.1016/j.jaci.2010.09.043

Depner M, Ege MJ, Cox MJ, Dwyer S, Walker AW, Birzele LT, Genuneit J, Horak E, Braun-Fahrlander C, Danielewicz H, Maier RM, Moffatt MF, Cookson WO, Heederik D, von Mutius E, Legatzki A (2017) Bacterial microbiota of the upper respiratory tract and childhood asthma. J Allergy Clin Immunol 139(3):826–834. e813. https://doi.org/10.1016/j.jaci.2016.05.050

Dixon AE, Pratley RE, Forgione PM, Kaminsky DA, Whittaker-Leclair LA, Griffes LA, Garudathri J, Raymond D, Poynter ME, Bunn JY, Irvin CG (2011) Effects of obesity and bariatric surgery on airway hyperresponsiveness, asthma control, and inflammation. J Allergy Clin Immunol 128(3):508–515.e501–502. doi:https://doi.org/10.1016/j.jaci.2011.06.009

Dweik RA, Sorkness RL, Wenzel S, Hammel J, Curran-Everett D, Comhair SA, Bleecker E, Busse W, Calhoun WJ, Castro M, Chung KF, Israel E, Jarjour N, Moore W, Peters S, Teague G, Gaston B, Erzurum SC, National Heart, Lung, and Blood Institute Severe Asthma Research Program (2010) Use of exhaled nitric oxide measurement to identify a reactive, at-risk phenotype among patients with asthma. Am J Respir Crit Care Med 181 (10):1033–1041. https://doi.org/10.1164/rccm.200905-0695OC

Erin EM, Leaker BR, Nicholson GC, Tan AJ, Green LM, Neighbour H, Zacharasiewicz AS, Turner J, Barnathan ES, Kon OM, Barnes PJ, Hansel TT (2006) The effects of a monoclonal antibody directed against tumor necrosis factor-alpha in asthma. Am J Respir Crit Care Med 174(7):753–762. https://doi.org/10.1164/rccm.200601-072OC

Fahy JV, Fleming HE, Wong HH, Liu JT, Su JQ, Reimann J, Fick RB Jr, Boushey HA (1997) The effect of an anti-IgE monoclonal antibody on the early- and late-phase responses to allergen inhalation in asthmatic subjects. Am J Respir Crit Care Med 155(6):1828–1834. https://doi. org/10.1164/ajrccm.155.6.9196082

FitzGerald JM, Bleecker ER, Nair P, Korn S, Ohta K, Lommatzsch M, Ferguson GT, Busse WW, Barker P, Sproule S, Gilmartin G, Werkstrom V, Aurivillius M, Goldman M, CALIMA Study Investigators (2016) Benralizumab, an anti-interleukin-5 receptor alpha monoclonal antibody, as add-on treatment for patients with severe, uncontrolled, eosinophilic asthma (CALIMA): a randomised, double-blind, placebo-controlled phase 3 trial. Lancet 388(10056):2128–2141. https://doi.org/10.1016/S0140-6736(16)31322-8

Fitzpatrick AM, Teague WG, Meyers DA, Peters SP, Li X, Li H, Wenzel SE, Aujla S, Castro M, Bacharier LB, Gaston BM, Bleecker ER, Moore WC, National Institutes of Health/National Heart, Lung, and Blood Institute Severe Asthma Research Program (2011) Heterogeneity of severe asthma in childhood: confirmation by cluster analysis of children in the National Institutes of Health/National Heart, Lung, and Blood Institute Severe Asthma Research Program. J Allergy Clin Immunol 127(2):382–389.e381–313. doi:https://doi.org/10.1016/j. jaci.2010.11.015

Flood-Page PT, Menzies-Gow AN, Kay AB, Robinson DS (2003) Eosinophil's role remains uncertain as anti-interleukin-5 only partially depletes numbers in asthmatic airway. Am J Respir Crit Care Med 167(2):199–204. https://doi.org/10.1164/rccm.200208-789OC

Flood-Page P, Swenson C, Faiferman I, Matthews J, Williams M, Brannick L, Robinson D, Wenzel S, Busse W, Hansel TT, Barnes NC, International Mepolizumab Study Group (2007) A study to evaluate safety and efficacy of mepolizumab in patients with moderate persistent asthma. Am J Respir Crit Care Med 176(11):1062–1071. https://doi.org/10.1164/ rccm.200701-085OC

Galli SJ, Tsai M (2012) IgE and mast cells in allergic disease. Nat Med 18(5):693–704. https://doi. org/10.1038/nm.2755

Gauvreau GM, Boulet LP, Cockcroft DW, Fitzgerald JM, Carlsten C, Davis BE, Deschesnes F, Duong M, Durn BL, Howie KJ, Hui L, Kasaian MT, Killian KJ, Strinich TX, Watson RM, Nathalie Y, Zhou S, Raible D, O'Byrne PM (2011) Effects of interleukin-13 blockade on allergen-induced airway responses in mild atopic asthma. Am J Respir Crit Care Med 183 (8):1007–1014. https://doi.org/10.1164/rccm.201008-1210OC

Gauvreau GM, Boulet LP, Cockcroft DW, FitzGerald JM, Mayers I, Carlsten C, Laviolette M, Killian KJ, Davis BE, Larche M, Kipling C, Dua B, Mosesova S, Putnam W, Zheng Y, Scheerens H, McClintock D, Matthews JG, O'Byrne PM (2014a) OX40L blockade and allergen-induced airway responses in subjects with mild asthma. Clin Exp Allergy 44 (1):29–37. https://doi.org/10.1111/cea.12235

Gauvreau GM, O'Byrne PM, Boulet LP, Wang Y, Cockcroft D, Bigler J, FitzGerald JM, Boedigheimer M, Davis BE, Dias C, Gorski KS, Smith L, Bautista E, Comeau MR, Leigh R,

Parnes JR (2014b) Effects of an anti-TSLP antibody on allergen-induced asthmatic responses. N Engl J Med 370(22):2102–2110. https://doi.org/10.1056/NEJMoa1402895

Gauvreau GM, Arm JP, Boulet LP, Leigh R, Cockcroft DW, Davis BE, Mayers I, FitzGerald JM, Dahlen B, Killian KJ, Laviolette M, Carlsten C, Lazarinis N, Watson RM, Milot J, Swystun V, Bowen M, Hui L, Lantz AS, Meiser K, Maahs S, Lowe PJ, Skerjanec A, Drollmann A, O'Byrne PM (2016) Efficacy and safety of multiple doses of QGE031 (ligelizumab) versus omalizumab and placebo in inhibiting allergen-induced early asthmatic responses. J Allergy Clin Immunol 138(4):1051–1059. https://doi.org/10.1016/j.jaci.2016.02.027

Grunig G, Warnock M, Wakil AE, Venkayya R, Brombacher F, Rennick DM, Sheppard D, Mohrs M, Donaldson DD, Locksley RM, Corry DB (1998) Requirement for IL-13 independently of IL-4 in experimental asthma. Science 282(5397):2261–2263

Haldar P, Pavord ID, Shaw DE, Berry MA, Thomas M, Brightling CE, Wardlaw AJ, Green RH (2008) Cluster analysis and clinical asthma phenotypes. Am J Respir Crit Care Med 178 (3):218–224. https://doi.org/10.1164/rccm.200711-1754OC

Haldar P, Brightling CE, Hargadon B, Gupta S, Monteiro W, Sousa A, Marshall RP, Bradding P, Green RH, Wardlaw AJ, Pavord ID (2009) Mepolizumab and exacerbations of refractory eosinophilic asthma. N Engl J Med 360(10):973–984. https://doi.org/10.1056/NEJMoa0808991

Hanania NA, Alpan O, Hamilos DL, Condemi JJ, Reyes-Rivera I, Zhu J, Rosen KE, Eisner MD, Wong DA, Busse W (2011) Omalizumab in severe allergic asthma inadequately controlled with standard therapy: a randomized trial. Ann Intern Med 154(9):573–582. https://doi.org/10.7326/0003-4819-154-9-201105030-00002

Hanania NA, Wenzel S, Rosen K, Hsieh HJ, Mosesova S, Choy DF, Lal P, Arron JR, Harris JM, Busse W (2013) Exploring the effects of omalizumab in allergic asthma: an analysis of biomarkers in the EXTRA study. Am J Respir Crit Care Med 187(8):804–811. https://doi.org/10.1164/rccm.201208-1414OC

Hanania NA, Korenblat P, Chapman KR, Bateman ED, Kopecky P, Paggiaro P, Yokoyama A, Olsson J, Gray S, Holweg CT, Eisner M, Asare C, Fischer SK, Peng K, Putnam WS, Matthews JG (2016) Efficacy and safety of lebrikizumab in patients with uncontrolled asthma (LAVOLTA I and LAVOLTA II): replicate, phase 3, randomised, double-blind, placebo-controlled trials. Lancet Respir Med 4(10):781–796. https://doi.org/10.1016/S2213-2600(16)30265-X

Harris JM, Maciuca R, Bradley MS, Cabanski CR, Scheerens H, Lim J, Cai F, Kishnani M, Liao XC, Samineni D, Zhu R, Cochran C, Soong W, Diaz JD, Perin P, Tsukayama M, Dimov D, Agache I, Kelsen SG (2016) A randomized trial of the efficacy and safety of quilizumab in adults with inadequately controlled allergic asthma. Respir Res 17:29. https://doi.org/10.1186/s12931-016-0347-2

Hastie AT, Moore WC, Meyers DA, Vestal PL, Li H, Peters SP, Bleecker ER, National Heart, Lung, Blood Institute Severe Asthma Research Program (2010) Analyses of asthma severity phenotypes and inflammatory proteins in subjects stratified by sputum granulocytes. J Allergy Clin Immunol 125(5):1028–1036. e1013. https://doi.org/10.1016/j.jaci.2010.02.008

Holgate ST, Chuchalin AG, Hebert J, Lotvall J, Persson GB, Chung KF, Bousquet J, Kerstjens HA, Fox H, Thirlwell J, Cioppa GD, Omalizumab 011 International Study Group (2004) Efficacy and safety of a recombinant anti-immunoglobulin E antibody (omalizumab) in severe allergic asthma. Clin Exp Allergy 34(4):632–638. https://doi.org/10.1111/j.1365-2222.2004.1916.x

Homburg U, Renz H, Timmer W, Hohlfeld JM, Seitz F, Luer K, Mayer A, Wacker A, Schmidt O, Kuhlmann J, Turowska A, Roller J, Kutz K, Schluter G, Krug N, Garn H (2015) Safety and tolerability of a novel inhaled GATA3 mRNA targeting DNAzyme in patients with TH2-driven asthma. J Allergy Clin Immunol 136(3):797–800. https://doi.org/10.1016/j.jaci.2015.02.018

Hooper LV, Littman DR, Macpherson AJ (2012) Interactions between the microbiota and the immune system. Science 336(6086):1268–1273. https://doi.org/10.1126/science.1223490

Howarth PH, Babu KS, Arshad HS, Lau L, Buckley M, McConnell W, Beckett P, Al Ali M, Chauhan A, Wilson SJ, Reynolds A, Davies DE, Holgate ST (2005) Tumour necrosis factor (TNFalpha) as a novel therapeutic target in symptomatic corticosteroid dependent asthma. Thorax 60(12):1012–1018. https://doi.org/10.1136/thx.2005.045260

Humbert M, Beasley R, Ayres J, Slavin R, Hebert J, Bousquet J, Beeh KM, Ramos S, Canonica GW, Hedgecock S, Fox H, Blogg M, Surrey K (2005) Benefits of omalizumab as add-on therapy in patients with severe persistent asthma who are inadequately controlled despite best available therapy (GINA 2002 step 4 treatment): INNOVATE. Allergy 60(3):309–316. https://doi.org/10.1111/j.1398-9995.2004.00772.x

Iida S, Misaka H, Inoue M, Shibata M, Nakano R, Yamane-Ohnuki N, Wakitani M, Yano K, Shitara K, Satoh M (2006) Nonfucosylated therapeutic IgG1 antibody can evade the inhibitory effect of serum immunoglobulin G on antibody-dependent cellular cytotoxicity through its high binding to FcgammaRIIIa. Clin Cancer Res 12(9):2879–2887. https://doi.org/10.1158/1078-0432.CCR-05-2619

Jatakanon A, Uasuf C, Maziak W, Lim S, Chung KF, Barnes PJ (1999) Neutrophilic inflammation in severe persistent asthma. Am J Respir Crit Care Med 160(5 Pt 1):1532–1539. https://doi.org/10.1164/ajrccm.160.5.9806170

Jia G, Erickson RW, Choy DF, Mosesova S, Wu LC, Solberg OD, Shikotra A, Carter R, Audusseau S, Hamid Q, Bradding P, Fahy JV, Woodruff PG, Harris JM, Arron JR, Bronchoscopic Exploratory Research Study of Biomarkers in Corticosteroid-refractory Asthma Study Group (2012) Periostin is a systemic biomarker of eosinophilic airway inflammation in asthmatic patients. J Allergy Clin Immunol 130 (3):647-654.e610. doi:https://doi.org/10.1016/j.jaci.2012.06.025

Kips JC, Tavernier J, Pauwels RA (1992) Tumor necrosis factor causes bronchial hyperresponsiveness in rats. Am Rev Respir Dis 145(2 Pt 1):332–336. https://doi.org/10.1164/ajrccm/145.2_Pt_1.332

Kolbeck R, Kozhich A, Koike M, Peng L, Andersson CK, Damschroder MM, Reed JL, Woods R, Dall'acqua WW, Stephens GL, Erjefalt JS, Bjermer L, Humbles AA, Gossage D, Wu H, Kiener PA, Spitalny GL, Mackay CR, Molfino NA, Coyle AJ (2010) MEDI-563, a humanized anti-IL-5 receptor alpha mAb with enhanced antibody-dependent cell-mediated cytotoxicity function. J Allergy Clin Immunol 125(6):1344–1353. e1342. https://doi.org/10.1016/j.jaci.2010.04.004

Kolwijck E, van de Veerdonk FL (2014) The potential impact of the pulmonary microbiome on immunopathogenesis of Aspergillus-related lung disease. Eur J Immunol 44(11):3156–3165. https://doi.org/10.1002/eji.201344404

Krug N, Hohlfeld JM, Kirsten AM, Kornmann O, Beeh KM, Kappeler D, Korn S, Ignatenko S, Timmer W, Rogon C, Zeitvogel J, Zhang N, Bille J, Homburg U, Turowska A, Bachert C, Werfel T, Buhl R, Renz J, Garn H, Renz H (2015) Allergen-induced asthmatic responses modified by a GATA3-specific DNAzyme. N Engl J Med 372(21):1987–1995. https://doi.org/10.1056/NEJMoa1411776

Lanier B, Bridges T, Kulus M, Taylor AF, Berhane I, Vidaurre CF (2009) Omalizumab for the treatment of exacerbations in children with inadequately controlled allergic (IgE-mediated) asthma. J Allergy Clin Immunol 124(6):1210–1216. https://doi.org/10.1016/j.jaci.2009.09.021

Laviolette M, Gossage DL, Gauvreau G, Leigh R, Olivenstein R, Katial R, Busse WW, Wenzel S, Wu Y, Datta V, Kolbeck R, Molfino NA (2013) Effects of benralizumab on airway eosinophils in asthmatic patients with sputum eosinophilia. J Allergy Clin Immunol 132 (5):1086-1096. e1085. doi:https://doi.org/10.1016/j.jaci.2013.05.020

Leckie MJ, ten Brinke A, Khan J, Diamant Z, O'Connor BJ, Walls CM, Mathur AK, Cowley HC, Chung KF, Djukanovic R, Hansel TT, Holgate ST, Sterk PJ, Barnes PJ (2000) Effects of an interleukin-5 blocking monoclonal antibody on eosinophils, airway hyper-responsiveness, and the late asthmatic response. Lancet 356(9248):2144–2148

Leung E, Al Efraij K, FitzGerald JM (2017) The safety of mepolizumab for the treatment of asthma. Expert Opin Drug Saf:1–8. https://doi.org/10.1080/14740338.2017.1286327

Lin TH, Cheng CC, Su HH, Huang NC, Chen JJ, Kang HY, Chang TH (2016) Lipopolysaccharide attenuates induction of proallergic cytokines, thymic stromal lymphopoietin, and interleukin 33 in respiratory epithelial cells stimulated with PolyI:C and human parechovirus. Front Immunol 7:440. https://doi.org/10.3389/fimmu.2016.00440

Lotvall J, Akdis CA, Bacharier LB, Bjermer L, Casale TB, Custovic A, Lemanske RF Jr, Wardlaw AJ, Wenzel SE, Greenberger PA (2011) Asthma endotypes: a new approach to classification of disease entities within the asthma syndrome. J Allergy Clin Immunol 127(2):355–360. https://doi.org/10.1016/j.jaci.2010.11.037

Mansbach JM, Hasegawa K, Henke DM, Ajami NJ, Petrosino JF, Shaw CA, Piedra PA, Sullivan AF, Espinola JA, Camargo CA, Jr (2016) Respiratory syncytial virus and rhinovirus severe bronchiolitis are associated with distinct nasopharyngeal microbiota. J Allergy Clin Immunol 137 (6):1909-1913.e1904. doi:https://doi.org/10.1016/j.jaci.2016.01.036

Marsland BJ (2013) Influences of the microbiome on the early origins of allergic asthma. Ann Am Thorac Soc 10(Suppl):S165–S169. https://doi.org/10.1513/AnnalsATS.201305-118AW

Marsland BJ, Gollwitzer ES (2014) Host-microorganism interactions in lung diseases. Nat Rev Immunol 14(12):827–835. https://doi.org/10.1038/nri3769

Martinez-Gonzalez I, Steer CA, Takei F (2015) Lung ILC2s link innate and adaptive responses in allergic inflammation. Trends Immunol 36(3):189–195. https://doi.org/10.1016/j.it.2015.01.005

Massoud AH, Charbonnier LM, Lopez D, Pellegrini M, Phipatanakul W, Chatila TA (2016) An asthma-associated IL4R variant exacerbates airway inflammation by promoting conversion of regulatory T cells to TH17-like cells. Nat Med 22(9):1013–1022. https://doi.org/10.1038/nm.4147

McGrath KW, Icitovic N, Boushey HA, Lazarus SC, Sutherland ER, Chinchilli VM, Fahy JV, Asthma Clinical Research Network of the National Heart, Lung, Blood Institute (2012) A large subgroup of mild-to-moderate asthma is persistently noneosinophilic. Am J Respir Crit Care Med 185(6):612–619. https://doi.org/10.1164/rccm.201109-1640OC

Milgrom H, Fick RB Jr, Su JQ, Reimann JD, Bush RK, Watrous ML, Metzger WJ (1999) Treatment of allergic asthma with monoclonal anti-IgE antibody. rhuMAb-E25 Study Group. N Engl J Med 341(26):1966–1973. https://doi.org/10.1056/NEJM199912233412603

Milgrom H, Berger W, Nayak A, Gupta N, Pollard S, McAlary M, Taylor AF, Rohane P (2001) Treatment of childhood asthma with anti-immunoglobulin E antibody (omalizumab). Pediatrics 108(2):E36

Miranda C, Busacker A, Balzar S, Trudeau J, Wenzel SE (2004) Distinguishing severe asthma phenotypes: role of age at onset and eosinophilic inflammation. J Allergy Clin Immunol 113 (1):101–108. https://doi.org/10.1016/j.jaci.2003.10.041

Mochizuki H, Kusuda S, Okada K, Yoshihara S, Furuya H, Simoes EA (2017) Palivizumab prophylaxis in preterm infants and subsequent recurrent wheezing: 6 year follow up study. Am J Respir Crit Care Med 196(1):–29, 38. https://doi.org/10.1164/rccm.201609-1812OC

Modena BD, Tedrow JR, Milosevic J, Bleecker ER, Meyers DA, Wu W, Bar-Joseph Z, Erzurum SC, Gaston BM, Busse WW, Jarjour NN, Kaminski N, Wenzel SE (2014) Gene expression in relation to exhaled nitric oxide identifies novel asthma phenotypes with unique biomolecular pathways. Am J Respir Crit Care Med 190(12):1363–1372. https://doi.org/10.1164/rccm.201406-1099OC

Moore WC, Meyers DA, Wenzel SE, Teague WG, Li H, Li X, D'Agostino R Jr, Castro M, Curran-Everett D, Fitzpatrick AM, Gaston B, Jarjour NN, Sorkness R, Calhoun WJ, Chung KF, Comhair SA, Dweik RA, Israel E, Peters SP, Busse WW, Erzurum SC, Bleecker ER, National Heart, Lung, Blood Institute's Severe Asthma Research Program (2010) Identification of asthma phenotypes using cluster analysis in the Severe Asthma Research Program. Am J Respir Crit Care Med 181(4):315–323. https://doi.org/10.1164/rccm.200906-0896OC

Moore WC, Hastie AT, Li X, Li H, Busse WW, Jarjour NN, Wenzel SE, Peters SP, Meyers DA, Bleecker ER, National Heart, Lung, and Blood Institute's Severe Asthma Research Program (2014) Sputum neutrophil counts are associated with more severe asthma phenotypes using cluster analysis. J Allergy Clin Immunol 133(6):1557–1563.e1555. https://doi.org/10.1016/j.jaci.2013.10.011

Morjaria JB, Chauhan AJ, Babu KS, Polosa R, Davies DE, Holgate ST (2008) The role of a soluble TNFalpha receptor fusion protein (etanercept) in corticosteroid refractory asthma: a double blind, randomised, placebo controlled trial. Thorax 63(7):584–591. https://doi.org/10.1136/thx.2007.086314

Nair P, Pizzichini MM, Kjarsgaard M, Inman MD, Efthimiadis A, Pizzichini E, Hargreave FE, O'Byrne PM (2009) Mepolizumab for prednisone-dependent asthma with sputum eosinophilia. N Engl J Med 360(10):985–993. https://doi.org/10.1056/NEJMoa0805435

Nair P, Gaga M, Zervas E, Alagha K, Hargreave FE, O'Byrne PM, Stryszak P, Gann L, Sadeh J, Chanez P, Study Investigators (2012) Safety and efficacy of a CXCR2 antagonist in patients with severe asthma and sputum neutrophils: a randomized, placebo-controlled clinical trial. Clin Exp Allergy 42(7):1097–1103. https://doi.org/10.1111/j.1365-2222.2012.04014.x

National Asthma Education and Prevention Program (2007) Expert panel report 3 (EPR-3): guidelines for the diagnosis and management of asthma-summary report 2007. J Allergy Clin Immunol 120(5 Suppl):S94–138. https://doi.org/10.1016/j.jaci.2007.09.043

Ober C, Yao TC (2011) The genetics of asthma and allergic disease: a 21st century perspective. Immunol Rev 242(1):10–30. https://doi.org/10.1111/j.1600-065X.2011.01029.x

Oh CK, Leigh R, McLaurin KK, Kim K, Hultquist M, Molfino NA (2013) A randomized, controlled trial to evaluate the effect of an anti-interleukin-9 monoclonal antibody in adults with uncontrolled asthma. Respir Res 14:93. https://doi.org/10.1186/1465-9921-14-93

Ortega HG, Yancey SW, Mayer B, Gunsoy NB, Keene ON, Bleecker ER, Brightling CE, Pavord ID (2016) Severe eosinophilic asthma treated with mepolizumab stratified by baseline eosinophil thresholds: a secondary analysis of the DREAM and MENSA studies. Lancet Respir Med 4(7):549–556. https://doi.org/10.1016/S2213-2600(16)30031-5

Parker JM, Oh CK, LaForce C, Miller SD, Pearlman DS, Le C, Robbie GJ, White WI, White B, Molfino NA, MEDI-528 Clinical Trials Group (2011) Safety profile and clinical activity of multiple subcutaneous doses of MEDI-528, a humanized anti-interleukin-9 monoclonal antibody, in two randomized phase 2a studies in subjects with asthma. BMC Pulm Med 11:14. https://doi.org/10.1186/1471-2466-11-14

Pavord ID, Brightling CE, Woltmann G, Wardlaw AJ (1999) Non-eosinophilic corticosteroid unresponsive asthma. Lancet 353(9171):2213–2214. https://doi.org/10.1016/S0140-6736(99)01813-9

Pavord ID, Korn S, Howarth P, Bleecker ER, Buhl R, Keene ON, Ortega H, Chanez P (2012) Mepolizumab for severe eosinophilic asthma (DREAM): a multicentre, double-blind, placebo-controlled trial. Lancet 380(9842):651–659. https://doi.org/10.1016/S0140-6736(12)60988-X

Piper E, Brightling C, Niven R, Oh C, Faggioni R, Poon K, She D, Kell C, May RD, Geba GP, Molfino NA (2013) A phase II placebo-controlled study of tralokinumab in moderate-to-severe asthma. Eur Respir J 41(2):330–338. https://doi.org/10.1183/09031936.00223411

Prazma CM, Wenzel S, Barnes N, Douglass JA, Hartley BF, Ortega H (2014) Characterisation of an OCS-dependent severe asthma population treated with mepolizumab. Thorax 69 (12):1141–1142. https://doi.org/10.1136/thoraxjnl-2014-205581

Queen C, Schneider WP, Selick HE, Payne PW, Landolfi NF, Duncan JF, Avdalovic NM, Levitt M, Junghans RP, Waldmann TA (1989) A humanized antibody that binds to the interleukin 2 receptor. Proc Natl Acad Sci USA 86(24):10029–10033

Rickel EA, Siegel LA, Yoon BR, Rottman JB, Kugler DG, Swart DA, Anders PM, Tocker JE, Comeau MR, Budelsky AL (2008) Identification of functional roles for both IL-17RB and IL-17RA in mediating IL-25-induced activities. J Immunol 181(6):4299–4310

Robinson DS, Hamid Q, Ying S, Tsicopoulos A, Barkans J, Bentley AM, Corrigan C, Durham SR, Kay AB (1992) Predominant TH2-like bronchoalveolar T-lymphocyte population in atopic asthma. N Engl J Med 326(5):298–304. https://doi.org/10.1056/NEJM199201303260504

Schmidt-Weber CB, Akdis M, Akdis CA (2007) TH17 cells in the big picture of immunology. J Allergy Clin Immunol 120(2):247–254. https://doi.org/10.1016/j.jaci.2007.06.039

Schmitt E, Germann T, Goedert S, Hoehn P, Huels C, Koelsch S, Kühn R, Müller W, Palm N, Rüde E (1994) IL-9 production of naive CD4+ T cells depends on IL-2, is synergistically enhanced by a combination of TGF-beta and IL-4, and is inhibited by IFN-gamma. J Immunol 153(9):3989–3996

Segal LN, Blaser MJ (2014) A brave new world: the lung microbiota in an era of change. Ann Am Thorac Soc 11(Suppl 1):S21–S27. https://doi.org/10.1513/AnnalsATS.201306-189MG

Sel S, Wegmann M, Dicke T, Sel S, Henke W, Yildirim AO, Renz H, Garn H (2008) Effective prevention and therapy of experimental allergic asthma using a GATA-3-specific DNAzyme. J Allergy Clin Immunol 121(4):910–916.e915. https://doi.org/10.1016/j.jaci.2007.12.1175

Sheehan WJ, Mauger DT, Paul IM, Moy JN, Boehmer SJ, Szefler SJ, Fitzpatrick AM, Jackson DJ, Bacharier LB, Cabana MD, Covar R, Holguin F, Lemanske RF Jr, Martinez FD, Pongracic JA, Beigelman A, Baxi SN, Benson M, Blake K, Chmiel JF, Daines CL, Daines MO, Gaffin JM, Gentile DA, Gower WA, Israel E, Kumar HV, Lang JE, Lazarus SC, Lima JJ, Ly N, Marbin J, Morgan WJ, Myers RE, Olin JT, Peters SP, Raissy HH, Robison RG, Ross K, Sorkness CA, Thyne SM, Wechsler ME, Phipatanakul W, AsthmaNet NN (2016) Acetaminophen versus Ibuprofen in Young Children with Mild Persistent Asthma. N Engl J Med 375(7):619–630. https://doi.org/10.1056/NEJMoa1515990

Shook BC, Lin K (2017) Recent advances in developing antiviral therapies for respiratory syncytial virus. Top Curr Chem 375(2):40. https://doi.org/10.1007/s41061-017-0129-4

Slager RE, Hawkins GA, Ampleford EJ, Bowden A, Stevens LE, Morton MT, Tomkinson A, Wenzel SE, Longphre M, Bleecker ER, Meyers DA (2010) IL-4 receptor alpha polymorphisms are predictors of a pharmacogenetic response to a novel IL-4/IL-13 antagonist. J Allergy Clin Immunol 126(4):875–878. https://doi.org/10.1016/j.jaci.2010.08.001

Slager RE, Otulana BA, Hawkins GA, Yen YP, Peters SP, Wenzel SE, Meyers DA, Bleecker ER (2012) IL-4 receptor polymorphisms predict reduction in asthma exacerbations during response to an anti-IL-4 receptor alpha antagonist. J Allergy Clin Immunol 130(2):516–522. e514. https://doi.org/10.1016/j.jaci.2012.03.030

Soler M, Matz J, Townley R, Buhl R, O'Brien J, Fox H, Thirlwell J, Gupta N, Della Cioppa G (2001) The anti-IgE antibody omalizumab reduces exacerbations and steroid requirement in allergic asthmatics. Eur Respir J 18(2):254–261

Tan WC (2005) Viruses in asthma exacerbations. Curr Opin Pulm Med 11(1):21–26

Tan LD, Bratt JM, Godor D, Louie S, Kenyon NJ (2016) Benralizumab: a unique IL-5 inhibitor for severe asthma. J Asthma Allergy 9:71–81. https://doi.org/10.2147/JAA.S78049

Teo SM, Mok D, Pham K, Kusel M, Serralha M, Troy N, Holt BJ, Hales BJ, Walker ML, Hollams E, Bochkov YA, Grindle K, Johnston SL, Gern JE, Sly PD, Holt PG, Holt KE, Inouye M (2015) The infant nasopharyngeal microbiome impacts severity of lower respiratory infection and risk of asthma development. Cell Host Microbe 17(5):704–715. https://doi.org/10.1016/j.chom.2015.03.008

Tomosada Y, Chiba E, Zelaya H, Takahashi T, Tsukida K, Kitazawa H, Alvarez S, Villena J (2013) Nasally administered Lactobacillus rhamnosus strains differentially modulate respiratory antiviral immune responses and induce protection against respiratory syncytial virus infection. BMC Immunol 14:40. https://doi.org/10.1186/1471-2172-14-40

van Rensen EL, Evertse CE, van Schadewijk WA, van Wijngaarden S, Ayre G, Mauad T, Hiemstra PS, Sterk PJ, Rabe KF (2009) Eosinophils in bronchial mucosa of asthmatics after allergen challenge: effect of anti-IgE treatment. Allergy 64(1):72–80. https://doi.org/10.1111/j.1398-9995.2008.01881.x

Vignola AM, Humbert M, Bousquet J, Boulet LP, Hedgecock S, Blogg M, Fox H, Surrey K (2004) Efficacy and tolerability of anti-immunoglobulin E therapy with omalizumab in patients with concomitant allergic asthma and persistent allergic rhinitis: SOLAR. Allergy 59(7):709–717. https://doi.org/10.1111/j.1398-9995.2004.00550.x

Wang Y, Misumi I, Gu AD, Curtis TA, Su L, Whitmire JK, Wan YY (2013) GATA-3 controls the maintenance and proliferation of T cells downstream of TCR and cytokine signaling. Nat Immunol 14(7):714–722. https://doi.org/10.1038/ni.2623

Wenzel SE (2006) Asthma: defining of the persistent adult phenotypes. Lancet 368 (9537):804–813. https://doi.org/10.1016/S0140-6736(06)69290-8

Wenzel SE (2013) Complex phenotypes in asthma: current definitions. Pulm Pharmacol Ther 26 (6):710–715. https://doi.org/10.1016/j.pupt.2013.07.003

Wenzel SE, Szefler SJ, Leung DY, Sloan SI, Rex MD, Martin RJ (1997) Bronchoscopic evaluation of severe asthma. Persistent inflammation associated with high dose glucocorticoids. Am J Respir Crit Care Med 156(3 Pt 1):737–743. https://doi.org/10.1164/ajrccm.156.3.9610046

Wenzel SE, Schwartz LB, Langmack EL, Halliday JL, Trudeau JB, Gibbs RL, Chu HW (1999) Evidence that severe asthma can be divided pathologically into two inflammatory subtypes with distinct physiologic and clinical characteristics. Am J Respir Crit Care Med 160 (3):1001–1008. https://doi.org/10.1164/ajrccm.160.3.9812110

Wenzel S, Wilbraham D, Fuller R, Getz EB, Longphre M (2007) Effect of an interleukin-4 variant on late phase asthmatic response to allergen challenge in asthmatic patients: results of two phase 2a studies. Lancet 370(9596):1422–1431. https://doi.org/10.1016/S0140-6736(07) 61600-6

Wenzel SE, Barnes PJ, Bleecker ER, Bousquet J, Busse W, Dahlen SE, Holgate ST, Meyers DA, Rabe KF, Antczak A, Baker J, Horvath I, Mark Z, Bernstein D, Kerwin E, Schlenker-Herceg R, Lo KH, Watt R, Barnathan ES, Chanez P, Investigators TA (2009) A randomized, double-blind, placebo-controlled study of tumor necrosis factor-alpha blockade in severe persistent asthma. Am J Respir Crit Care Med 179(7):549–558. https://doi.org/10.1164/rccm.200809-1512OC

Wenzel S, Ford L, Pearlman D, Spector S, Sher L, Skobieranda F, Wang L, Kirkesseli S, Rocklin R, Bock B, Hamilton J, Ming JE, Radin A, Stahl N, Yancopoulos GD, Graham N, Pirozzi G (2013) Dupilumab in persistent asthma with elevated eosinophil levels. N Engl J Med 368(26):2455–2466. https://doi.org/10.1056/NEJMoa1304048

Wenzel S, Castro M, Corren J, Maspero J, Wang L, Zhang B, Pirozzi G, Sutherland ER, Evans RR, Joish VN, Eckert L, Graham NM, Stahl N, Yancopoulos GD, Louis-Tisserand M, Teper A (2016) Dupilumab efficacy and safety in adults with uncontrolled persistent asthma despite use of medium-to-high-dose inhaled corticosteroids plus a long-acting beta2 agonist: a randomised double-blind placebo-controlled pivotal phase 2b dose-ranging trial. Lancet 388 (10039):31–44. https://doi.org/10.1016/S0140-6736(16)30307-5

White B, Leon F, White W, Robbie G (2009) Two first-in-human, open-label, phase I dose-escalation safety trials of MEDI-528, a monoclonal antibody against interleukin-9, in healthy adult volunteers. Clin Ther 31(4):728–740. https://doi.org/10.1016/j.clinthera.2009.04.019

Wills-Karp M, Luyimbazi J, Xu X, Schofield B, Neben TY, Karp CL, Donaldson DD (1998) Interleukin-13: central mediator of allergic asthma. Science 282(5397):2258–2261

Woodruff PG, Boushey HA, Dolganov GM, Barker CS, Yang YH, Donnelly S, Ellwanger A, Sidhu SS, Dao-Pick TP, Pantoja C, Erle DJ, Yamamoto KR, Fahy JV (2007) Genome-wide profiling identifies epithelial cell genes associated with asthma and with treatment response to corticosteroids. Proc Natl Acad Sci USA 104(40):15858–15863. https://doi.org/10.1073/pnas.0707413104

Woodruff PG, Modrek B, Choy DF, Jia G, Abbas AR, Ellwanger A, Koth LL, Arron JR, Fahy JV (2009) T-helper type 2-driven inflammation defines major subphenotypes of asthma. Am J Respir Crit Care Med 180(5):388–395. https://doi.org/10.1164/rccm.200903-0392OC

Wu W, Bleecker E, Moore W, Busse WW, Castro M, Chung KF, Calhoun WJ, Erzurum S, Gaston B, Israel E, Curran-Everett D, Wenzel SE (2014) Unsupervised phenotyping of Severe Asthma Research Program participants using expanded lung data. J Allergy Clin Immunol 133 (5):1280–1288. https://doi.org/10.1016/j.jaci.2013.11.042

Ye P, Rodriguez FH, Kanaly S, Stocking KL, Schurr J, Schwarzenberger P, Oliver P, Huang W, Zhang P, Zhang J, Shellito JE, Bagby GJ, Nelson S, Charrier K, Peschon JJ, Kolls JK (2001) Requirement of interleukin 17 receptor signaling for lung CXC chemokine and granulocyte colony-stimulating factor expression, neutrophil recruitment, and host defense. J Exp Med 194 (4):519–527

Ying S, O'Connor B, Ratoff J, Meng Q, Mallett K, Cousins D, Robinson D, Zhang G, Zhao J, Lee TH, Corrigan C (2005) Thymic stromal lymphopoietin expression is increased in asthmatic airways and correlates with expression of Th2-attracting chemokines and disease severity. J Immunol 174(12):8183–8190

Ying S, O'Connor B, Ratoff J, Meng Q, Fang C, Cousins D, Zhang G, Gu S, Gao Z, Shamji B, Edwards MJ, Lee TH, Corrigan CJ (2008) Expression and cellular provenance of thymic stromal lymphopoietin and chemokines in patients with severe asthma and chronic obstructive pulmonary disease. J Immunol 181(4):2790–2798

Zhu J (2015) T helper 2 (Th2) cell differentiation, type 2 innate lymphoid cell (ILC2) development and regulation of interleukin-4 (IL-4) and IL-13 production. Cytokine 75(1):14–24. https://doi.org/10.1016/j.cyto.2015.05.010

Zissler UM, Chaker AM, Effner R, Ulrich M, Guerth F, Piontek G, Dietz K, Regn M, Knapp B, Theis FJ, Heine H, Suttner K, Schmidt-Weber CB (2016a) Interleukin-4 and interferon-gamma orchestrate an epithelial polarization in the airways. Mucosal Immunol 9(4):917–926. https://doi.org/10.1038/mi.2015.110

Zissler UM, Esser-von Bieren J, Jakwerth CA, Chaker AM, Schmidt-Weber CB (2016b) Current and future biomarkers in allergic asthma. Allergy 71(4):475–494. https://doi.org/10.1111/all.12828

Index